Introductio[n to]
Abstract Al[gebra]

Third Edition

T. A. WHITELAW
Department of Mathematics
University of Glasgow

BLACKIE ACADEMIC & PROFESSIONAL
An Imprint of Chapman & Hall

London · Glasgow · Weinheim · New York · Tokyo · Melbourne · Madras

Published by
Blackie Academic and Professional, an imprint of Chapman & Hall,
Wester Cleddens Road, Bishopbriggs, Glasgow G64 2NZ

Chapman & Hall, 2–6 Boundary Row, London SE1 8HN, UK

Blackie Academic & Professional, Wester Cleddens Road, Bishopbriggs, Glasgow G64 2NZ, UK

Chapman & Hall GmbH, Pappelallee 3, 69469 Weinheim, Germany

Chapman & Hall USA, One Penn Plaza, 41st Floor, New York NY 10119, USA

Chapman & Hall Japan, ITP-Japan, Kyowa Building, 3F, 2-2-1 Hirakawacho, Chiyoda-ku, Tokyo 102, Japan

DA Book (Aust.) Pty Ltd, 648 Whitehorse Road, Mitcham 3132, Victoria, Australia

Chapman & Hall India, R. Seshadri, 32 Second Main Road, CIT East, Madras 600 035, India

First edition 1978

Second edition 1988

This edition 1995

© 1995 Chapman & Hall

Typeset in 10/12pt Times by AFS Image Setters Ltd, Glasgow

Printed in Great Britain by Hartnolls Ltd, Bodmin, Cornwall

ISBN   0 7514 0147 1 (PB)

A catalogue record for this book is available from the British Library

Library of Congress Catalog Card Number: 94-78425

∞ Printed on acid-free text paper, manufactured in accordance with ANSI/NISO Z39.48-1992 (Permanence of Paper)

Preface

The first and second editions of *Introduction to Abstract Algebra* have proved very helpful to many students encountering abstract algebra for the first time. In this third edition, as in the previous editions, my aim throughout has been to present the ideas of abstract algebra in a manner that is, *judged from the undergraduate student's point of view*, as clear as possible. So, for example, in considering mappings, relations, etc., I have avoided presentations which, though satisfyingly neat and sophisticated to those who have long understood the concepts, do little to communicate them to the uninitiated. In the same spirit, I have sought to ensure that there is no hidden agenda of unspoken elementary points that students must pick up for themselves if they are to make headway in digesting and using definitions, theorems, etc. My policy has been to remark explicitly on all that is vital, even when it is simple.

The book demands little in the way of prerequisites, though an elementary knowledge of complex numbers and matrices is assumed in some sections. In just one section (the penultimate section of the last chapter) is anything presented that calls for a knowledge of vector space theory: however, the chapter is organized in such a way that readers who know nothing of vector spaces can omit that section and still gain a worthwhile education in the field theory presented in Chapter 11.

This is not an advanced treatise on abstract algebra, but, in contrast to many introductory textbooks, it does penetrate far enough into the subject to show students that abstract algebra has worthwhile insights to offer and to introduce them to some of the distinctive ways of thinking that produce interesting results. For example, in group theory the material covered includes cyclic groups, Lagrange's theorem, homomorphisms, normal subgroups, quotient groups, the partition of the symmetric group of degree n into conjugacy classes, and an introduction to the alternating group of degree n. In the chapters on rings and fields a wide variety of ideas is covered, and students will find it illuminating to see that both the construction of $\mathbb{Q}$ from $\mathbb{Z}$ and the construction of $\mathbb{C}$ from $\mathbb{R}$ can be usefully generalized.

Experience shows that students find abstract algebra difficult to grasp if they start with an inadequate understanding of basic topics such as equivalence relations, mappings, and the notations and vocabulary of set theory. Accordingly, as a preliminary, the first four chapters of this book deal with all the relevant details of these basic topics, along with certain properties of integers. This last item is included to establish machinery (e.g.

iii

division algorithm) used in group theory and elsewhere, and also to provide material that will serve to motivate and illustrate some of the important concepts of abstract algebra. Chapter 1 also includes some material that comes under the heading of *Logic* and consists of explanations and advice about such fundamental matters as quantifiers, the use of the "implies" sign, and proof by contradiction.

In this new edition greater space is given to consideration of rings and fields. Chapter 10 gives an account of the fundamentals of elementary ring theory (homomorphisms, ideals, factor rings, etc.), while Chapter 11 gives a full treatment of polynomials over a field and discusses the beginnings of the theory of field extensions and the theory of finite fields. The study of rings and fields is, therefore, better represented than in the second edition. I expect the appearance of the newly included material to be widely welcomed, both because of its undoubted intrinsic interest and because of its relevance to other important parts of mathematics and its applications.

The text includes worked examples, and these have been chosen to reveal important methods of argument and show how certain kinds of proof may be set down. At the end of each chapter there is an ample supply of varied exercises, and partial solutions to these are at the back of the book. This adjunct to the exercises will, if sensibly used, prove beneficial to students.

I should like to record openly how much the merits of this book owe to the tradition of care and enthusiasm in the teaching of mathematics that has long existed in the Mathematics Department of the University of Glasgow. The tradition I refer to nourished me as an undergraduate; and more recently, in working and conversing with my colleagues, I have learned much that has enriched my ideas about how mathematics can and should be taught. Special thanks go to my colleague Dr J.B. Hickey for the help he has given me with the correction of the proofs of this edition: his kindness and his attention to detail have far exceeded all that I could have asked for. I must thank also my colleague Dr I. Anderson for his extremely helpful advice concerning some of the new material incorporated in this third edition.

T.A.W.

Contents

CHAPTER ONE

SETS AND LOGIC

1. Some very general remarks

This opening section is intended as an introduction to the entire book and not merely to the first chapter. It begins with an attempt to give brief answers to the questions: What is Abstract Algebra? and Why is it a worthwhile subject of study?

The thoughtful student may have realized that he has had to learn the properties of several different algebraic systems—the system of integers, the system of real numbers, the system of complex numbers, systems of matrices, systems of vectors, etc. He would be correct to guess that the number of algebraic systems he has to deal with will multiply rapidly as he proceeds further in the study of mathematics. This being the case, it is desirable to try to gain general insights about algebraic systems and to prove theorems that apply to broad classes of algebraic systems, instead of devoting energy to the study of a seemingly endless sequence of individual systems. If we can achieve such general insights, we stand to gain a deeper and more unified understanding of numerous elementary matters such as properties of integers or properties of matrices; and we shall be equipped to think intelligently about new algebraic systems when we encounter them. Thus there are ample reasons for believing it worthwhile (perhaps indeed essential) to attempt the study of algebraic systems in general, as opposed to the separate study of many individual algebraic systems. It is such study of algebraic systems in general that is called Abstract Algebra, and this is the part of mathematics to which this textbook provides an introduction.

The word *abstract* must not be taken to mean either difficult or unrelated to the more tangible objects of mathematical study (such as numbers and matrices). Instead, *abstract* indicates that the discussion will not be tied to one particular concrete algebraic system (since generality of insight is the paramount aim), while every general result obtained will have applications to a variety of different concrete situations.

1

In this book the study of abstract algebra proper begins at chapter five. Before that, attention is given to preliminary material of fundamental importance, including sets, equivalence relations, and mappings.

This first chapter contains a survey of the main ideas used continually in the discussion of sets. Interwoven with that are some remarks on elementary logical matters such as quantifiers, the implies sign, and proof by contradiction. It is hoped that these remarks will help students both to digest logical arguments that they read and to execute their own logical arguments. Even so, a clear warning must be given that mastery of the art of logical argument (as required in abstract algebra and comparable mathematical disciplines) depends less on knowledge of any core of facts about logic than on commitment to complete precision in the use of words and symbols.

2. Introductory remarks on sets

(a) Very often in mathematics we discuss sets or collections of objects and think of each set as a single thing that may be denoted by a single letter (S or A or ...).

(b) Standard notations for familiar sets of numbers are:

$\mathbb{N}$ denoting the set of all natural numbers (i.e. positive integers);

$\mathbb{Z}$ denoting the set of all integers;

$\mathbb{Q}$ denoting the set of all rational numbers;

$\mathbb{R}$ denoting the set of all real numbers;

$\mathbb{C}$ denoting the set of all complex numbers.

(c) Suppose that S is a set and x is an object. If x is one of the objects in the set S, we say that x is a **member** of S (or x is an **element** of S, or x **belongs to** S) and we indicate this by writing

$$x \in S.$$

In the contrary case, i.e. if x is not a member of S, we write

$$x \notin S.$$

We also use the $\in$ symbol as an abbreviation for *belonging to*, as in the phrase *for every* $x \in S$ (read *for every x belonging to S*).

(d) The set whose members are all the objects appearing in the list $x_1, x_2, \ldots, x_n$ and no others is denoted by

$$\{x_1, x_2, \ldots, x_n\}.$$

(e) A set S is called **finite** if only finitely many different objects are members of S. Otherwise S is described as an infinite set.

If S is a finite set, the number of different objects in S is called the **order** (or **cardinality**) of S and denoted by $|S|$; e.g. if $S = \{1, 3, 5, 7\}$, then $|S| = 4$.

(f) It is important to have a clear criterion for deciding when two sets S_1, S_2 should be called *equal*. We adopt the natural definition that "$S_1 = S_2$" shall mean that every member of S_1 is a member of S_2 and every member of S_2 is a member of S_1.

(g) It is already apparent that we are basing our discussion of sets on the simplistic assumption that we can recognize a "set" or "collection of objects" when we meet one. In fact, some far-reaching objections can be raised to this approach. Most seriously, it can be shown that contradictions arise when we apply our simple-minded notions about sets to collections of objects so vast that we cannot really discern whether they fit in with the simple-minded notions. The circumvention of these difficulties lies beyond the scope of this book. Here our attitude is to assume provisionally that the simple-minded concept of a set is a reasonable and useful one in the situations we are exploring, and to concentrate on explaining facts and notations that are frequently used in these situations. (Students who would like to know more about the problems we have hinted at should read the earlier chapters of *Naive Set Theory* by P. R. Halmos [1].)

3. Statements and conditions; quantifiers

In thinking generally about the sentences that occur in mathematics, it is valuable to draw a distinction between statements and conditions.

A *statement* means a sentence about specific numbers, sets, or other objects, the sentence being, in its own right, either true or false (even if it is difficult to say which). So examples of statements include:

$$3 \times 4 = 11$$

$$2^{19937} - 1 \text{ is a prime number.}$$

A *condition*, on the other hand, is a sentence (such as "$x^3 > 3x$" or "$y^2 = 4x$") that involves one or more variables (symbols like $x, y, \ldots$) and becomes a statement when each variable is given a specific value (i.e. is replaced by a definite object). There should always be an indication of the set of values each variable can take (i.e. be replaced by); and for each variable this set is called the **domain** of the variable.

For illustrative purposes, consider for a moment one particular example of a condition, namely

$$y^2 = 4x \qquad (x, y \in \mathbb{R}).$$

The bracketed "$x, y \in \mathbb{R}$" indicates that both x and y have $\mathbb{R}$ as domain. From this condition, true statements can be produced by giving x, y certain specific values (e.g. giving x the value 9 and y the value 6 produces the true statement $6^2 = 4 \times 9$); and false statements can be produced by giving x, y other values.

There is a wide variety of accepted ways of expressing the fact that a true statement is produced from the condition on replacing x by 9 and y by 6: for example, one may say that "the condition is true for $x = 9$, $y = 6$" or "the condition holds when $x = 9$, $y = 6$" or "the condition is satisfied by $x = 9$, $y = 6$", etc.

The commonest kind of condition is the one-variable condition, i.e. condition involving just one variable. Such a condition may conveniently be denoted by $P(x)$, x being the variable with, let us suppose, domain S. We denote by

$$\{x \in S : P(x)\}$$

the set of elements in S that satisfy the condition $P(x)$, i.e. the set of values of x in S for which $P(x)$ is true. For example,

$$\{x \in \mathbb{Z} : x^2 = 4\} = \{-2, 2\}.$$

The universal quantifier
Suppose here that $P(x)$ is a condition involving the variable x, which has domain S.

The symbol $\forall$ is shorthand for "for all" or "for every" and may be used to build the statement

$$\forall x \in S, P(x), \tag{1}$$

which is the assertion that $P(x)$ is true for all values of x in the set S. It should be noticed that this new sentence is not a condition but a statement, true or false in its own right—true if $P(x)$ is true for all values of x in S, false otherwise. A statement of the form (1) is called a **universal statement**, and the symbol $\forall$ is called the **universal quantifier.**

As specific illustrations of universal statements, consider:

$$\forall x \in \mathbb{R}, \quad (x+1)^2 = x^2 + 2x + 1,$$
$$\forall x \in \mathbb{R}, \quad (x+1)^2 = x^2 + 1.$$

Elementary algebra tells us that the former is true. But the latter is false, since the condition "$(x+1)^2 = x^2 + 1$" is false for certain values of x (e.g. $x = 1$).

Remarks. (*a*) The assertion "$\forall x \in S, P(x)$" is equivalent to the assertion that the set $\{x \in S : P(x)\}$ is the whole of S.

(*b*) Universal statements can also be made from conditions involving more than one variable. If $P(x, y, z, \ldots)$ is a condition involving the variables $x, y, z, \ldots$, with respective domains $A, B, C, \ldots$, then the statement

$$\forall x \in A, \quad \forall y \in B, \quad \forall z \in C, \ldots, \quad P(x, y, z, \ldots)$$

is the assertion that $P(x, y, z, \ldots)$ is true for all relevant values of $x, y, z, \ldots$ (i.e. all values of x in A, etc.). For instance, a familiar index law is:

$$\forall x \in \mathbb{R}, \quad \forall m \in \mathbb{N}, \quad \forall n \in \mathbb{N}, \quad x^m . x^n = x^{m+n}.$$

(c) It should be observed that all of the following are synonymous:

$$\forall x \in \mathbb{R}, \; x^2 . x^3 = x^5; \quad \forall y \in \mathbb{R}, \; y^2 . y^3 = y^5; \quad \forall \alpha \in \mathbb{R}, \; \alpha^2 . \alpha^3 = \alpha^5; \quad \text{etc.}$$

Each of them says, in formal language, that the product of the square of a real number and its cube is always equal to the fifth power of that number. More generally, the fact is that, in a universal statement, one can change the symbol used for the variable quite arbitrarily without altering the meaning of the statement. This is often expressed by saying that the variable is a *dummy*.

(d) A well-established convention in mathematicians' use of language is that where an asserted fact takes the form of a condition (accompanied by an indication of the domain(s) of the variable(s)), it is to be understood that what is being asserted is the truth of the condition for all values of the variables. For example, one may find asserted as a theorem

$$x^m . x^n = x^{m+n} \qquad (x \in \mathbb{R}, m, n \in \mathbb{N}).$$

This must be interpreted as meaning

$$\forall x \in \mathbb{R}, \quad \forall m \in \mathbb{N}, \quad \forall n \in \mathbb{N}, \quad x^m . x^n = x^{m+n}.$$

The existential quantifier
Suppose again that $P(x)$ is a condition involving the variable x, which has domain S.

The symbol $\exists$ is shorthand for "there exists" and may be used to build the statement

$$\exists x \in S \text{ s.t. } P(x)$$

(read "there exists x belonging to S such that $P(x)$"), which is the assertion that there is at least one value of x (in S) for which $P(x)$ is true.

Thus, for instance,

$$\exists x \in \mathbb{R} \text{ s.t. } x^2 = 4$$

is true: for the condition $x^2 = 4$ is satisfied by at least one value (in $\mathbb{R}$) of x (e.g. $x = 2$). But

$$\exists x \in \mathbb{R} \text{ s.t. } x^2 = -1$$

is false: for there is no real value of x satisfying $x^2 = -1$.

It should be noted that a sentence of the form "$\exists x \in S \text{ s.t. } P(x)$" is a statement, not a condition, as it is true or false in its own right. Such a statement is called an **existential statement**, and the symbol ∃ is called the **existential quantifier.**

Remarks. (*a*) The assertion "∃ $x \in S$ s.t. $P(x)$" is equivalent to the assertion that the set $\{x \in S : P(x)\}$ has at least one member.

(*b*) It should be observed that the following are synonymous:

$$\exists x \in \mathbb{R} \text{ s.t. } x^2 = 4; \quad \exists y \in \mathbb{R} \text{ s.t. } y^2 = 4; \quad \exists \alpha \in \mathbb{R} \text{ s.t. } \alpha^2 = 4; \quad \text{etc.}$$

Once again, we describe this situation by saying that the variables are dummies.

Negation of a statement

For any given statement p, the **negation** of p is the statement "not p" (denoted $\neg p$) produced from p by inserting the word "not" at the appropriate point so as to produce a statement denying precisely that which p asserts. For example, if p is "$2^{100} > 3^{63}$", then $\neg p$ is "2^{100} is not greater than 3^{63}" (or, synonymously, $2^{100} \leqslant 3^{63}$). Obviously $\neg(\neg p)$ is always synonymous with p.

Likewise, from any condition $P(x)$ we can produce its negation, $\neg P(x)$, by inserting "not" at the crucial point, so that if $P(x)$ is true for a certain value of x, then $\neg P(x)$ is false for that value of x, and vice versa.

The following general rules tell us that the negation of a universal statement can be re-expressed as an existential statement, and vice versa.

3.1 (i) $\neg [\forall x \in S, P(x)]$ is synonymous with $\exists x \in S$ s.t. $\neg P(x)$.

(ii) $\neg [\exists x \in S \text{ s.t. } P(x)]$ is synonymous with $\forall x \in S, \neg P(x)$.

Proof. (i) Let α stand for "$\neg [\forall x \in S, P(x)]$" and β stand for "$\exists x \in S$ s.t. $\neg P(x)$".

The crux of the matter is that "not everyone does" says the same as "at least one doesn't". Therefore α, which asserts that not every value of x (in S) satisfies $P(x)$, is equivalent to the assertion that at least one value of x (in S) does not satisfy $P(x)$, i.e. that at least one value of x (in S) satisfies $\neg P(x)$: and this is precisely what β asserts.

(ii) can be justified similarly by observing that "no one does" says the same as "everyone does not".

To give a very simple illustration of 3.1, let us suppose that S is a class of students and that $P(x)$ stands for "student x is aged 21 or over" (so that $\neg P(x)$ means "student x is aged under 21"). Applied in this case, 3.1(i) tells us that the negation of "every student in the class is aged 21 or over" is "there exists a student in the class aged under 21"; and 3.1(ii) tells us that the negation of "there exists a student in the class aged 21 or over" is "every student in the class is aged under 21".

These deductions from 3.1 may rightly be considered to be simple common sense: the point of 3.1 is that it brings out the general principle vaguely discerned in such "common sense" and equips us to apply the principle in more complicated contexts.

Proofs of universal and existential statements

Notice first that to prove that an existential statement is true, we need only consider one (appropriately chosen) value of the variable. For example, to prove true "$\exists x \in \mathbb{R}$ s.t. $x^2 = 4$", we need only produce one value of x (in $\mathbb{R}$) satisfying $x^2 = 4$; and thus it suffices to consider $x = 2$.

An existential statement being the negation of a universal statement, it is also evident that a universal statement may be proved false by citing just one (appropriately chosen) value of the variable. For example, to prove that "$\forall x \in \mathbb{R}, (x+1)^2 = x^2 + 1$" is false, we need only exhibit one value of x (in $\mathbb{R}$) that does not satisfy the condition $(x+1)^2 = x^2 + 1$ (and clearly 1 is such a value of x).

A value of x for which $P(x)$ is false, exhibited to disprove the universal statement "$\forall x \in S, P(x)$", is called a **counterexample** to this universal statement.

Obviously, however, one cannot hope to prove the truth of a true universal statement merely by considering a single value of the variable! The basic ideas generally used for tackling and setting down proof of a universal statement are perhaps most easily absorbed by study of a very straightforward illustration. For this purpose a proof will now be given of the universal statement

$$\forall x \in E, \quad x+4 \text{ is even,} \tag{2}$$

where E denotes the set of all even integers. [Here and elsewhere in this chapter, it must be explained, we shall work from the following assumptions about even and odd integers: (i) an even integer is one expressible as $2 \times$ (an integer); (ii) an odd integer is one expressible as $2 \times$ (an integer) $+ 1$; (iii) every integer is either even or odd (and not both).]

Proof of (2). Let x be an arbitrary member of E.
Then $x = 2k$ for some integer k, and hence

$$x+4 = 2k+4 = 2(k+2).$$

Therefore, since $k+2$ is an integer, $x+4$ is even.
It follows that $\forall x \in E, x+4$ is even.

In the above proof, note particularly the device of opening by saying, "Let x be an arbitrary member of...". Some may be unhappy about a more pedantic point—the double use of the symbol x (as dummy variable in the universal statement and as arbitrary element of a set in the proof); but an assurance can be given that this very widespread habit does not in practice cause confusion.

To conclude the section, notice that disproof of an existential statement, which has not been separately discussed, is (by 3.1(ii)) equivalent to proof of a universal statement.

4. The implies sign ($\Rightarrow$)

Here let P, Q stand either for statements or for conditions involving the same variables x, y,.... The compound sentence

$$P \Rightarrow Q$$

(read "P implies Q" or, alternatively, "if P then Q") is the assertion that in every admissible case when P is true, Q is true too, i.e. that in no allowed case do we have P true and Q false. For example,

$$x > 2 \Rightarrow x^2 > 4 \qquad (x \in \mathbb{R})$$

is true because every value of x (in $\mathbb{R}$) that satisfies the condition "$x > 2$" satisfies also the condition "$x^2 > 4$". But

$$x^2 > 4 \Rightarrow x > 2 \qquad (x \in \mathbb{R})$$

is false, as we can show by exhibiting a value of x (in $\mathbb{R}$) that satisfies "$x^2 > 4$" but does not satisfy "$x > 2$" (and $x = -3$ is such a counterexample).

Remarks. (*a*) If $P(x)$, $Q(x)$ are conditions involving the variable x with domain S, the sentence

$$P(x) \Rightarrow Q(x) \qquad (x \in S) \tag{3}$$

is really a universal statement, because it says that a certain condition is satisfied for every value (in S) that might be given to x (namely that the combination $P(x)$ true, $Q(x)$ false does not arise). Purists might therefore prefer that (3) should be re-written

$$\forall x \in S, \quad P(x) \Rightarrow Q(x)$$

though in fact (3) is typical of the way mathematicians usually express themselves.

Notice also that the universal statement "$\forall x \in S, P(x)$" can be re-written as

$$x \in S \Rightarrow P(x) \qquad (x \in S).$$

(*b*) "$P \Rightarrow Q$" is held to be true (by default, as it were) if there is no case in which P is true; e.g. the statement

$$x \neq x \Rightarrow x = 7 \qquad (x \in \mathbb{R})$$

is classified as true, because in no allowed case do we have $x \neq x$ true and $x = 7$ false.

(c) It is essential to realize that the statement "$P \Rightarrow Q$" says something about the *relationship between P and Q*. It does not say anything about the truth of P in any or all cases; nor does it say anything about the truth of Q in any or all cases. In particular "$\Rightarrow$", which comes in the middle of a compound sentence, is most certainly not a synonym for "hence" or "therefore", which are appropriate words at the beginning of a new sentence. If, in the course of a proof, we wish to assert that, because of what has just been said, the truth of Q is apparent, then we should say "Hence Q" or "Therefore Q" or "It follows that Q" (but not "$\Rightarrow Q$").

Formats for the proof of "$P \Rightarrow Q$"
A straightforward way of giving such a proof is to begin by saying "Suppose P" or "Assume P", and then to proceed to demonstrate that Q is a consequence of this supposition. Thus, by temporarily regarding P as true (for the purpose of the argument), we show that we cannot have P true without also having Q true—which is precisely what "$P \Rightarrow Q$" means.

As an illustration, here is a proof of the simple result

$$n \text{ is odd} \Rightarrow n^2 \text{ is odd} \qquad (n \in \mathbb{Z}).$$

Proof. Suppose that n is odd ($n \in \mathbb{Z}$). Then $n = 2k+1$ for some $k \in \mathbb{Z}$, and hence

$$n^2 = (2k+1)^2 = 4k^2 + 4k + 1 = 2(2k^2 + 2k) + 1.$$

Therefore, since $2k^2 + 2k \in \mathbb{Z}$, n^2 is odd.
The stated result follows.

Another common form of proof of "$P \Rightarrow Q$" is a sequence

$$P \Rightarrow \alpha_1 \Rightarrow \alpha_2 \Rightarrow \alpha_3 \Rightarrow \ldots \Rightarrow \alpha_n \Rightarrow Q. \qquad (4)$$

In interpreting such a proof, it is essential to appreciate that (4) is an abbreviation for:

$$\left\{ \begin{array}{l} P \Rightarrow \alpha_1, \ \alpha_1 \Rightarrow \alpha_2, \ \alpha_2 \Rightarrow \alpha_3, \ldots, \alpha_n \Rightarrow Q \text{ are all true.} \\ \text{Consequently, } P \Rightarrow Q \text{ is true.} \end{array} \right\}$$

The logic underlying the word "consequently" is not hard to discern.

As a simple illustration of the format (4), here is a proof of this type of the result

$$n \text{ is odd} \Rightarrow n^2 \text{ is odd} \qquad (n \in \mathbb{Z}).$$

Proof. For $n \in \mathbb{Z}$,

$$n \text{ is odd} \Rightarrow n = 2k+1 \ (k \text{ being an integer})$$
$$\Rightarrow n^2 = (2k+1)^2 = 2(2k^2+2k)+1$$
$$\Rightarrow n^2 \text{ is odd (since } 2k^2+2k \text{ is an integer)}.$$

The law of contraposition

4.1 "$P \Rightarrow Q$" is synonymous with "$(\neg Q) \Rightarrow (\neg P)$".

Proof. "$P \Rightarrow Q$" says that it is not possible to have P true and Q false.

"$(\neg Q) \Rightarrow (\neg P)$" says that it is not possible to have $\neg Q$ true and $\neg P$ false, i.e. to have Q false and P true.

Thus the two are synonymous.

We call "$(\neg Q) \Rightarrow (\neg P)$" the **contrapositive** of "$P \Rightarrow Q$", and we call the very useful law 4.1 (the fact that "$P \Rightarrow Q$" is synonymous with its contrapositive) the *law of contraposition*.

In contrast, the sentence "$Q \Rightarrow P$" is called the **converse** of "$P \Rightarrow Q$". A common and bad error is to confuse "$P \Rightarrow Q$" and its converse, or to believe that they are synonymous. That, let it be emphasized, is certainly not so. Indeed we observed early in the section that, while

$$x > 2 \Rightarrow x^2 > 4 \qquad (x \in \mathbb{R})$$

is true, its converse is false.

"If and only if" $(\Leftrightarrow)$

The compound sentence "$P \Leftrightarrow Q$" (read "P if and only if Q" or "P is equivalent to Q") means "both $P \Rightarrow Q$ and $Q \Rightarrow P$". Thus, when we assert the truth of "$P \Leftrightarrow Q$", we are asserting the truth both of "$P \Rightarrow Q$" and of its converse "$Q \Rightarrow P$", and we are therefore claiming that, in every admissible case, P and Q are either both true or both false.

Sometimes, especially in definitions, it is convenient to employ the abbreviation "iff" (rather than the symbol $\Leftrightarrow$) for the words "if and only if".

The proof of "$P \Leftrightarrow Q$" must often be split into two parts—the first a proof of "$P \Rightarrow Q$" and the second a proof of "$Q \Rightarrow P$". Sometimes we encounter a proof of "$P \Leftrightarrow Q$" taking the form of a sequence

$$P \Leftrightarrow \alpha_1 \Leftrightarrow \alpha_2 \Leftrightarrow \alpha_3 \Leftrightarrow \ldots \Leftrightarrow \alpha_n \Leftrightarrow Q.$$

This (cf. corresponding remarks on proving "$P \Rightarrow Q$") must be understood as an abbreviation for:

$$\left\{ \begin{array}{l} P \Leftrightarrow \alpha_1, \ \alpha_1 \Leftrightarrow \alpha_2, \ \alpha_2 \Leftrightarrow \alpha_3, \ldots, \alpha_n \Leftrightarrow Q \text{ are all true.} \\ \text{Consequently, } P \Leftrightarrow Q \text{ is true.} \end{array} \right\}$$

5. Proof by contradiction

In many contexts a powerful method of proof is *proof by contradiction*, which, in effect, consists of eliminating the possibility that the statement to be proved is false. Those unfamiliar with the method are recommended to study the next two paragraphs of general explanation in conjunction with the simple example which follows them.

When using proof by contradiction, we proceed as follows.

(*a*) We begin by assuming, for the sake of argument, that the statement to be proved is false.

(*b*) We deduce from that assumption a contradiction, i.e. something that is manifestly false because it is self-contradictory (e.g. "*n* is both zero and nonzero") or contrary to known facts (e.g. "2 = 1") or contrary to what was given or assumed at stage (*a*).

(*c*) We infer that the assumption made at stage (*a*) was wrong—which is to say that the statement to be proved is in fact true.

The thinking behind the vital step (*c*) is that the obtaining through accurate logical deduction of a contradiction reveals that at some point we have made a wrong assumption (since valid assumptions lead only to true conclusions): so, since the only questionable assumption made was the one in stage (*a*), that assumption must be wrong.

Example. It is given that m, n are integers and that the product mn is odd. Prove that m is odd.

Proof. Suppose the contrary, i.e. that m is even. Then $m = 2k$ for some integer k, and hence

$$mn = 2kn = 2 \times (\text{an integer}).$$

So mn is even—a contradiction [i.e. something contrary to what is given].

From this contradiction it follows that m is odd.

6. Subsets

Let A, B denote sets. We say that A is a **subset** of B (or that A is **contained in** B) iff every member of A is also a member of B.

In this book, our notation for "A is a subset of B" will be

$$A \subseteq B.$$

When $A \subseteq B$ and $A \neq B$, we say that A is a **proper subset** of B (or that A is **strictly contained** in B) and we write $A \subset B$. [The student is advised that mathematicians (and the textbooks they write) vary slightly in their usages of the symbols $\subseteq$ and $\subset$.]

Clearly:

6.1 For every set A, $A \subseteq A$.
6.2 For sets A, B, C, $(A \subseteq B$ and $B \subseteq C) \Rightarrow A \subseteq C$.
6.3 For sets A, B, $A = B \Leftrightarrow (A \subseteq B$ and $B \subseteq A)$.

Remarks. (*a*) If A, B are subsets of the set S, the statement "$A \subseteq B$" is equivalent to

$$x \in A \Rightarrow x \in B \qquad (x \in S),$$

and the statement "$A = B$" is equivalent to

$$x \in A \Leftrightarrow x \in B \qquad (x \in S).$$

These simple observations are the basis of many proofs of statements of the form "$A \subseteq B$" or "$A = B$".

(*b*) The notation $\{x \in S : P(x)\}$ for specifying a subset of the set S was introduced in §3. Notice that in this notation x is a dummy variable. For example,

$$\{x \in \mathbb{R} . x^2 - 4\}, \quad \{y \in \mathbb{R} . y^2 = 4\}, \quad \{u \in \mathbb{R} . u^2 = 4\}, \quad \text{etc.}$$

all mean the same thing, viz. $\{-2, 2\}$.

(*c*) In illustrations we shall often refer to **intervals**, which are "connected" subsets of $\mathbb{R}$. Specifically, if $a, b \in \mathbb{R}$ and $a < b$, the intervals with a and b as endpoints are the four subsets

$$(a, b) = \{x \in \mathbb{R} : a < x < b\}, \qquad [a, b] = \{x \in \mathbb{R} : a \leqslant x \leqslant b\},$$
$$(a, b] = \{x \in \mathbb{R} : a < x \leqslant b\}, \qquad [a, b) = \{x \in \mathbb{R} : a \leqslant x < b\};$$

and, for each $a \in \mathbb{R}$, we write

$$(-\infty, a) = \{x \in \mathbb{R} : x < a\}, \qquad (-\infty, a] = \{x \in \mathbb{R} : x \leqslant a\},$$
$$(a, \infty) = \{x \in \mathbb{R} : x > a\}, \qquad [a, \infty) = \{x \in \mathbb{R} : x \geqslant a\}.$$

(*d*) Another notation used in specifying a subset of a set S takes the form

$$\{f(x) : x \in A\}$$

where A is a given set and $f(x)$ stands for something constructible from x (e.g. x^2). It is to be understood that if x is given a definite value in A, then correspondingly $f(x)$ will take a definite value in S. And $\{f(x) : x \in A\}$ means the set of all values $f(x)$ can take if x takes a value in A. For example,

$$\{x^2 : x \in \mathbb{N}\} = \{1, 4, 9, 16, 25, \ldots\};$$

and, if $A = \{1, -1, 2, -2, 3, -3\}$, then $\left\{\dfrac{1}{x^2 + 1} : x \in A\right\} = \{\tfrac{1}{2}, \tfrac{1}{5}, \tfrac{1}{10}\}$. In formal terms, $\{f(x):x\in A\}$ can be defined as

$$\{y\in S:\exists x\in A \text{ s.t. } y=f(x)\}.$$

More complicated variants of this notation abound but are to some extent self-explanatory. When one occurs in the pages of this book, there will always be accompanying words aimed at making the meaning fully clear.

The empty set
An **empty set** is a set with no members: e.g. $\{x\in\mathbb{R}:x^2=-1\}$.

If θ is an empty set and A is any set whatever, then it is true by default that every member of θ is a member of A, i.e. that

$$x\in\theta \Rightarrow x\in A \qquad (x\in A).$$

Hence:

6.4 If θ is an empty set and A is any set whatever, then $\theta\subseteq A$.

From this it can immediately be deduced that:
6.5 There is only one empty set.

Proof. If θ_1 and θ_2 are empty sets, then (by 6.4) $\theta_1\subseteq\theta_2$ and $\theta_2\subseteq\theta_1$, and hence (see 6.3) $\theta_1=\theta_2$. The result follows.

Because of 6.5, it makes sense to speak of *the* empty set, which we always denote by θ.

Complement of a subset
Definition. Let A be a subset of a given set S. The **complement** of A in S is the subset $\{x\in S:x\notin A\}$. It is denoted by $\mathscr{C}_S A$.

For example, if E is the set of all even integers, then $\mathscr{C}_\mathbb{Z} E$ is the set of all odd integers.

When it is understood that a particular set S is our "universal set" (i.e. that all the sets being considered are subsets of S and that "complement of A" means "complement of A in S"), it is feasible to use simpler notations for $\mathscr{C}_S A$, e.g. $\mathscr{C}A$ or A'.

Clearly, when S is the universal set:

6.6 $\mathscr{C}\theta=S$; $\mathscr{C}S=\theta$; and, for every $A\subseteq S$, $\mathscr{C}(\mathscr{C}A)=A$.

Another definition. If A, B are sets, the **relative complement** of B in A (denoted by $A-B$) is the set $\{x\in A:x\notin B\}$. [Here, notice, we are not insisting that $B\subseteq A$. Obviously, when $B\subseteq A$, $A-B$ coincides with $\mathscr{C}_A B$.]

For example, if $A=\{1,2,3\}$ and $B=\{2,4,5,6\}$, then $A-B=\{1,3\}$.

Venn diagrams
When subsets of a set S are discussed, it is often helpful to draw pictures in which S is represented by a rectangular region (the elements of S being

represented by points of the region) and subsets $A, B, \ldots$ of S are represented by subregions of the rectangle. Such a pictorial representation is called a Venn diagram.

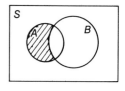

(i)   (ii) $\mathscr{C}A$ shaded   (iii) $A - B$ shaded

Parts (ii) and (iii) are Venn diagrams which, in an obvious way, illustrate how $\mathscr{C}A$ is related to A and how $A - B$ is related to A and B.

To be useful, a Venn diagram must represent the most general situation under discussion. Thus, for example, for considering two arbitrary subsets A, B of the set S, the Venn diagram (ii) below is a satisfactory visual aid while (i) is not, as it suggests that A, B have no members in common—which of course need not be the case.

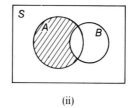

(i)   (ii)

Typical of the sort of proposition whose truth can, so to speak, be seen from a Venn diagram is:

6.7 For subsets A, B of a set S, $A \subseteq B \Leftrightarrow \mathscr{C}B \subseteq \mathscr{C}A$.

Justification by Venn diagram. We refer to diagram (ii). Both $A \subseteq B$ and $\mathscr{C}B \subseteq \mathscr{C}A$ are equivalent to the assertion that the set represented by the shaded region (i.e. $A - B$) is empty.

Formal proof. $A \subseteq B$ is equivalent to

$$x \in A \Rightarrow x \in B \qquad (x \in S).$$

By the law of contraposition (4.1), this is equivalent to

$$x \notin B \Rightarrow x \notin A \qquad (x \in S),$$

i.e.

$$x \in \mathscr{C}B \Rightarrow x \in \mathscr{C}A \qquad (x \in S),$$

and that is precisely what $\mathscr{C}B \subseteq \mathscr{C}A$ says.

Power set of a given set

Definition. For a given set S, the **power set** of S (denoted by $\mathscr{P}(S)$) is the set whose members are the subsets of S. (So $A \in \mathscr{P}(S)$ means $A \subseteq S$.)

For example, if $S = \{a, b\}$ (where $a \neq b$), then $\mathscr{P}(S)$ has 4 members, viz. $\emptyset$, S, $\{a\}$, $\{b\}$; i.e. $\mathscr{P}(S) = \{\emptyset, S, \{a\}, \{b\}\}$.

6.8 Let S be a finite set of order n. Then $|\mathscr{P}(S)| = 2^n$.

Sketch of proof. In choosing a subset of S, we have, for each of the n elements of S, 2 possibilities—inclusion in the subset or exclusion from it. Hence, in the choosing of a subset of S, the total number of possibilities is $2 \times 2 \times 2 \times \ldots$ to n factors, i.e. 2^n.

7. Unions and intersections

Consider first a sequence of n sets $A_1, A_2, \ldots, A_n$ (where $n \geqslant 2$). For clarity, we suppose that all of $A_1, A_2, \ldots, A_n$ are subsets of a universal set S, but we do not insist that the sets $A_1, A_2, \ldots, A_n$ must all be different from one another.

Definitions. (1) The **intersection** of the sets $A_1, A_2, \ldots, A_n$ is the subset of S comprising those elements that belong to all of $A_1, A_2, \ldots, A_n$. It is denoted by

$$A_1 \cap A_2 \cap \ldots \cap A_n \quad \text{or} \quad \bigcap_{r=1}^{n} A_r.$$

(2) The **union** of the sets $A_1, A_2, \ldots, A_n$ is the subset of S comprising those elements that belong to at least one of the sets $A_1, A_2, \ldots, A_n$. It is denoted by

$$A_1 \cup A_2 \cup \ldots \cup A_n \quad \text{or} \quad \bigcup_{r=1}^{n} A_r.$$

As an illustration, consider the following subsets of $\mathbb{N}$:

$$A_1 = \{1, 2, 3, 4\}, \quad A_2 = \{1, 2, 5\}, \quad A_3 = \{2, 4, 6, 8, 12\}.$$

In this case, as the student should verify to his satisfaction,

$$A_1 \cap A_2 \cap A_3 = \{2\}, \quad \text{and} \quad A_1 \cup A_2 \cup A_3 = \{1, 2, 3, 4, 5, 6, 8, 12\}.$$

It should be observed that a more formal phrasing of the above definitions brings in the quantifiers:

$$\bigcap_{r=1}^{n} A_r = \{x \in S : \forall r \in \{1, 2, \ldots, n\}, x \in A_r\},$$

$$\bigcup_{r=1}^{n} A_r = \{x \in S : \exists r \in \{1, 2, \ldots, n\} \text{ s.t. } x \in A_r\}.$$

The simplest case is the intersection or union of two sets A, B ($\subseteq S$), illustrated by the Venn diagrams below. Notice that

$$A \cap B = \{x \in S : x \in A \text{ and } x \in B\}$$
$$A \cup B = \{x \in S : x \in A \text{ or } x \in B\}$$

where, in the second line (as always in mathematics), we use the word "or" in the *inclusive* sense, i.e. we hold "P or Q" to be true whenever at least one of P, Q is true.

(i) $A \cap B$ shaded   (ii) $A \cup B$ shaded

Sometimes one has to deal with infinite families of sets. For example, one might have an infinite sequence of sets $A_1, A_2, A_3, \ldots$, or one might have a set A_x corresponding to each real number x. In all such cases the intersection and union of the family of sets are defined in the natural way: the intersection is the set consisting of the objects that belong to all of the sets in the given family; the union is the set consisting of the objects that belong to at least one of the sets in the given family.

Two further definitions. (1) Sets A, B are called **disjoint** iff $A \cap B = \emptyset$.

(2) If $\mathcal{T}$ is a collection of subsets of the set S (i.e. $\mathcal{T} \subseteq \mathcal{P}(S)$), we call $\mathcal{T}$ a **partition** of S iff the following conditions are all satisfied:

(i) the union of all the subsets belonging to $\mathcal{T}$ is the whole of S;

(ii) unequal members of $\mathcal{T}$ are disjoint;

(iii) each subset belonging to $\mathcal{T}$ is non-empty.

(Note that conditions (i) and (ii) are equivalent to the single condition that every element of S belongs to precisely one of the subsets in the collection $\mathcal{T}$.)

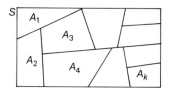

The Venn diagram illustrates the situation when $A_1, A_2, \ldots, A_k$ are different subsets forming a partition of S (i.e. the collection $\{A_1, A_2, \ldots, A_k\}$ is a partition of S).

There exist many interesting results about complements, unions and intersections of subsets, but to devote much space to them would be at variance with the basic purpose of this book. Accordingly, we shall here discuss only a few of these results, which have widespread uses or have proofs which instructively illustrate logical points described earlier in the chapter.

We consider first the *de Morgan laws* (7.1 below), which we state here for a finite family of sets, though the perceptive student will realize that, by the same approach, they can be proved true for a perfectly arbitrary family of sets.

7.1 For sets $A_1, A_2, \ldots, A_n$ (all subsets of universal set S),

$$\text{(i)} \quad \mathscr{C}\left(\bigcap_{r=1}^{n} A_r\right) = \bigcup_{r=1}^{n} (\mathscr{C}A_r),$$

$$\text{(ii)} \quad \mathscr{C}\left(\bigcup_{r=1}^{n} A_r\right) = \bigcap_{r=1}^{n} (\mathscr{C}A_r).$$

Proof. (i) We show (cf. remark (*a*) in §6) that, for $x \in S$,

$$x \in \text{left-hand side} \Leftrightarrow x \in \text{right-hand side}.$$

$$x \in \mathscr{C}\left(\bigcap_{r=1}^{n} A_r\right) \Leftrightarrow x \notin \bigcap_{r=1}^{n} A_r$$

$$\Leftrightarrow \neg \left[x \in \bigcap_{r=1}^{n} A_r \right]$$

$$\Leftrightarrow \neg \left[\forall r \in \{1, 2, \ldots, n\}, x \in A_r \right]$$

$$\Leftrightarrow \exists r \in \{1, 2, \ldots, n\} \text{ s.t. } \neg (x \in A_r) \qquad \text{(by 3.1(i))}$$

$$\Leftrightarrow \exists r \in \{1, 2, \ldots, n\} \text{ s.t. } x \in \mathscr{C}A_r$$

$$\Leftrightarrow x \in \bigcup_{r=1}^{n} \mathscr{C}A_r.$$

This establishes part (i).

Part (ii) may be proved similarly (using 3.1(ii)); or it may be deduced from (i) as follows.

Let $B_r = \mathscr{C}A_r$ $(r = 1, 2, \ldots, n)$. Then, by (i),

$$\mathscr{C}\left(\bigcap_{r=1}^{n} B_r\right) = \bigcup_{r=1}^{n} \mathscr{C}B_r, \quad \text{i.e.} \quad \mathscr{C}\left(\bigcap_{r=1}^{n} B_r\right) = \bigcup_{r=1}^{n} A_r.$$

The result now follows on taking the complement of each side.

It is clear (as the Venn diagrams on page 16 illustrate) that, for every pair of sets A, B, $A \cap B$ is a subset of A and of B, while both A and B are subsets of $A \cup B$. We now take note of a sharper result in this vein.

7.2 Let A, B be sets (subsets of universal set S). Then:

(i) $A \cap B$ is the largest subset of S contained in both A and B, in the sense that, for $C \subseteq S$, $(C \subseteq A$ and $C \subseteq B) \Rightarrow C \subseteq A \cap B$;

(ii) $A \cup B$ is the smallest subset of S containing both A and B, in the sense that, for $C \subseteq S$, $(A \subseteq C$ and $B \subseteq C) \Rightarrow A \cup B \subseteq C$.

Proof of part (ii). Suppose that $A \subseteq C$ and $B \subseteq C$ $(C \subseteq S)$.

Let x be an arbitrary element of $A \cup B$. Then x belongs to at least one of A, B. If $x \in A$, then also $x \in C$ (since $A \subseteq C$). Otherwise, $x \in B$ and so again $x \in C$ (since $B \subseteq C$). So in all cases $x \in C$.

This shows that: $x \in A \cup B \Rightarrow x \in C$, i.e. that $A \cup B \subseteq C$.

Part (ii) is now proved.

It is possible (and slightly easier) to prove part (i) by a similar argument.

Although neither part of 7.2 can be regarded as a profound result, the format of the simple proof given of part (ii) is very worthy of study by newcomers to the art of rigorous argument about sets.

A further pair of easy results that can be deduced from 7.2 is:

7.3 For sets A, B,

(i) $A \cup B = B \Leftrightarrow A \subseteq B$,   (ii) $A \cap B = B \Leftrightarrow B \subseteq A$.

Finally, we mention a simple version of the *distributive laws* for sets, viz:

7.4 For sets A, B, C,

(i) $A \cap (B \cup C) = (A \cap B) \cup (A \cap C)$,

(ii) $A \cup (B \cap C) = (A \cup B) \cap (A \cup C)$.

Recently it has become fashionable to give prominence in elementary mathematics courses to the technique of proving by Venn diagram results (such as 7.4) about three sets or subsets. An account of the technique may be found in, for example, *Algebra and Number Systems* by J. Hunter *et al.* [2]. The drawback in "proofs by Venn diagram" is that they involve an appeal (of dubious logical status) to a picture. It is left as a challenge to the student to find a proof of 7.4 that involves no such appeal to a picture.

8. Cartesian product of sets

For any positive integer n, an **n-tuple** means something of the form $(x_1, x_2, \ldots, x_n)$, i.e. a sequence of n objects $x_1, x_2, \ldots, x_n$ listed in a definite order. In particular, a 2-tuple is something of the form (x, y) and is usually called an **ordered pair.**

It must be emphasized that the order matters in an n-tuple; e.g. we look upon $(1, 2, 3)$ and $(2, 1, 3)$ as different 3-tuples. In fact, we consider two n-tuples $(x_1, x_2, \ldots, x_n)$, $(y_1, y_2, \ldots, y_n)$ to be equal if and only if $x_i = y_i$ for every i from 1 to n.

Note also that repetitions are allowed in an n-tuple; e.g. $(1, 2, 1)$ and $(1, 1, 1)$ are perfectly good 3-tuples.

Definition. Let $A_1, A_2, \ldots, A_n$ be sets. The **cartesian product** $A_1 \times A_2 \times \ldots$ $\times A_n$ of the given sets (in their given order) is the set of all n-tuples with first object belonging to A_1, second object belonging to $A_2, \ldots$, and nth object belonging to A_n. This may be expressed concisely by writing

$$A_1 \times A_2 \times \ldots \times A_n = \{(x_1, x_2, \ldots, x_n) : x_i \in A_i \quad \text{for} \quad i = 1, 2, \ldots, n\}.$$

In particular, for sets A, B, the cartesian product $A \times B$ is the set of all ordered pairs with first object in A, second object in B.

Remarks. (a) For a given set A, we usually denote the cartesian product $A \times A \times \ldots \times A (n$ factors) by A^n. For example, $\mathbb{R}^2$ is the set of all ordered pairs of real numbers.

(b) It is natural to identify $\mathbb{R}^2$ with the coordinate plane: each element (a, b) of $\mathbb{R}^2$ is thought of as a point of the coordinate plane (viz. the point with coordinates (a, b)), and $\mathbb{R}^2$, being the set of all such objects, is thought of as the entire coordinate plane.

(c) Observe that $(\frac{1}{2}, 1) \in \mathbb{R} \times \mathbb{Z}$, while $(\frac{1}{2}, 1) \notin \mathbb{Z} \times \mathbb{R}$, and therefore $\mathbb{R} \times \mathbb{Z} \neq \mathbb{Z} \times \mathbb{R}$. Thus $A \times B$ and $B \times A$ certainly may be different; and indeed it is not hard to prove that $A \times B \neq B \times A$ whenever A, B are unequal non-empty sets.

(d) Clearly, for every set A, $A \times \emptyset = \emptyset \times A = \emptyset$.

(e) It is left as an exercise for the reader to prove that:

8.1 If A, B are finite sets, then $|A \times B| = |A| \times |B|$.

Worked example. Let A, B, C be sets with $B \neq \emptyset$. Given that $A \times B \subseteq C \times C$, show that $A \subseteq C$.

Solution. Since $B \neq \emptyset$, we can introduce into the argument an element $b \in B$. We have:

$$x \in A \Rightarrow (x, b) \in A \times B$$
$$\Rightarrow (x, b) \in C \times C \quad \text{(since } A \times B \subseteq C \times C)$$
$$\Rightarrow x \in C.$$

It follows that $A \subseteq C$.

EXERCISES ON CHAPTER ONE

1. Suppose that S is a given subset of $\mathbb{R}$. Write down in simplest form the negations of the statements (i) $\forall x \in S,\ x \leqslant 3$, (ii) $\exists x \in S$ s.t. $x \leqslant 3$.

State whether each of (i), (ii) is true or false (a) when $S = \{2, 3, 4\}$, (b) when $S = [0, 3]$. Which of these four answers can be justified by citing a single value of x in S?

2. Of the statements

(i) $x < 5 \Rightarrow x \leqslant 5$ $(x \in \mathbb{R})$,   (ii) $x \leqslant 5 \Rightarrow x < 5$ $(x \in \mathbb{R})$,

(iii) $x > 5 \Rightarrow x \geqslant 5$ $(x \in \mathbb{R})$,   (iv) $x \geqslant 5 \Rightarrow x > 5$ $(x \in \mathbb{R})$,

(a) which is the converse of (i), (b) which is the contrapositive of (i), and (c) which are true and which false?

3. Suppose that x, y, z are nonzero real numbers, not all equal to each other. Show that x/y, y/z, z/x are not all equal.

4. What is the order (or cardinality) of each of the following sets:

$$S_1 = \{-1, 0, 1\}, \quad S_2 = \{x \in \mathbb{Z} : 0 < x < 6\}, \quad S_3 = \{x^2 - x : x \in S_1\},$$
$$S_4 = \{X \in \mathscr{P}(S_2) : |X| = 3\}, \quad S_5 = \mathscr{P}(\emptyset)?$$

5. Let A, B be subsets of the set S. Prove that $A - B = A \cap \mathscr{C}_S B$, and deduce (without appealing to a Venn diagram) that

$$A - (A - B) = B - (B - A) = A \cap B.$$

6. Let A, B denote subsets of the set S. Write down a simple form of the negation of "$\forall x \in A,\ x \in \mathscr{C}_S B$". Hence prove that, for $A, B \subseteq S$,

$$A \cap B = \emptyset \Leftrightarrow A \subseteq \mathscr{C}_S B.$$

Prove further that

$$\exists X \in \mathscr{P}(S) \text{ s.t. } (A \cap X) \cup (B \cap X') = \emptyset \Leftrightarrow A \cap B = \emptyset$$

(where X' denotes $\mathscr{C}_S X$).

7. For sets A, B, prove that (i) $\mathscr{P}(A) \cup \mathscr{P}(B) \subseteq \mathscr{P}(A \cup B)$ and (ii) $\mathscr{P}(A) \cap \mathscr{P}(B) = \mathscr{P}(A \cap B)$. Give an example of a pair of sets C, D such that $\mathscr{P}(C) \cup \mathscr{P}(D) \neq \mathscr{P}(C \cup D)$.

8. Let A, B be finite sets. What can you say about the relation between $|A \cup B|$ and $|A| + |B|$ (a) when A, B are disjoint, (b) when A, B are not disjoint?

9. The *symmetric difference* $A + B$ of two subsets A, B of the set S is defined by $A + B = (A - B) \cup (B - A)$.

(i) Simplify $A + \emptyset$, $A + S$, $A + A$, $A + \mathscr{C}_S A$ $(A \subseteq S)$.

(ii) Prove that, for $A, B, C \subseteq S$, $(A + B) + C = A + (B + C)$.

10. For each $n \in \mathbb{N}$, let $A_n = [0, 1/n]$. Show that the intersection of the infinite family of sets $A_1, A_2, A_3, \ldots$ is $\{0\}$.

11. Let A, B, C, D be subsets of a set S, and let $P = A \cap B \cap (C \cup D)$, $Q = (A \cap D) \cup (B \cap C)$. By using 7.4 to re-express P, prove that $P \subseteq Q$. Prove also that $P = Q$ iff both $B \cap C \subseteq A$ and $A \cap D \subseteq B$.

12. Let A, B, C, D be subsets of a set S. Show that if there is a subset X of S satisfying the simultaneous equations

$$A \cup (X \cap B) = C, \quad (A \cup X) \cap B = D, \tag{1}$$

then

$$A \cap B \subseteq D \subseteq B \quad \text{and} \quad A \cup D = C. \tag{2}$$

Prove conversely that if the conditions (2) are satisfied, then the equations (1) hold if $X = D - A$.

13. Let A, B be non-empty sets. Prove that if $A \times B = B \times A$, then $A = B$.

14. For sets A, B, C, D, prove that $(A \times B) \cap (C \times D) = (A \cap C) \times (B \cap D)$.

15. Show that, for sets P, Q, R, $P \times (Q \cup R) = (P \times Q) \cup (P \times R)$.
Deduce that, for sets A, B, C, D, $(A \times B) \cup (C \times D) \subseteq (A \cup C) \times (B \cup D)$, and give an example to show that equality need not occur.

16. How many partitions are there of the set S (i) if $|S| = 2$, (ii) if $|S| = 3$, (iii) if $|S| = 4$?

CHAPTER TWO

SOME PROPERTIES OF $\mathbb{Z}$

9. Introduction

One of our aims in this chapter is to establish certain elementary facts about the system of integers that we need in order to develop and illustrate the material of later chapters. These facts include simple results on congruences (introduced in §14) and, on a more fundamental level, the well-ordering principle (an absolutely basic property of $\mathbb{Z}$, explained in detail in §10) and the division algorithm (§11). The last of these is the fact, suggested by our experience of elementary arithmetic, that, given an integer a and a divisor d ($\in \mathbb{N}$), we can, in exactly one way, express a in the form

$$a = (\text{QUOTIENT}) \times d + (\text{REMAINDER})$$

where the quotient and remainder are integers and the remainder lies in the range 0 to $d-1$, inclusive.

This chapter also includes proof of the very straightforward factorization properties of the integers (see §13). The student is advised to note particularly how the proof given of these properties depends crucially on the important result 12.3—the fact that the highest common factor of two positive integers a, b can be expressed in the form $x_1 a + x_2 b$ (with $x_1, x_2 \in \mathbb{Z}$). In abstract algebra, this material about $\mathbb{Z}$ motivates fruitful exploration of factorization in a more general framework, where one discovers, in particular, that many other algebraic systems have nice factorization properties like those of $\mathbb{Z}$. We return to this topic in the final chapter of the book.

10. The well-ordering principle

We first introduce some terminology. In the definitions which follow, S will denote a subset of $\mathbb{R}$.

22

Definition 1. We say that S is **bounded below** iff $\exists\, k \in \mathbb{R}$ such that

$$\forall x \in S, \quad k \leqslant x. \tag{α}$$

And when S is bounded below, each number k that satisfies (α) is called a **lower bound** of S.

   [For example, the interval $U = (1, \infty)$ is bounded below; -7, 1, $\frac{1}{2}$ are examples of lower bounds of U; the set of all lower bounds of U is $(-\infty, 1]$.]

Definition 2. We say that S is **bounded above** iff $\exists\, K \in \mathbb{R}$ such that

$$\forall x \in S, \quad K \geqslant x. \tag{β}$$

And when S is bounded above, each number K satisfying (β) is called an **upper bound** of S.

   [For example, the interval $V = (-\infty, 2]$ is bounded above; the set of all the upper bounds of V is $[2, \infty).$]

   We are now in a position to state one form of the well-ordering principle, namely:

10.1 Every non-empty set of integers that is bounded below has a least member, i.e. a member which is less than or equal to every member of the set.

   A dual version is:

10.2 Every non-empty set of integers that is bounded above has a greatest member, i.e. a member which is greater than or equal to every member of the set.

   It should be noted that a non-empty subset of $\mathbb{R}$ that is bounded below need not have a least member: e.g. the interval $(1, \infty)$, though bounded below, clearly does not have a least member. This observation highlights the fact that 10.1 says something distinctive and non-trivial about the integers (as does 10.2).

   It is also true that the well-ordering principle is a fundamental property of $\mathbb{Z}$, in the sense that no description of $\mathbb{Z}$ is adequate unless it incorporates the well-ordering principle or something equivalent to it (e.g. the principle of induction). Accordingly, any so-called proof of the well-ordering principle is bound to be misleading to the extent that it must somewhere appeal to an assumption that is no more fundamental than the well-ordering principle. [Students familiar with the principle of induction can gain some sense of perspective in these matters by thinking out how the well-ordering principle can be proved from the principle of induction and vice versa.]

   In a number of proofs we shall make use of the following consequence of 10.1 which may be termed the principle of the least counterexample.

10.3 Suppose that $P(n)$ is a condition involving the variable n, whose domain is the set $\{n \in \mathbb{Z} : n \geqslant l\}$, l being a fixed integer. Suppose further that the universal statement "$\forall n \geqslant l$, $P(n)$" is false. Then there is a least counter-example, i.e. a lowest integral value ($\geqslant l$) of n for which $P(n)$ is false.

Proof. Let S be the set of all counterexamples to the universal statement, i.e. the set of all values ($\geqslant l$) of n for which $P(n)$ is false. Then, since the universal statement is false, $S \neq \emptyset$. Moreover, S consists of integers and is bounded below (all its members being $\geqslant l$). By 10.1, it follows that S has a least member; and this proves the result.

Integral part of a real number

Let $x \in \mathbb{R}$. We shall take it for granted that there exist integers which are $\leqslant x$. So the set of all such integers is non-empty and is bounded above (x being an upper bound). Accordingly (by 10.2) there is a greatest integer less than or equal to x: this integer is called the **integral part** of x and is denoted by $[x]$, e.g.

$$[\tfrac{5}{2}] = 2, \quad [-\tfrac{5}{2}] = -3, \quad [14] = 14, \quad [\sqrt{10}] = 3.$$

Certain things follow immediately from the definition:

10.4 Let $x \in \mathbb{R}$. Then:
  (i) $[x] \leqslant x < [x]+1$;
  (ii) $x - [x] \in [0,1)$;
  (iii) if $x = n+t$, where $n \in \mathbb{Z}$ and $t \in [0,1)$, then $n = [x]$.

11. The division algorithm

The formal statement of this result, briefly previewed in §9, is:

11.1 Let a, d be integers, with $d > 0$. Then there exist unique integers q, r such that

$$a = qd + r \quad \text{and} \quad 0 \leqslant r \leqslant d-1. \tag{*}$$

Proof. Part (i)—the existence of q, r. Here we shall produce integers q, r satisfying (*).

Let $q = \left[\dfrac{a}{d}\right]$ and $t = \dfrac{a}{d} - \left[\dfrac{a}{d}\right]$, so that $q \in \mathbb{Z}$ and $t \in [0,1)$. Since $a/d = q+t$, we have

$$a = qd + r \tag{**}$$

where $r = td$. Since a, q, $d \in \mathbb{Z}$, it follows from (**) that r ($= a - qd$) is an integer too. Further, since $0 \leqslant t < 1$ and $d > 0$, it follows that $0 \leqslant td < d$, i.e. $0 \leqslant r < d$; and hence, since $r \in \mathbb{Z}$, $0 \leqslant r \leqslant d-1$.

We have thus produced q, r with the stated properties.

Part (ii)—the uniqueness of q, r.

Suppose that $a = qd + r$, where $q, r \in \mathbb{Z}$ and $0 \leqslant r \leqslant d - 1$. We have to prove that these suppositions uniquely determine q, r in terms of a and d. We have

$$\frac{a}{d} = q + \frac{r}{d} \quad \text{and} \quad 0 \leqslant \frac{r}{d} \leqslant \frac{d-1}{d} < 1.$$

Hence (cf. 10.4(iii)) $q = \left[\dfrac{a}{d}\right]$, and so $r = a - d\left[\dfrac{a}{d}\right]$.

The result follows.

In the above result, we call r the **principal remainder** on division of a by d.

If and only if $r = 0$, i.e. iff a is an integral multiple of d, we say that d **divides** a (or, alternatively, d is a **factor** of a). We often use a short vertical line to stand for the verb "divides", so that

$$d|a, \quad d \nmid a$$

are the standard shorthands for, respectively, "d divides a" and "d does not divide a". So, for example, $5|(-30)$ and $4 \nmid 19$.

Simple results on this theme include:

11.2 If $d|a$, where $d \in \mathbb{N}$ and a is a nonzero integer, then $|a| \geqslant d$.

11.3 If $d|e$ and $e|f$, where $d, e, f \in \mathbb{N}$, then $d|f$.

11.4 If $d|a$ and $d|b$, where $d, a, b \in \mathbb{Z}$ and $d > 0$, then $d|(a+b)$ and $d|(a-b)$ and, more generally, $d|(x_1 a + x_2 b)$ for every $x_1, x_2 \in \mathbb{Z}$.

The proofs of these results are straightforward. As an indication of method of attack, here is a proof of 11.4.

Proof. Suppose that $d|a$ and $d|b$ $(d, a, b \in \mathbb{Z}, d > 0)$. Let x_1, x_2 be arbitrary integers.

Since $d|a$ and $d|b$, $a = kd$ and $b = ld$ for some integers k, l. Hence

$$x_1 a + x_2 b = x_1 kd + x_2 ld = (x_1 k + x_2 l)d.$$

Since $x_1 k + x_2 l \in \mathbb{Z}$, it follows that $d|(x_1 a + x_2 b)$.

This proves 11.4.

12. Highest common factors and Euclid's algorithm

Let $a_1, a_2, \ldots, a_n$ be a finite sequence of positive integers. Each positive integer which divides all of $a_1, a_2, \ldots, a_n$ is called a **common factor** of $a_1, a_2, \ldots, a_n$ (e.g. 3 is a common factor of 12, 18, 30). In the general case there is certainly at least one common factor of $a_1, a_2, \ldots, a_n$, viz. 1. On the other hand, every such common factor is, in particular, a factor of a_1 and so (cf. 11.2) is not greater than a_1. Clearly, therefore, the set of common factors of $a_1, a_2, \ldots, a_n$ has a greatest member, which is termed the **highest common factor** of $a_1, a_2, \ldots, a_n$

and denoted by

$$(a_1, a_2, \ldots, a_n) \quad \text{or} \quad \text{hcf}(a_1, a_2, \ldots, a_n).$$

We call two positive integers a, b **relatively prime** iff their highest common factor is 1, e.g. the integers 6 and 25 are relatively prime (even though neither is a prime number). [The student should know that a **prime number** means an integer $p \geqslant 2$ whose only factors (in $\mathbb{N}$) are 1 and p. An integer greater than 1 that is not prime is termed **composite.**]

It is easily seen that:

12.1 If p is a prime number and a is a positive integer such that $p \nmid a$, then p and a are relatively prime.

We now focus attention on the determination of $\text{hcf}(a_1, a_2)$, where a_1, a_2 are given positive integers and where, let us suppose, $a_1 \geqslant a_2$. The highest common factor may be found systematically by a process called *Euclid's algorithm*, exhibited below. This process, it will be seen, gives a very useful by-product—the fact that the highest common factor of a_1 and a_2 can be expressed as an integral linear combination of a_1 and a_2, i.e. in the form $x_1 a_1 + x_2 a_2$ with x_1, x_2 integers.

Starting from the given positive integers a_1, a_2, we can make repeated use of the division algorithm to obtain (all symbols denoting integers):

$$a_1 = q_1 a_2 + a_3, \quad \text{where} \quad 0 \leqslant a_3 < a_2 (\leqslant a_1);$$

$$\text{and, if } a_3 \neq 0, \quad a_2 = q_2 a_3 + a_4, \quad \text{where} \quad 0 \leqslant a_4 < a_3 < a_2 \leqslant a_1;$$

$$\text{and, if } a_4 \neq 0, \quad a_3 = q_3 a_4 + a_5, \quad \text{where} \quad 0 \leqslant a_5 < a_4 < a_3 < a_2 \leqslant a_1;$$

$$\text{etc.}$$

Since there is only a finite number of integers in the interval $[0, a_1]$, we must, after some finite number of steps in the process, obtain zero remainder. Thus we obtain a finite sequence of equations

$$a_1 = q_1 a_2 + a_3 \tag{1}$$

$$a_2 = q_2 a_3 + a_4 \tag{2}$$

$$a_3 = q_3 a_4 + a_5 \tag{3}$$

$$\cdots$$

$$a_{r-3} = q_{r-3} a_{r-2} + a_{r-1} \tag{$r-3$}$$

$$a_{r-2} = q_{r-2} a_{r-1} + a_r \tag{$r-2$}$$

$$a_{r-1} = q_{r-1} a_r \tag{$r-1$}$$

12.2 In this process a_r (which is the last nonzero remainder, except in the trivial case $a_2 | a_1$) equals $\text{hcf}(a_1, a_2)$.

Proof. The result is trivial if $a_2|a_1$ (i.e. $r = 2$). Henceforth we deal with the non-trivial case $a_2 \nmid a_1$ (so that $r \geqslant 3$).

Let $d = \mathrm{hcf}(a_1, a_2)$. Using the general version of 11.4, we reason progressively as follows.

By (1), since $d|a_1$ and $d|a_2$, d also divides a_3 $(= a_1 - q_1 a_2)$; hence, by (2), since $d|a_2$ and $d|a_3$, d also divides a_4 $(= a_2 - q_2 a_3)$; etc; and eventually, at the $(r-2)$th such step: hence, by $(r-2)$, since $d|a_{r-2}$ and $d|a_{r-1}$, d also divides a_r.

Consequently, by 11.2, $a_r \geqslant d$.

But, using the equations in reverse order, we can argue thus:

by $(r-1)$, $a_r|a_{r-1}$;

hence, by $(r-2)$, a_r also divides a_{r-2};

hence, by $(r-3)$, since $a_r|a_{r-1}$ and $a_r|a_{r-2}$, a_r also divides a_{r-3};

etc; and eventually:

hence, by (2), since $a_r|a_4$ and $a_r|a_3$, a_r also divides a_2;

hence, by (1), since $a_r|a_3$ and $a_r|a_2$, a_r also divides a_1.

By the last two conclusions, a_r is a common factor of a_1 and a_2, and so $a_r \leqslant d$ (d being the highest such common factor).

Since it has already been proved that $a_r \geqslant d$, it follows that $a_r = d$, which is the stated result.

At this stage it is instructive to work through a specific numerical example. Let us consider $\mathrm{hcf}(2190, 465)$.

Following the procedure described, we successively obtain

$$2190 = 4.465 + 330 \tag{1}$$
$$465 = 1.330 + 135 \tag{2}$$
$$330 = 2.135 + 60 \tag{3}$$
$$135 = 2.60 + 15 \tag{4}$$
$$60 = 4.15 \tag{5}$$

By 12.2, we conclude that $\mathrm{hcf}(2190, 465) = 15$.

More information lies close at hand. By using the equations (4), (3), (2), (1) in that order, we can express the highest common factor 15 as an integral linear combination of 2190 and 465. We proceed as follows.

$15 = 135 - 2.60$ (by (4))

   $= 135 - 2(330 - 2.135)$ (using (3) to replace the smaller of 60, 135 in the previous line)

   $= 5.135 - 2.330$

   $= 5(465 - 330) - 2.330$ (using (2) to replace the smaller of 135, 330 in the previous line)

   $= 5.465 - 7.330$

   $= 5.465 - 7(2190 - 4.465)$ (using (1) to replace the smaller of 330, 465 in the previous line)

   $= 33.465 - 7.2190.$

Now consider again an arbitrary pair of positive integers a_1, a_2 with $a_1 \geqslant a_2$. Provided $a_2 \nmid a_1$, it is evident that the process demonstrated above can be carried out to yield an expression for $\mathrm{hcf}(a_1, a_2)$ as an integral linear combination of a_1, a_2; and in the trivial case $a_2 | a_1$ we of course have

$$\mathrm{hcf}(a_1, a_2) = a_2 = 0a_1 + 1a_2$$

Therefore:

12.3 If $a, b \in \mathbb{N}$ and $d = \mathrm{hcf}(a, b)$, then $\exists x_1, x_2 \in \mathbb{Z}$ such that

$$d = x_1 a + x_2 b.$$

This theorem has important consequences, including the results 12.4, 12.5 below.

12.4 Let $a, b \in \mathbb{N}$, let $d = \mathrm{hcf}(a, b)$, and let c be any common factor of a and b. Then $c | d$.

Proof. By 12.3, $d = x_1 a + x_2 b$ for some $x_1, x_2 \in \mathbb{Z}$. Since $c | a$ and $c | b$, it follows by 11.4 that $c | d$.

It is of interest that 12.4 shows that the highest common factor of two integers is supreme among the common factors of the two integers in two senses. As well as being supreme in the sense of numerically greatest, it is also supreme in the sense of being divisible by every common factor. This observation gives a clue as to how one can introduce the concept of "highest common factor" in other algebraic systems where "numerically greatest" would be meaningless.

12.5 Suppose that the prime number p divides the product ab, where $a, b \in \mathbb{N}$. Then $p | a$ or $p | b$ (possibly both).

Proof. Suppose that $p \nmid a$. (It will suffice to prove that $p | b$ follows from this assumption.)

By 12.1, $\mathrm{hcf}(p, a) = 1$ and hence, by 12.3, there are integers x_1, x_2 such that $x_1 p + x_2 a = 1$. Hence we have

$$b = b \times 1 = b(x_1 p + x_2 a) = (bx_1)p + x_2(ab),$$

which (by 11.4) is divisible by p (since $p | ab$).

Thus $p | b$, and the result follows.

A generalization of 12.5 is obtainable by repeated application:

12.6 If the prime number p divides the product $a_1 a_2 \ldots a_k$ (where each $a_i \in \mathbb{N}$), then p divides at least one of $a_1, a_2, \ldots, a_k$.

Proof. $p|a_1 a_2 a_3 \ldots a_k \Rightarrow p|a_1$ or $p|a_2 a_3 \ldots a_k$ (by 12.5)

                      $\Rightarrow p|a_1$ or $p|a_2$ or $p|a_3 \ldots a_k$ (by 12.5)

                      . . .

                      $\Rightarrow p|a_1$ or $p|a_2$ or $p|a_3$ or $\ldots$ or $p|a_k$,

after altogether $k-1$ applications of 12.5.

The basic results 12.3 and 12.4 are also capable of generalization. We state the more general results without proof.

12.7 Let $a_1, a_2, \ldots, a_n$ be positive integers, and let d be their highest common factor. Then:
    (i) d can be expressed as $x_1 a_1 + x_2 a_2 + \ldots + x_n a_n$ for some $x_1, x_2, \ldots, x_n \in \mathbb{Z}$;
    (ii) every common factor of $a_1, a_2, \ldots, a_n$ is a divisor of d.

13. The fundamental theorem of arithmetic

The *fundamental theorem of arithmetic* is the name given to the twofold assertion that, for every integer $n \geqslant 2$,
    (i) n can be expressed as a product of one or more primes, and
    (ii) there is essentially only one expression for n as a product of primes, i.e. any two such expressions for n can differ only in the order in which the prime factors appear.
    (Note that we allow the simple case in which there is just one factor in the "product". Indeed if p is a prime number, the one and only way to express p as a product of primes is to write it as the one-factor product p.)

The unique expression for the integer n ($\geqslant 2$) as a product of primes is called the **prime decomposition** of n. Examples of prime decompositions include:

$$504 = 2 \times 2 \times 2 \times 3 \times 3 \times 7, \quad 16 = 2 \times 2 \times 2 \times 2, \quad 17 = 17,$$
$$1978 = 2 \times 23 \times 43, \quad 1988 = 2 \times 2 \times 7 \times 71.$$

The results 13.1 and 13.2, which follow, establish respectively parts (i) and (ii) of the fundamental theorem (as stated above).

13.1 For every integer $n \geqslant 2$, n can be expressed as a product of primes.

Proof. With a view to obtaining a contradiction, suppose the result false. Then, by 10.3, there is a least counterexample integer m ($\geqslant 2$) which cannot be expressed as a product of primes. As a prime number is, in a trivial way, expressible as a product of primes, it is clear that m cannot be prime. Hence $m = kl$ for some $k, l \in \mathbb{N}$ satisfying $1 < k, l < m$. Because m is the least counterexample, both k and l are expressible as products of primes: say $k = p_1 p_2 \ldots p_r$, $l = q_1 q_2 \ldots q_s$ (all of $p_1, \ldots, q_s$ being primes). Hence we have

$$m = p_1 p_2 \ldots p_r q_1 q_2 \ldots q_s,$$

giving an expression for the supposed counterexample m as a product of primes. This is a contradiction, and from it the result follows.

13.2 For every integer $n \geqslant 2$, any two expressions for n as products of primes differ only in the order of the factors.

Proof. Suppose, with a view to obtaining a contradiction, that the result is false. Then, by 10.3, there is a least counterexample integer m which can be expressed as a product of primes in two genuinely different ways, say

$$m = p_1 p_2 \ldots p_r = q_1 q_2 \ldots q_s,$$

where $p_1, \ldots, p_r, q_1, \ldots, q_s$ are all primes. Clearly m itself is not prime and so r, s $\geqslant 2$. Without loss of generality we can insist that, in each of the prime products equal to m, the prime factors are arranged in ascending order, i.e. $p_1 \leqslant p_2 \leqslant \ldots$ $\leqslant p_r$ and $q_1 \leqslant q_2 \leqslant \ldots \leqslant q_s$; and also that $p_1 \leqslant q_1$.

We note that $p_1 \neq q_1$; for, if p_1 were equal to q_1, we would have

$$\frac{m}{p_1} = p_2 p_3 \ldots p_r = q_2 q_3 \ldots q_s$$

giving two genuinely different prime product expressions for the integer m/p_1, which, however, is less than the least counterexample m.

Hence $p_1 < q_1$, and so, since $q_1 \leqslant q_2 \leqslant \ldots \leqslant q_s$, p_1 is less than q_j for each j. Now each q_j is prime and so, since $1 < p_1 < q_j$, we deduce that, for each j, $p_1 \nmid q_j$.

But p_1 does divide m, which equals $q_1 q_2 \ldots q_s$, and so, by 12.6, p_1 must divide at least one of $q_1, q_2, \ldots, q_s$.

We thus have a contradiction, and the stated result follows.

14. Congruence modulo m ($m \in \mathbb{N}$)

When we face a question like, "If it is now 8 o'clock, what time will it be 30 hours from now?", we find ourselves wanting to ignore integral multiples of 12. This is a somewhat trivial example of a host of reasons why it is profitable to have a special phraseology and notation for indicating that two integers differ by an integral multiple of some given integer m. That is basically what this section is about. Throughout the general discussion, m will denote a fixed positive integer.

For $a, b \in \mathbb{Z}$, we write

$$a \equiv b \,(\text{mod } m)$$

(read "a is *congruent* to b modulo m") to mean $m|(b-a)$, i.e. $b-a$ is an integral multiple of m, i.e. $b = a + km$ for some integer k.

So, for example, it is true that $13 \equiv 5 \,(\text{mod } 4)$ and $9 \equiv -1 \,(\text{mod } 5)$, while, equally clearly, $10 \not\equiv 2 \pmod 6$, i.e. 10 is not congruent to 2 modulo 6.

It is useful at the outset to note the simple information in the following results 14.1 and 14.2.

14.1 (i) For every integer a, $a \equiv a \pmod m$.

(ii) $a \equiv b \pmod m \Rightarrow b \equiv a \pmod m$ $(a, b \in \mathbb{Z})$.

(iii) $a \equiv b \pmod m$ and $b \equiv c \pmod m \Rightarrow a \equiv c \pmod m$ $(a, b, c \in \mathbb{Z})$.

Proof. (i) is true because, for every $a \in \mathbb{Z}$, $a - a$ is an integral multiple of m (viz. $0 \times m$).

(ii) For $a, b \in \mathbb{Z}$,

$$a \equiv b \pmod m \Rightarrow b = a + km \quad (k \text{ being an integer})$$
$$\Rightarrow a - b = (-k)m$$
$$\Rightarrow m|(a-b) \quad (\text{since } -k \text{ is an integer})$$
$$\Rightarrow b \equiv a \pmod m.$$

(iii) For $a, b, c \in \mathbb{Z}$,

$$a \equiv b \pmod m \text{ and } b \equiv c \pmod m \Rightarrow m|(b-a) \text{ and } m|(c-b)$$
$$\Rightarrow m|\{(b-a)+(c-b)\} \quad (\text{by } 11.4)$$
$$\Rightarrow m|(c-a)$$
$$\Rightarrow a \equiv c \pmod m.$$

14.2 (i) If the integer a has principal remainder r on division by m, then $a \equiv r \pmod m$.

(ii) Each integer is congruent (modulo m) to precisely one of the integers $0, 1, 2, \ldots, m-1$.

Proof. (i) If a has principal remainder r on division by m, then

$$a = qm + r \text{ for some integer } q,$$

and hence $a \equiv r \pmod m$.

(ii) Let a be an arbitrary integer. It follows at once from (i) that a is congruent (modulo m) to at least one of the integers in the interval $[0, m-1]$. If a were congruent (modulo m) to two different integers r_1, r_2 in this interval, we would have

$$a = q_1 m + r_1 = q_2 m + r_2 \text{ for some } q_1, q_2 \in \mathbb{Z}$$

—an impossibility in view of the uniqueness clause in the division algorithm (11.1). The result follows.

One of the advantages of handling information like "$m|(b-a)$" in the form "$a \equiv b \pmod m$" is that sentences of the latter form can be treated to a great extent like equations. The next theorem amplifies this claim.

14.3 Suppose that $a \equiv b \pmod{m}$ and that $c \equiv d \pmod{m}$, where $a, b, c, d \in \mathbb{Z}$. Then:

   (i) $a + c \equiv b + d \pmod{m}$;

   (ii) $a - c \equiv b - d \pmod{m}$;

   (iii) $ac \equiv bd \pmod{m}$.

  Proof. Since $a \equiv b \pmod{m}$ and $c \equiv d \pmod{m}$,

$$b = a + km \quad \text{and} \quad d = c + lm, \quad \text{for some } k, l \in \mathbb{Z}.$$

  (i) $(b + d) - (a + c) = (a + km + c + lm) - (a + c) = (k + l)m$. Since $k + l \in \mathbb{Z}$, it follows that $a + c \equiv b + d \pmod{m}$.

  (ii) is proved similarly by considering $(b - d) - (a - c)$.

  (iii) $bd = (a + km)(c + lm) = ac + (kc + al + klm)m$. Therefore, since $kc + al + klm \in \mathbb{Z}$, $ac \equiv bd \pmod{m}$.

  The conclusions (i), (ii), (iii) in 14.3 may be described as the results of (respectively) adding, subtracting, and multiplying the given congruences (i.e. the given statements of the form $x \equiv y \pmod{m}$).

  Notice that certain things can be deduced via 14.3(iii) from one given congruence $a \equiv b \pmod{m}$. By multiplying the given congruence by the clearly true congruence $c \equiv c \pmod{m}$ [c being an arbitrary integer], we deduce that $ca \equiv cb \pmod{m}$. On another line of attack, by multiplying the given congruence by itself, we deduce that $a^2 \equiv b^2 \pmod{m}$; then, multiplying this by the given congruence, we deduce $a^3 \equiv b^3 \pmod{m}$; and, by continuing in this way, we can deduce $a^n \equiv b^n \pmod{m}$ for any specified positive integer n. To sum up:

14.4 If $a, b \in \mathbb{Z}$ and $a \equiv b \pmod{m}$, then

   (i) $\forall c \in \mathbb{Z}, ca \equiv cb \pmod{m}$,

   (ii) $\forall n \in \mathbb{N}, a^n \equiv b^n \pmod{m}$.

Worked examples

  1. Show that, for every $n \in \mathbb{Z}, n^2 \not\equiv 2$ and $n^2 \not\equiv 3 \pmod{5}$. [Note: this is a fact familiar to the observant—that (in the customary notation for integers) the square of an integer never ends in 2, 3, 7, or 8.]

  Solution. Let n be an arbitrary integer. By 14.2(ii),

$$n \equiv 0 \text{ or } n \equiv 1 \text{ or } n \equiv 2 \text{ or } n \equiv 3 \text{ or } n \equiv 4 \pmod{5}.$$

Hence, by 14.4(ii),

$$n^2 \equiv 0 \text{ or } n^2 \equiv 1 \text{ or } n^2 \equiv 4 \text{ or } n^2 \equiv 9 \text{ or } n^2 \equiv 16 \pmod{5}.$$

Since $9 \equiv 4$ and $16 \equiv 1 \pmod{5}$, it follows [by 14.1(iii)] that, in every case, $n^2 \equiv 0$ or 1 or 4 $\pmod{5}$.

By 14.2(ii), the result follows.

2. Using the result of example 1, show that there do not exist integers x, y satisfying $7x^2 - 15y^2 = 1$.

Solution. Suppose the contrary, so that $7x^2 - 15y^2 = 1$ for some integers x, y. Then (since $15y^2$ is an integral multiple of 5)

$$7x^2 \equiv 1 \pmod 5.$$

Since $7 \equiv 2 \pmod 5$, it follows (by 14.4(i)) that $7x^2 \equiv 2x^2 \pmod 5$ and hence (by 14.1) that $2x^2 \equiv 1 \pmod 5$.

But, by example 1, $x^2 \equiv 0$ or 1 or 4 (mod 5) and hence $2x^2 \equiv 0$ or 2 or 8 (mod 5), i.e. $2x^2 \equiv 0$ or 2 or 3 (mod 5).

In view of 14.2(ii), we now have contradictory information about the integer $2x^2$. From this contradiction the result follows.

EXERCISES ON CHAPTER TWO

1. Let $x \in \mathbb{R}$. Prove that $[x] + [-x]$ equals 0 if $x \in \mathbb{Z}$ and equals -1 if $x \notin \mathbb{Z}$.

2. Suppose that a, b are relatively prime positive integers and that the positive integer c is a factor of $a + b$. Prove that $\mathrm{hcf}(a, c) = \mathrm{hcf}(b, c) = 1$.

3. Show that, $\forall n \in \mathbb{N}$, $\mathrm{hcf}(9n + 8, 6n + 5) = 1$.

4. Let a, b, c denote positive integers, and let d denote $\mathrm{hcf}(a, b)$. Prove that
$$\exists x, y \in \mathbb{Z} \text{ s.t. } ax + by = c \Leftrightarrow d | c.$$

5. Suppose that d, a, b are positive integers, that a and d are relatively prime, and that $d | ab$. Prove that $d | b$.

6. Suppose that d_1, d_2 are relatively prime positive integers such that $d_1 | a$ and $d_2 | a$, a being a positive integer. Use 12.3 to prove that $d_1 d_2 | a$.

7. Show that $n \equiv 7 \pmod{12} \Rightarrow n \equiv 3 \pmod{4}$ $(n \in \mathbb{Z})$.

8. Show that every positive integer (when written in the customary notation) is congruent modulo 9 to the sum of its digits. Deduce a simple test for divisibility by 9.

9. Show that, $\forall n \in \mathbb{Z}$, $n^2 \equiv 0$ or $1 \pmod 3$. Show further that
$$3 | (x^2 + y^2) \Rightarrow 3 | x \text{ and } 3 | y \qquad (x, y \in \mathbb{Z}).$$
Deduce that there is no solution in integers of the equation
$$x^2 + y^2 = 3z^2$$
except $x = y = z = 0$.

10. Show that, $\forall n \in \mathbb{N}$, $3^{3n+1} \equiv 3.5^n \pmod{11}$. Obtain a corresponding result about 2^{4n+3} and deduce that, $\forall n \in \mathbb{N}$, $11 | (3^{3n+1} + 2^{4n+3})$.

11. Use Euclid's algorithm to find an integer x such that $31x \equiv 1 \pmod{56}$.

12. Show that: $a \equiv b \pmod m \Rightarrow \mathrm{hcf}(a, m) = \mathrm{hcf}(b, m)$ $(a, b, m \in \mathbb{N})$.

13. An integer m is called a *perfect square* iff $\exists n \in \mathbb{Z}$ such that $m = n^2$, i.e. iff $m \in \{1, 4, 9, 16, 25, \ldots\}$. Suppose that the integer r is a perfect square and that $r = st$, where s, t are relatively prime positive integers. Prove that s and t are perfect squares.

14. Prove that: $\sqrt{n} \in \mathbb{Q} \Rightarrow \sqrt{n} \in \mathbb{N}$ $(n \in \mathbb{N})$.

15. Prove that the set of prime numbers is infinite. [Suggested method: Suppose the contrary, and let $p_1, p_2, \ldots, p_n$ be a complete list of all the primes. Obtain a contradiction by showing that no prime divides the integer $p_1 p_2 \ldots p_n + 1$.]
Prove by a similar approach that there are infinitely many primes that are congruent to 3 (mod 4).

CHAPTER THREE

EQUIVALENCE RELATIONS AND-
EQUIVALENCE CLASSES

15. Relations in general

Underlying the discussion which we now begin is the widespread occurrence in mathematics of sentences of the form

x is related [in some particular way] to y

in which x, y denote elements of some given set S. Examples in the case $S = \mathbb{Z}$ include:

$$x = y; \quad x < y; \quad x \equiv y \,(\text{mod } 5).$$

All such sentences can be written in mathematical shorthand as

$$xRy$$

where R stands for "is related [in some particular way] to": e.g. R could stand for "equals" or "is less than" or "is congruent modulo 5 to".

Naturally, if in any specific context we are going to employ such a shorthand, we must make clear what exactly R stands for. When this has been made fully clear, i.e. when, for any given a, b belonging to the set S, it is possible to decide whether "aRb" is true or false, we say that R is a **relation** on S. So, for instance,

$$= [\text{equality}]$$
$$<$$
$$\equiv (\text{mod } 5) \quad [\text{congruence modulo 5}]$$

are all examples of relations on $\mathbb{Z}$.

When R is a relation on S and x, $y \in S$, a useful notation for the negation of "xRy" is

$$xR'y.$$

Of course, symbols other than R (e.g. ρ, $\sim$) may be used for a relation in general discussions.

It is possible to give a more formal definition of "a relation on S", S being a set. Newcomers to the topic are advised, for the time being at least, to ignore this possibility. However, some students will wish to know that in the formal approach one begins by defining "a relation R on S" to be a subset of $S \times S$; then one introduces "xRy" as an alternative notation for $(x, y) \in R$. It is left to interested students to think out why this approach makes sense and what advantages it has over the informal approach.

The next two sections of the chapter are devoted to the study of equivalence relations, i.e. relations with certain special properties. Knowledge gained in these sections is applied later in the chapter to develop a very significant seam of theory that arises from the relation of congruence (modulo a given positive integer m) on $\mathbb{Z}$.

16. Equivalence relations

Throughout this section, S will denote a set and R will denote a relation on S.

We begin by detailing some properties that R may possess.

(*a*) We say that R is **reflexive** (or R possesses the reflexive property) iff $\forall x \in S$, xRx.

(*b*) We say that R is **symmetric** (or R possesses the symmetric property) iff

$$xRy \Rightarrow yRx \qquad (x, y \in S).$$

(*c*) We say that R is **transitive** (or R possesses the transitive property) iff

$$xRy \text{ and } yRz \Rightarrow xRz \qquad (x, y, z \in S).$$

[Note: when R is symmetric, it is true that

$$xRy \Leftrightarrow yRx \qquad (x, y \in S);$$

for the symmetric property tells us that, if a, $b \in S$ and one of the statements "aRb", "bRa" is true, then so must be the other.]

The main definition of this chapter is:

The relation R on the set S is called an **equivalence relation** iff it is reflexive, symmetric, and transitive.

Obviously "$=$" (equality) is an equivalence relation on S (whatever the set S is). Many less trivial examples are also encountered in elementary mathematics: e.g. in geometry, congruence is an equivalence relation on the set of triangles in a plane. Further, as the result 14.1 tells us:

16.1 For each $m \in \mathbb{N}$, congruence modulo m is an equivalence relation on $\mathbb{Z}$.

Worked example. Let S be the set of all $n \times n$ matrices with real entries. A relation $\sim$ (called *similarity*) is defined on S by

$$X \sim Y \Leftrightarrow \exists \text{ nonsingular } P \in S \text{ s.t. } Y = P^{-1}XP \qquad (X, Y \in S).$$

Prove that $\sim$ is an equivalence relation on S.

Remark. Students must beware of mis-interpreting "$\exists$" here. It is important to understand that, if $A, B \in S$, "$A \sim B$" means that a true statement can be made by inserting some nonsingular matrix in S (depending on what exactly A, B are) into both blanks in the sentence "$B = (..)^{-1}A(..)$"; but the definition of $\sim$ does *not* say that there is a special matrix P that can be used to fill the blanks in all cases.

Solution. (1) For each $X \in S$, $X \sim X$ because $X = I^{-1}XI$ (and I is a nonsingular member of S). Thus $\sim$ is reflexive.
(2) For $X, Y \in S$,

$$X \sim Y \Rightarrow Y = P^{-1}XP \qquad (P \text{ being a nonsingular member of } S)$$
$$\Rightarrow X = PYP^{-1}$$
$$\Rightarrow X = Q^{-1}YQ \qquad (\text{where } Q = P^{-1})$$
$$\Rightarrow Y \sim X \qquad (\text{since } Q \text{ is a nonsingular member of } S).$$

Thus $\sim$ is symmetric.
(3) For $X, Y, Z \in S$,

$$X \sim Y \text{ and } Y \sim Z \Rightarrow Y = P^{-1}XP \quad \text{and} \quad Z = Q^{-1}YQ$$
$$(P, Q \text{ being nonsingular members of } S)$$
$$\Rightarrow Z = Q^{-1}P^{-1}XPQ$$
$$\Rightarrow Z = T^{-1}XT \qquad (\text{where } T = PQ)$$
$$\Rightarrow X \sim Z \qquad (\text{since } T \text{ is a nonsingular member of } S).$$

Thus $\sim$ is transitive.
It now follows that $\sim$ is an equivalence relation on S.

17. Equivalence classes

Throughout this section, R will denote an equivalence relation on the set S.

Definition. Let x be an arbitrary element of S. Each element t of S that satisfies tRx (or equivalently xRt) is called an **R-relative** of x. And the set of all R-relatives of x is called the **R-class** of x (or the **equivalence class** of x *under R*). We shall denote it by C_x. Thus

$$C_x = \{t \in S : tRx\}.$$

For example, if $S = \mathbb{Z}$ and R is congruence modulo 5, then

$$C_0 = \{0, 5, -5, 10, -10, 15, -15, 20, \ldots\}.$$

It is essential to have a thorough understanding of the basic properties of equivalence classes before attempting to study in depth abstract algebra (or indeed any other part of advanced pure mathematics). The rest of this section is mainly devoted to establishing these basic properties, including (17.5 below) the very important fact that the collection of subsets of S that occur as R-classes is a partition of S (or, as some may prefer to express it, the different R-classes form a partition of S). In trying to digest these results, students may find it helpful to think through what each tells us in a very simple case. Consider for example the case where S is a set of coloured beads and the relation R is defined on S by: xRy iff x and y have the same colour ($x, y \in S$). Clearly this relation R is an equivalence relation, and, for each $x \in S$, the R-class C_x is the set of all beads in S with the same colour as x. So, in this case, theorem 17.5, mentioned above, tells us the fairly obvious fact that the subsets of S each comprising all the beads of a particular colour form a partition of S.

Starting now our exploration of the general case (where R is an arbitrary equivalence relation on the set S), we note first that, because R is reflexive, each element x of S is an R-relative of itself. Therefore:

17.1 For every $x \in S$, C_x is a non-empty subset of S, x being a member of C_x.

The next few results concern the relationship between the subsets C_x and C_y ($x, y \in S$) (i) when xRy, (ii) when $xR'y$.

17.2 $xRy \Rightarrow C_x = C_y$ ($x, y \in S$).

Proof. Suppose that xRy.
Then we have, for $t \in S$,

$$t \in C_x \Rightarrow tRx \qquad \text{(by definition of } C_x)$$
$$\Rightarrow tRy \qquad \text{(since } xRy \text{ and } R \text{ is transitive)}$$
$$\Rightarrow t \in C_y \qquad \text{(by meaning of } C_y).$$

Hence $C_x \subseteq C_y$.

Just as we have deduced $C_x \subseteq C_y$ from xRy, so from yRx (true also by the symmetric property of R) we can deduce $C_y \subseteq C_x$. Therefore $C_x = C_y$.
The result follows.

It is not difficult to improve 17.2 to:

17.3 $C_x = C_y \Leftrightarrow xRy$ ($x, y \in S$).

Proof. We know from 17.2 that $xRy \Rightarrow C_x = C_y$ ($x, y \in S$).
To prove the converse, suppose that $C_x = C_y$ ($x, y \in S$). Then since (by 17.1) $x \in C_x$, it follows that $x \in C_y$, i.e. xRy. The result now follows.

Remark. It is now clear that C_a and C_b may be the same set when $a \neq b$ $(a, b \in S)$; and it is often important to bear in mind that an equivalence class is liable to have several possible labels of the form C_x.

17.4 If $xR'y$, then $C_x \cap C_y = \emptyset$ $(x, y \in S)$.

Proof. Suppose that $xR'y$ $(x, y \in S)$.

If $C_x \cap C_y$ were non-empty, there would be an element z belonging to both C_x and C_y, so that we would have both xRz and zRy and hence (by the transitive property of R) xRy—a contradiction.

The result follows.

We reach now the climax of this sequence of theorems.

17.5 The set of R-classes is a partition of S.

Proof. By the definition of "partition" (see §7), we have to prove:
  (i) the union of the R-classes is the whole of S;
  (ii) unequal R-classes are disjoint;
  (iii) each R-class is non-empty.
Of these, (iii) and (i) follow from 17.1, which tells us both that every R-class is non-empty and that every element of S belongs to at least one R-class (since $x \in C_x$). And (ii) is true since, for $x, y \in S$,

$$C_x \neq C_y \Rightarrow xR'y \qquad \text{(by 17.3)}$$
$$\Rightarrow C_x \cap C_y = \emptyset \qquad \text{(by 17.4)}.$$

Clarificatory note. Suppose that $x_1, x_2, x_3, \ldots$ is a complete repetition-free list of the elements of S. Then certainly

$$C_{x_1}, \quad C_{x_2}, \quad C_{x_3}, \quad \ldots \qquad (*)$$

is an exhaustive list of R-classes, but (cf. 17.3) it is not in general a repetition-free list of them. Thus it would be wrong to imagine that 17.5 says that each element of S belongs to C_{x_i} for only one value of i. What 17.5 does say is that the collection of subsets that appear (whether once or repeatedly) in the list (*) is a partition of S.

For a mathematical illustration of 17.5, consider the case where S is $\mathbb{Z}$ and R is congruence modulo 5. It follows at once from 14.2(ii) that each integer belongs to precisely one of the classes C_0, C_1, C_2, C_3, C_4. So in this case a complete repetition-free list of the R-classes (forming a partition of $\mathbb{Z}$) is:

$$C_0 = \{\ldots, -10, -5, 0, 5, 10, 15, 20, \ldots\},$$
$$C_1 = \{\ldots, -9, -4, 1, 6, 11, 16, 21, \ldots\},$$
$$C_2 = \{\ldots, -8, -3, 2, 7, 12, 17, 22, \ldots\},$$
$$C_3 = \{\ldots, -7, -2, 3, 8, 13, 18, 23, \ldots\},$$
$$C_4 = \{\ldots, -6, -1, 4, 9, 14, 19, 24, \ldots\}.$$

Turning again to the general discussion of an equivalence relation R on a set S, we round off the section by noting two further simple facts.

17.6 For each $x \in S$, the one and only R-class to which x belongs is C_x (and therefore, for $x, y \in S$, $x \in C_y \Rightarrow C_x = C_y$).

Proof. Let $x \in S$. Because the set of R-classes is a partition of S, there is precisely one R-class to which x belongs; and this R-class is identified, by 17.1, as C_x.

17.7 The following are all equivalent to one another:
    (i) x, y belong to the same R-class;
    (ii) $C_x = C_y$; $(x, y \in S)$
    (iii) xRy.

Proof. (i) and (ii) are equivalent because (by 17.6) C_x is the (unique) R-class to which x belongs and C_y is the (unique) R-class to which y belongs. And (ii) and (iii) are equivalent by 17.3. Therefore (i), (ii), (iii) are all equivalent to one another.

18. Congruence classes

Throughout this section, m will denote a fixed positive integer, and, for each integer x, C_x will denote the congruence class of $x \pmod{m}$, i.e. the R-class of x in the case where R is congruence modulo m. Thus, for each $x \in \mathbb{Z}$, we have

$$C_x = \{\ldots, x - 2m, x - m, x, x + m, x + 2m, \ldots\}.$$

From earlier sections we can obtain immediately certain simple facts about congruence classes $\pmod{m}$, e.g. from 17.3, we have:

18.1 $C_x = C_y \Leftrightarrow x \equiv y \pmod{m}$ $(x, y \in \mathbb{Z})$.

By 17.5, we know that:

18.2 The collection of congruence classes $\pmod{m}$ is a partition of $\mathbb{Z}$.
    And thirdly, it follows from 14.2(ii) that:

18.3 There are m different congruence classes $\pmod{m}$, and $C_0, C_1, C_2, \ldots, C_{m-1}$ is a complete repetition-free list of them.

The set of congruence classes $\pmod{m}$, i.e. $\{C_0, C_1, C_2, \ldots, C_{m-1}\}$, is denoted by $\mathbb{Z}_m$. Thus $\mathbb{Z}_m$ is a set of order m (see remark (e) of §2). While each of its member objects C_j is an infinite subset of $\mathbb{Z}$, we must be willing, when discussing $\mathbb{Z}_m$, to think of each C_j as a single object.
    We can turn $\mathbb{Z}_m$ into an algebraic system by giving meanings to the sum and product of two of its member objects. This algebraic system turns out to be of very great interest, both because calculations in it provide a powerful and

efficient means of gaining information about integers, and because, as will be seen in later chapters, this is just one example of many algebraic systems that can be produced by following, in a more general context, the pattern we are about to see in the case of $\mathbb{Z}_m$.

Tentatively, let us propose the following definitions of addition and multiplication of congruence classes (mod m):

$$C_x + C_y = C_{x+y} \qquad (x, y \in \mathbb{Z}) \tag{1}$$

$$C_x \cdot C_y = C_{xy} \qquad (x, y \in \mathbb{Z}). \tag{2}$$

Here a logical point of some subtlety arises because each member object of $\mathbb{Z}_m$ has many possible labels (e.g. $C_0 = C_m = C_{2m} = \ldots$), and serious attention must be given to the objection that, apparently at least, the meaning given by the equations (1), (2) to the sum and product of two congruence classes (mod m) depends on the particular labels C_x, C_y that we use for the objects being added or multiplied: if we were to use alternative labels for the same objects (say $C_{x'}$, $C_{y'}$), might we not come (via equations (1) and (2)) to a different conclusion about what the sum and product of these objects are? Happily we can, by the following short lemma, show that the attempt to define the sum and product of two congruence classes (mod m) by the equations (1), (2) survives these objections.

18.4 The equations (1) and (2) provide *consistent* definitions of $\xi + \eta$ and $\xi\eta$ (where ξ, η are members of $\mathbb{Z}_m$) in the sense that the meanings they give to $\xi + \eta$ and $\xi\eta$ are independent of the choice of labels of ξ, η.

Proof. Let ξ, η be two arbitrary member objects of $\mathbb{Z}_m$. Let C_x, $C_{x'}$ be two possible labels for ξ and C_y, $C_{y'}$ be two possible labels for η. Use of the former label in each case would lead one to the conclusions

$$\xi + \eta = C_{x+y} \quad \text{and} \quad \xi\eta = C_{xy}$$

while use of the latter label in each case would lead to

$$\xi + \eta = C_{x'+y'} \quad \text{and} \quad \xi\eta = C_{x'y'}$$

Our task, therefore, is to show that $C_{x+y} = C_{x'+y'}$ and $C_{xy} = C_{x'y'}$.

Since $C_x = C_{x'} (= \xi)$ and $C_y = C_{y'} (= \eta)$, it follows (by 18.1) that

$$x \equiv x' \,(\text{mod } m) \quad \text{and} \quad y \equiv y' \,(\text{mod } m).$$

Hence, by 14.3, $x + y \equiv x' + y' \,(\text{mod } m)$ and $xy \equiv x'y' \,(\text{mod } m)$. Therefore, by 18.1,

$$C_{x+y} = C_{x'+y'} \quad \text{and} \quad C_{xy} = C_{x'y'}$$

and our task is achieved.

It should be noted that the problem of consistency of definition will arise whenever we attempt to give a meaning to the sum, product, etc., of equivalence classes.

At this point we end §18, knowing that we have satisfactorily turned $\mathbb{Z}_m$ into an algebraic system. In §19 below, we shall consider properties of this system against the background of a very general preliminary discussion.

19. Properties of $\mathbb{Z}_m$ as an algebraic system

To begin with, suppose simply that we have a set S on which addition and multiplication are defined: i.e. for all x, $y \in S$, each of $x+y$ and xy has a meaning. Here is a list of interesting basic properties that *may* be possessed by the system $(S, +, .)$, i.e. the set S together with the addition and multiplication defined on it:

A0: $\forall x, y \in S, x+y$ is a member of S;

A1: (associative property of $+$) $\forall x, y, z \in S, (x+y)+z = x+(y+z)$;

A2: (commutative property of $+$) $\forall x, y \in S, x+y = y+x$;

A3: (existence of a zero) S contains a special element $\circ$ with the property that $\forall x \in S, x+\circ = \circ+x = x$;

A4: (existence of negatives [applicable only when A3 holds]): corresponding to each element $x \in S$, there is an element x' (also in S) such that $x+x' = x'+x = \circ$;

M0: $\forall x, y \in S, xy$ is a member of S;

M1: (associative property of .) $\forall x, y, z \in S, (xy)z = x(yz)$;

M2: (commutative property of .) $\forall x, y \in S, xy = yx$;

M3: (existence of a unity) S contains a special element i (different from $\circ$ if A3 holds) with the property that $\forall x \in S, ix = xi = x$;

M4: (existence of inverses [applicable only when M3 holds]) corresponding to each $x \in S$, excepting $x = \circ$ if A3 holds, there is an element x^* (also in S) such that $xx^* = x^*x = i$;

D: (distributive property) $\forall x, y, z \in S,$
$$x(y+z) = xy+xz \quad \text{and} \quad (y+z)x = yx+zx.$$

When we construct or encounter for the first time a new system $(S, +, .)$, we may expect, in general, to find that it possesses some of the above properties but not others. We call the system $(S, +, .)$ a **field** iff it possesses *all* the properties in the above list. Thus a field is a system of a specially nice kind, lacking no basic property. After a little thought the student will realize that the systems of rational, real, and complex numbers are all fields.

After the above generalities, let us consider specifically the system $(\mathbb{Z}_m, +, .)$, where $m \in \mathbb{N}$.

19.1 For every integer $m \geq 2$, the system $(\mathbb{Z}_m, +, .)$ possesses all of the properties A0, A1, A2, A3, A4, M0, M1, M2, M3, D.

Proof. Let m be an arbitrary integer $\geqslant 2$.

From the definition of the sum and product of two congruence classes (mod m), it is clear that A0 and M0 hold for $(\mathbb{Z}_m, +, .)$.

To check A1, consider three arbitrary elements C_x, C_y, C_z of $\mathbb{Z}_m$ ($x, y, z \in \mathbb{Z}$). We have

$$\begin{aligned}
(C_x + C_y) + C_z &= C_{x+y} + C_z && \text{(by the definition of addition on } \mathbb{Z}_m) \\
&= C_{(x+y)+z} && \text{(by the definition of addition on } \mathbb{Z}_m) \\
&= C_{x+(y+z)} && \text{(by associative property of addition of integers)} \\
&= C_x + C_{y+z} && \text{(by definition of addition on } \mathbb{Z}_m) \\
&= C_x + (C_y + C_z) && \text{(by definition of addition on } \mathbb{Z}_m).
\end{aligned}$$

Hence A1 holds for the system $(\mathbb{Z}_m, +, .)$.

Similarly, for arbitrary elements C_x, C_y of $\mathbb{Z}_m$, we have

$$C_x + C_y = C_{x+y} = C_{y+x} = C_y + C_x$$

proving that A2 holds.

By the same kind of approach, we can show that M1, M2, D all hold too.

The congruence class C_0 acts as a zero, since

$$\forall x \in \mathbb{Z}, \quad C_x + C_0 = C_0 + C_x = C_x$$

And C_1 (which is different from C_0 since $1 \not\equiv 0 \pmod{m}$, m being $\geqslant 2$) acts as a unity, since

$$\forall x \in \mathbb{Z}, \quad C_1 C_x = C_x C_1 = C_x$$

Thus A3 and M3 hold.

Finally, we note that, for every $x \in \mathbb{Z}$

$$C_x + C_{(-x)} = C_{(-x)} + C_x = C_{x+(-x)} = C_0$$

and (C_0 being the "zero") this establishes that A4 holds.

We could re-phrase 19.1 by saying that, for $m \geqslant 2$, the system $(\mathbb{Z}_m, +, .)$ is a field iff it possesses the further property M4. The next theorem gives the interesting completion of this story.

19.2 For $m \geqslant 2$,

(i) if m is composite, $(\mathbb{Z}_m, +, .)$ is not a field,

(ii) if m is prime, $(\mathbb{Z}_m, +, .)$ is a field.

Proof. (i) Suppose that m ($\geqslant 2$) is composite. Then $m = kl$ for some $k, l \in \mathbb{N}$ satisfying $1 < k, l < m$, and we have

$$C_0 = C_m = C_{kl}$$

i.e.

$$C_0 = C_k C_l \qquad (\alpha)$$

Since $1 < k < m$, $k \not\equiv 0$ (*mod m*); and therefore $C_k \neq C_0$. Hence, if $(\mathbb{Z}_m, +, .)$ were a field, there would be, in $\mathbb{Z}_m$, an inverse of C_k, i.e. an element $C_u \in \mathbb{Z}_m$ ($u \in \mathbb{Z}$) such that $C_u C_k = C_1$; and from the above equation (α), we would deduce

$$C_u C_0 = C_u(C_k C_l) = (C_u C_k)C_l = C_1 C_l$$

i.e.

$$C_0 = C_l$$

which is a contradiction since (because $1 < l < m$) $l \not\equiv 0$ (mod m).

It follows that $(\mathbb{Z}_m, +, .)$ is not a field in this case.

(ii) Suppose that m is prime.

Consider an arbitrary nonzero member C_x of $\mathbb{Z}_m$ ($x \in \mathbb{Z}$). Because $C_x \neq C_0$, $x \not\equiv 0$ (mod m) and so $m \nmid x$. Therefore (by 12.1), since m is prime, hcf$(m, x) = 1$, and hence (by 12.3) there are integers u, v such that $ux + vm = 1$. From this it follows that $ux \equiv 1$ (mod m), i.e. $C_{ux} = C_1$, i.e.

$$C_u C_x = C_x C_u = C_1 \qquad (= \text{the unity of } \mathbb{Z}_m).$$

This shows that C_u is an inverse of C_x in $\mathbb{Z}_m$.

Thus the existence of an inverse of each nonzero member of $\mathbb{Z}_m$ is established, and so $(\mathbb{Z}_m, +, .)$ is a field in this case.

It is a useful easy exercise to make up addition and multiplication tables for $\mathbb{Z}_m$ for selected values of m. As an illustration the addition and multiplication tables for $\mathbb{Z}_5$ are given below.

+	C_0	C_1	C_2	C_3	C_4
C_0	C_0	C_1	C_2	C_3	C_4
C_1	C_1	C_2	C_3	C_4	C_0
C_2	C_2	C_3	C_4	C_0	C_1
C_3	C_3	C_4	C_0	C_1	C_2
C_4	C_4	C_0	C_1	C_2	C_3

×	C_0	C_1	C_2	C_3	C_4
C_0	C_0	C_0	C_0	C_0	C_0
C_1	C_0	C_1	C_2	C_3	C_4
C_2	C_0	C_2	C_4	C_1	C_3
C_3	C_0	C_3	C_1	C_4	C_2
C_4	C_0	C_4	C_3	C_2	C_1

(*Explanation*: the sum $C_x + C_y$ and the product $C_x C_y$ in $\mathbb{Z}_5$ are indicated in the row headed C_x and the column headed C_y in the left and right-hand tables, respectively.)

Were we interested primarily in number theory (i.e. properties of integers), we could now use our considerable knowledge of the systems $(\mathbb{Z}_m, +, .)$ to establish non-trivial theorems about integers, e.g. Fermat's theorem on congruences, which states that if p is a prime and n is an integer not divisible by p, then $n^{p-1} \equiv 1$ (mod p). However, it is more efficient and it gives more insight to pursue our stated aim in this textbook—the study of algebraic systems in general. In pursuing that aim, we shall obtain theorems that apply in particular to $\mathbb{Z}_m$ and enable many interesting results about integers like Fermat's theorem to be deduced in a relatively easy way.

EXERCISES ON CHAPTER THREE

1. A relation $\sim$ is defined on $\mathbb{R}$ by $x \sim y \Leftrightarrow [x] = [y]$ $(x, y \in \mathbb{R})$. Prove that $\sim$ is an equivalence relation on $\mathbb{R}$ and that, for each $n \in \mathbb{Z}$, the $\sim$-class of n is the interval $[n, n+1)$.

2. Let A, T be given sets, A being a subset of T, and let $S = \mathscr{P}(T)$. Let a relation α be defined on S by $X \alpha Y \Leftrightarrow X \cap A = Y \cap A$ $(X, Y \in S)$. (i) Prove that α is an equivalence relation. (ii) Prove that the α-class of $\emptyset$ is $\mathscr{P}(T - A)$. (iii) If $|A| = n$, where $n \in \mathbb{N}$, how many different α-classes are there?

3. Give examples of relations ρ, σ, τ on $\mathbb{Z}$ such that ρ is reflexive and symmetric but not transitive, σ is reflexive and transitive but not symmetric, and τ is symmetric and transitive but not reflexive. (The existence of these examples shows that no two of the reflexive, symmetric and transitive properties imply the third.)

4. A relation ρ is defined on $\mathbb{R}^2$ by

$$(a, b)\rho(c, d) \Leftrightarrow a + d = b + c \qquad (a, b, c, d \in \mathbb{R}).$$

Show that ρ is an equivalence relation. Interpreting $\mathbb{R}^2$ as the coordinate plane, give a geometrical description of the partition of $\mathbb{R}^2$ into ρ-classes.

5. A relation R is defined on $\mathbb{N}$ by

$$xRy \Leftrightarrow \exists n \in \mathbb{Z} \text{ s.t. } x = 2^n y \qquad (x, y \in \mathbb{N}).$$

Prove that R is an equivalence relation. Show that no R-class contains more than one prime. What is the smallest value of x $(\in \mathbb{N})$ for which the R-class of x contains no prime?

6. Let S be the set of all $n \times n$ matrices with real entries, and let a relation ρ be defined on S by

$$A\rho B \Leftrightarrow \text{there exist } r, s \in \mathbb{N} \text{ such that } A^r = B^s \quad (A, B \in S).$$

Prove that ρ is an equivalence relation on S.

7. Draw up a multiplication table for $\mathbb{Z}_6$. How many solutions are there in $\mathbb{Z}_6$ of the equation $x^2 = x$?

8. Taking a lead from the proof of 19.2(ii), find which of the classes $C_1, C_2, \ldots, C_{96}$ in the field $\mathbb{Z}_{97}$ is the inverse of C_{41}. Find the set of all integers satisfying $41x \equiv 2 \pmod{97}$.

9. An equivalence relation β is defined on $\mathbb{Z}$ by

$$x\beta y \Leftrightarrow x^2 \equiv y^2 \pmod 5 \qquad (x, y \in \mathbb{Z}).$$

How many different β-classes are there?

Let B_t stand for the β-class of t $(t \in \mathbb{Z})$. Suppose that the rules

$$\text{(i) } B_x + B_y = B_{x+y} \qquad \text{(ii) } B_x . B_y = B_{xy} \qquad (x, y \in \mathbb{Z})$$

are proposed as definitions of the sum and product of two β-classes. Investigate whether each of (i), (ii) should be accepted as a consistent definition.

CHAPTER FOUR

MAPPINGS

20. Introduction

The idea of a mapping is one of the most widely occurring in mathematics. This chapter is devoted to a survey of associated terminology, notations and theorems which will be frequently used in later chapters of the book (and are also in continual use in other parts of mathematics). Because this material plays such a key role, students will be well advised (whether or not they have any prior knowledge of mappings) to study this chapter closely until they have gained a detailed understanding of its contents.

Definition. Let S, T be non-empty sets (not necessarily different from each other). A **mapping** from S to T is a rule which associates with (or produces from) *each* element x of S a *single* element (depending on x) belonging to T.

Simple illustrations. (1) The "integral part function", which produces from each real number x its integral part, the integer $[x]$, is an example of a mapping from $\mathbb{R}$ to $\mathbb{Z}$.

(2) Let Γ be the coordinate plane. Then rotation anti-clockwise about the origin through $30°$ is a mapping from Γ to Γ: it is, in a certain sense, a rule which associates with (or produces from) each point X in Γ a single point X', also in Γ and related to X as shown in the diagram.

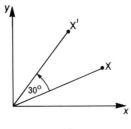

Remarks. (*a*) As the illustrations suggest, the word *rule* in the definition must be understood in a very general sense. Indeed it might be argued that the definition would convey its meaning more clearly if *rule* were replaced by a more general word or phrase such as *idea* or *mathematical notion.*

(*b*) The student should note particularly the force of the italicized words *each* and *single* in the definition.

(*c*) *Function, map, transformation* are among synonyms used for *mapping* in certain mathematical contexts.

(*d*) We usually denote mappings by single letters (such as $f, g, \ldots,$ $F, G, \ldots, \theta, \phi, \ldots$).

(*e*) In the case of a mapping f from S to T, the set S is called the **domain** of f, and the set T is called the **codomain** of f.

(*f*) The fact that f is a mapping from S to T is conveniently displayed symbolically as follows:

$$f : S \to T.$$

(*g*) Let f be a mapping from S to T. For each $x \in S$, the single element of T that f associates with (or produces from) x is denoted by $f(x)$. The correct technical term for $f(x)$ is the **image** of x under f. Another form of words often used is as follows: if $f(x) = y$, one says that "f maps x to y".

(*h*) It is sometimes helpful to picture a mapping $f : S \to T$ by means of the diagram below. The rectangular areas represent the sets S, T, and the dots within represent the elements of these sets. The "action" of f (i.e. the detail of how f associates a single element of T with each element of S) is indicated by drawing an arrow from x to $f(x)$ for each $x \in S$.

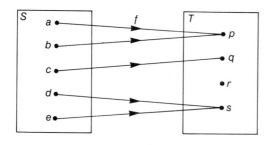

Here: $f(a) = f(b) = p$
      $f(c) = q$
      $f(d) = f(e) = s.$

(*i*) To specify a particular mapping $f : S \to T$, one has to say what $f(x)$ is for each element $x \in S$. Often this is most naturally achieved by means of a general formula telling (in terms of x) what $f(x)$ is for each $x \in S$, e.g. a mapping $f : \mathbb{R} \to \mathbb{R}$ may be specified by the formula

$$f(x) = x^2 + 1 \qquad (x \in \mathbb{R}).$$

A useful variant exists. One may specify a mapping from S to T by writing

$$x \mapsto \boxed{} \quad (x \in S)$$

filling in the shaded blank with a formula for the image of x (x^2+1, or whatever it might be).

(*j*) For far-reaching technical reasons, it is important to be clear about what we mean when two mappings f, g are called *equal*. In fact, we call the two mappings f, g equal iff all the following conditions hold:

(i) f, g have the same domain (S, say);

(ii) f, g have the same codomain;

(iii) $\forall x \in S, f(x) = g(x)$.

(*k*) This final remark is addressed only to students who are already to a great extent familiar with the idea of a mapping, and it may be ignored meantime by others. It is possible to give a more formal definition of a mapping—namely that a mapping is a 3-tuple (S, T, G) in which S and T are non-empty sets (the domain and codomain of the mapping) and G is a subset of $S \times T$ with the property that for each $x \in S$ there is precisely one member of G with first object x. As with the formal definition of a relation in chapter 3, it is left to the interested student to think out why this formal definition makes sense and what advantages it offers over the informal approach.

21. The image of a subset of the domain; surjections

Let f be a mapping from S to T, and let A be a subset of the domain S.

The set

$$\{f(x) : x \in A\}$$

(i.e. the subset of the codomain T comprising all the images under f of elements of A) is called the image of the set A under f and is denoted by $f(A)$. Thus:

21.1 For $y \in T$, $y \in f(A) \Leftrightarrow \exists x \in A$ s.t. $y = f(x)$.

A case of special interest is that in which A is the whole of S. We call the subset $f(S)$ of T the **image set** of f and denote it by im f.

Illustration. Let f be the mapping from $\{0, 1, 2, 3, 4, 5\}$ to $\{0, 1, 2, 3\}$ specified by

$$f(0) = 0, \quad f(1) = 0, \quad f(2) = 0, \quad f(3) = 1, \quad f(4) = 1, \quad f(5) = 3;$$

and let $A = \{0, 3\}$, $B = \{0, 1, 3\}$, $C = \{0, 1, 2\}$. Then, as the student should verify to his satisfaction,

$$f(A) = \{0, 1\}, \quad f(B) = \{0, 1\}, \quad f(C) = \{0\}, \quad \text{and} \quad \text{im} f = \{0, 1, 3\}.$$

It will be noted that we have now introduced three different usages of the word "image" in connection with a mapping $f : S \to T$, viz. (i) the image under f of an element of S, (ii) the image under f of a subset of S, (iii) the image set of f. It is important that the student should be able to avoid confusion among these three usages.

As the above illustration shows, the image set of a mapping need not be the whole of the codomain. This observation prompts a definition.

Definition. Let $f : S \to T$ be a mapping. We say that f is a **surjection** (or that f is a surjective mapping) iff im f is the whole of the codomain T.

The following simple result is the basis of the way we usually test whether a given mapping is surjective.

21.2 Let $f : S \to T$ be a mapping. Then f is surjective iff

$$\forall y \in T, \ \exists x \in S \text{ s.t. } y = f(x)$$

Proof. The condition

$$\exists x \in S \text{ s.t. } y = f(x) \qquad (y \in T)$$

holds if $y \in \text{im} f$ and does not hold if $y \in T - \text{im} f$ (see 21.1). Therefore the condition holds for all $y \in T$ iff $\text{im} f$ is the whole of T, i.e. iff f is surjective.

Remark. The word "onto" has traditionally been used, both as a preposition and as an adjective, to indicate surjections. For any mapping $f : S \to T$, one may say that f maps S to T. It is common to vary this by saying that f maps S onto T when f is surjective; and it is also common to speak of "an onto mapping", meaning simply "a surjection".

This section concludes with some interesting little results about images of subsets of the domain of a mapping.

As a preliminary, note that clearly:

21.3 If $f : S \to T$ is a mapping and A, B are subsets of S such that $A \subseteq B$, then $f(A) \subseteq f(B)$.

We employ this simple observation in proving the following theorem.

21.4 Let $f : S \to T$ be a mapping, and let A, B be subsets of S. Then:
  (i) $f(A \cap B) \subseteq f(A) \cap f(B)$;
  (ii) $f(A \cup B) = f(A) \cup f(B)$.

Proof. (i) Since $A \cap B \subseteq A$ and $A \cap B \subseteq B$, it follows from 21.3 that

$$f(A \cap B) \subseteq f(A) \quad \text{and} \quad f(A \cap B) \subseteq f(B).$$

Hence (by 7.2(i)) $f(A \cap B) \subseteq f(A) \cap f(B)$.

(ii) Similarly, it follows from 21.3 and 7.2(ii) that

$$f(A) \cup f(B) \subseteq f(A \cup B).$$

To establish the reverse containment, suppose that $y \in f(A \cup B)$. Then $y = f(x)$ for some $x \in A \cup B$. The element x must belong to A or to B, and so $y(=f(x))$ belongs to $f(A)$ or to $f(B)$. Therefore $y \in f(A) \cup f(B)$.

This shows that $y \in f(A \cup B) \Rightarrow y \in f(A) \cup f(B)$ $(y \in T)$

i.e. that

$$f(A \cup B) \subseteq f(A) \cup f(B).$$

The result now follows.

Post-script to 21.4(i). There are instances where $f(A \cap B)$ is not equal to $f(A) \cap f(B)$; e.g. if f is the mapping from $\mathbb{Z}$ to $\mathbb{Z}$ given by $x \mapsto x^2$ and if $A = \{0, 2\}$ and $B = \{-2, 0\}$, then, as is easily checked, $f(A) \cap f(B) = \{0, 4\}$, while $f(A \cap B) = \{0\}$.

22. Injections; bijections; inverse of a bijection

Definition. A mapping $f : S \to T$ is called an **injection** (or an injective mapping) iff in all cases distinct elements of S have distinct images under f. Thus f is injective iff

$$a \neq b \Rightarrow f(a) \neq f(b) \qquad (a, b \in S) \tag{α}$$

Illustration. Consider the mappings f_2 and f_3 from $\mathbb{R}$ to $\mathbb{R}$ given by $f_2(x) = x^2$, $f_3(x) = x^3$ $(x \in \mathbb{R})$. The mapping f_2 is not injective: for it is possible to find two distinct elements of $\mathbb{R}$ with the same image under f_2 (e.g. -5 and 5); but f_3 is injective because different real numbers have different cubes.

By applying the law of contraposition to the clause (α) in the definition, we see at once that:

22.1 The mapping $f : S \to T$ is injective iff

$$f(a) = f(b) \Rightarrow a = b \qquad (a, b \in S)$$

This result 22.1 is often used to prove that a given mapping is injective.

It is also clear from the definition of an injection that:

22.2 If $f : S \to T$ is an injection and A is a finite subset of S, then $|f(A)| = |A|$.

In the remainder of this section we are concerned with mappings which are both surjective and injective.

Definition 1. A mapping $f : S \to T$ is called a **bijection** (or a bijective mapping) iff f is both surjective and injective.

And, for the special case $S = T$:

Definition 2. A bijection $f : S \rightarrow S$ is called a **permutation** of S.

[*Note*: In certain elementary contexts, people tend to use the word *permutation* to mean arrangement in a definite order of some given objects: e.g. they may describe 53124 as a "permutation of the objects 1, 2, 3, 4, 5". But, at the level of mathematical education now reached, it is very much preferable to describe 53124 as a "*rearrangement* of 12345" and to reserve the word *permutation* to mean a mapping of the kind detailed in definition 2.]

Taking up the general discussion of bijective mappings, suppose that $f : S \rightarrow T$ is a bijection. For each $y \in T$, there is exactly one element $x \in S$ satisfying $y = f(x)$. (There is at least one since f is surjective, and there cannot be more than one since f is injective.) It follows that we can specify a mapping from T to S by

$$y \mapsto x \qquad (y \in T)$$

where x is the unique answer to the question "f(what in S) = y?". The mapping from T to S so specified is called the **inverse** of f and is denoted by f^{-1}.

So, to re-state the essential point in a less discursive way, when f is a bijection from S to T, f^{-1} is the mapping from T to S given by

$$f^{-1}(y) = \text{the element of } S \text{ that } f \text{ maps to } y \qquad (y \in T).$$

We note and record:

22.3 If $f : S \rightarrow T$ is a bijection, then

$$f^{-1}(y) = x \Leftrightarrow f(x) = y \qquad (x \in S, y \in T)$$

Illustration. Let E be the set of even integers and let f be the (clearly bijective) mapping from $\mathbb{Z}$ to E given by $f(x) = 2x$ $(x \in \mathbb{Z})$. Then f^{-1} is the mapping from E to $\mathbb{Z}$ given by $f^{-1}(y) = \frac{1}{2}y$ $(y \in E)$.

A significant observation is that whenever we have a bijection (f, say) from S to T, there is set up for us a "one-to-one correspondence" between the elements of S and the elements of T (see diagram): each element x of S is put in correspondence with the single element $f(x)$ of T, and each element y of T is put in correspondence with the single element $f^{-1}(y)$ of S.

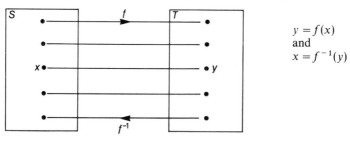

$$y = f(x)$$
and
$$x = f^{-1}(y)$$

Such thinking brings us close to the following result.

22.4 Let S, T be non-empty finite sets. Then there exists a bijection from S to T iff $|S| = |T|$.

Proof. (i) Suppose that there is a bijection f from S to T. Then

$$|S| = |f(S)| \qquad \text{(by 22.2, since } f \text{ is injective)}$$
$$= |T| \qquad \text{(since } f \text{ is surjective).}$$

(ii) Conversely, suppose that $|S| = |T|$ $(= n$, say). Let $x_1, x_2, \ldots, x_n$ and $y_1, y_2, \ldots, y_n$ be complete repetition-free lists of the elements of S and T, respectively. Then the mapping $f : S \to T$ given by

$$f(x_i) = y_i \qquad (i = 1, 2, \ldots, n)$$

is clearly a bijection from S to T.

Proof of the stated result is now complete.

23. Restriction of a mapping

Consider the following mappings f and g:

$$f : \mathbb{R} \to \mathbb{R} \quad \text{given by} \quad f(x) = x^2;$$
$$g : [0, \infty) \to [0, \infty) \quad \text{given by} \quad g(x) = x^2.$$

Although one might say that f and g have the same action (which could loosely be described as "squaring"), f and g are not equal (e.g. f, g do not have the same domain; moreover g^{-1} exists whereas f^{-1} does not). As the next definition makes clear, the relationship between these similar but unequal mappings is expressed by saying that g is a restriction of f.

Definition. Let $f : S \to T$ be a mapping, and let A, B be non-empty subsets of S, T, respectively, satisfying $f(A) \subseteq B$. Then we can specify a mapping $g : A \to B$ by

$$g(x) = f(x) \qquad (x \in A)$$

This mapping g is called the **restriction** of f to A as domain and B as codomain; and any mapping produced from f in this way is called a restriction of f.

When a non-bijective mapping is given, it is often possible and profitable to consider a bijective restriction of it. Given any mapping f, we can produce a surjective restriction of f just by changing the codomain to im$f$; and often a simple change of domain then produces a bijective restriction of f.

To illustrate what has just been said, consider the sine function, i.e. the mapping from $\mathbb{R}$ to $\mathbb{R}$ given by $x \mapsto \sin x$. This mapping is neither surjective nor injective. However, a surjective restriction is obtained on reducing the codomain to im(sin), i.e. to $[-1, 1]$; and then a bijective restriction is produced by reducing the domain to $[-\frac{1}{2}\pi, \frac{1}{2}\pi]$. (Consideration of a sketch of the curve $y = \sin x$ will make all this clear.) The bijective restriction described—a mapping from $[-\frac{1}{2}\pi, \frac{1}{2}\pi]$ to $[-1, 1]$—is important in calculus because its inverse is used there as the inverse sine function (denoted by $\sin^{-1}$ or arcsin).

Often one is concerned with a restriction of $f : S \rightarrow T$ to a smaller domain A ($\subset S$), but with the same codomain T as f itself. That restriction is denoted by $f|A$ and is usually called simply "the restriction of f to A".

24. Composition of mappings

Definition. Let $f : S \rightarrow T$ and $g : T \rightarrow U$ be mappings with the property that the domain of g is the same set as the codomain of f. Then a mapping from S to U may be specified by

$$x \mapsto g(f(x)) \qquad (x \in S)$$

This mapping is denoted by $g \circ f$ and is called the **composition** of f followed by g. (The diagram below may help with the assimilation of this definition.)

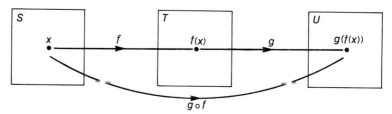

Note that (i) $g \circ f$ is not defined unless the domain of g equals the codomain of f, and (ii) when this condition is satisfied, i.e. when $g \circ f$ does exist,

domain of $g \circ f$ = domain of f,

codomain of $g \circ f$ = codomain of g.

Worked example. Let f, g be the mappings from $\mathbb{R}$ to $\mathbb{R}$ given by

$$f(x) = x^2, \quad g(x) = 2x + 1 \qquad (x \in \mathbb{R})$$

In this case both $g \circ f$ and $f \circ g$ exist. Find formulae specifying them.

Solution. For $x \in \mathbb{R}$, we have:

$$
\begin{aligned}
(g \circ f)(x) &= g(f(x)) && \text{(by definition of } g \circ f) \\
&= g(x^2) && \text{(by definition of } f) \\
&= 2x^2 + 1 && \text{(by definition of } g).
\end{aligned}
$$

Thus $g \circ f$ is the mapping from $\mathbb{R}$ to $\mathbb{R}$ given by $x \mapsto 2x^2 + 1$ $(x \in \mathbb{R})$. Similarly, for $x \in \mathbb{R}$,

$$(f \circ g)(x) = f(g(x)) = f(2x + 1) = (2x + 1)^2,$$

so that $f \circ g$ is the mapping from $\mathbb{R}$ to $\mathbb{R}$ given by $x \mapsto (2x + 1)^2$ $(x \in \mathbb{R})$.

In the above example, $(g \circ f)(1) = 3$ while $(f \circ g)(1) = 9$, which shows that

$g \circ f \neq f \circ g$. Thus, even in a case where $g \circ f$ and $f \circ g$ both exist and have the same domain and codomain, $g \circ f$ and $f \circ g$ may be unequal. We express this by saying that "mapping composition is not commutative".

In contrast, a very important positive property of mapping composition (called the "associative property") is that, for mappings f, g, h,

$$(f \circ g) \circ h = f \circ (g \circ h)$$

provided that the composite mappings involved exist. Here now is a formal statement of it.

24.1 Let $h: S \to T$, $g: T \to U$, $f: U \to V$ be mappings. Then

$$(f \circ g) \circ h = f \circ (g \circ h)$$

Proof. We note first that $f \circ g$ exists and is a mapping from T to V. Hence $(f \circ g) \circ h$ exists and is a mapping from S to V. Similarly, we find that $f \circ (g \circ h)$ exists and is a mapping from S to V. Thus $(f \circ g) \circ h$ and $f \circ (g \circ h)$ are mappings with the same domain and codomain. Further, for every $x \in S$,

$$
\begin{aligned}
[(f \circ g) \circ h](x) &= (f \circ g)[h(x)] && \text{(by meaning of } k \circ h\text{)} \\
&= f\{g[h(x)]\} && \text{(by meaning of } f \circ g\text{)} \\
&= f\{(g \circ h)(x)\} && \text{(by meaning of } g \circ h\text{)} \\
&= [f \circ (g \circ h)](x) && \text{(by meaning of } f \circ l\text{)}.
\end{aligned}
$$

It follows that $(f \circ g) \circ h = f \circ (g \circ h)$.

Identity mappings
For each non-empty set S, a mapping denoted by i_S from S to S is specified by the simple rule $x \mapsto x$ $(x \in S)$, i.e.

$$i_S(x) = x \qquad (x \in S)$$

This mapping i_S is called the **identity mapping** from S to itself.
It is easily seen that:

24.2 For every set S, i_S is a permutation of S.

An important property of identity mappings is that they behave in mapping composition like the number 1 in multiplication. That is:

24.3 Let $f: S \to T$ be a mapping. Then

$$\text{(i)} \ \ f \circ i_S = f, \quad \text{and} \quad \text{(ii)} \ \ i_T \circ f = f.$$

Proof. (i) $f \circ i_S$ and f are both mappings from S to T. Further, for every $x \in S$,

$$
\begin{aligned}
(f \circ i_S)(x) &= f(i_S(x)) && \text{(by meaning of } f \circ g\text{)} \\
&= f(x) && \text{(by definition of } i_S\text{)}.
\end{aligned}
$$

It follows that $f \circ i_S = f$.

Part (ii) is proved similarly.

The final result in this section is a formal expression of the fact that, when f is a bijection, f and f^{-1} are mappings which undo each other's effects (e.g. in an illustration in §22, the action of f was to double while the action of f^{-1} was to halve.)

24.4 Let $f : S \to T$ be a bijection, so that f^{-1} exists. Then

$$\text{(i) } f \circ f^{-1} = i_T \quad \text{and} \quad \text{(ii) } f^{-1} \circ f = i_S.$$

Proof. (i) It is easily checked that $f \circ f^{-1}$ exists and is a mapping from T to T. So $f \circ f^{-1}$ and i_T are mappings with the same domain and codomain.

Let y be an arbitrary element of T, and let $f^{-1}(y) = x$. Then (see 22.3) $y = f(x)$, and hence we have

$$(f \circ f^{-1})(y) = f(f^{-1}(y)) = f(x) = y = i_T(y).$$

It follows that $f \circ f^{-1} = i_T$.

Part (ii) can be proved in a fairly similar way, and so its proof is left as an exercise.

25. Some further results and examples on mappings

In 24.4 we showed that if f is a bijection from S to T and $g = f^{-1}$, then

$$g \circ f = i_S \quad \text{and} \quad f \circ g = i_T \tag{1}$$

We now prove (as a kind of converse) that the existence of a mapping g satisfying (1) implies that f is bijective and that a mapping g satisfying (1) can only be f^{-1}. More precisely:

25.1 Let $f : S \to T$ be a mapping. Suppose that a mapping $g : T \to S$ exists and is such that

$$g \circ f = i_S \quad \text{and} \quad f \circ g = i_T$$

Then f is bijective, and g is its inverse.

Proof. f is injective because (cf. 22.1), for $a, b \in S$,

$$
\begin{aligned}
f(a) = f(b) &\Rightarrow g(f(a)) = g(f(b)) \\
&\Rightarrow (g \circ f)(a) = (g \circ f)(b) \\
&\Rightarrow i_S(a) = i_S(b) \quad (\text{since } g \circ f = i_S) \\
&\Rightarrow a = b.
\end{aligned}
$$

Moreover, f is surjective: for (cf. 21.2) if y is an arbitrary element of T, y is the image under f of an element of S, viz. $g(y)$, since

$$f(g(y)) = (f \circ g)(y) = y \quad (f \circ g \text{ being equal to } i_T).$$

It follows that f is bijective, and therefore f^{-1} exists. Further:

$$f^{-1} = f^{-1} \circ i_T \quad \text{(by 24.3)}$$
$$= f^{-1} \circ (f \circ g) \quad \text{(since } f \circ g = i_T)$$
$$= (f^{-1} \circ f) \circ g \quad \text{(by 24.1)}$$
$$= i_S \circ g \quad \text{(by 24.4)}$$
$$= g \quad \text{(by 24.3)}.$$

The result is now completely proved.

The next two results are corollaries of 25.1.

25.2 Let $f : S \to T$ and $g : T \to U$ be bijections. Then $g \circ f$ is also a bijection, and its inverse equals $f^{-1} \circ g^{-1}$.

Proof. It is easily checked that $(g \circ f) \circ (f^{-1} \circ g^{-1})$ exists. Moreover:

$$(g \circ f) \circ (f^{-1} \circ g^{-1}) = g \circ [f \circ (f^{-1} \circ g^{-1})]$$
$$= g \circ [(f \circ f^{-1}) \circ g^{-1}] \quad \text{(by 24.1)}$$
$$= g \circ (i_T \circ g^{-1}) = g \circ g^{-1} = i_U \quad \text{(by 24.3 and 24.4)}.$$

Similarly, $(f^{-1} \circ g^{-1}) \circ (g \circ f)$ exists and equals i_S. So the equations

$$h \circ (g \circ f) = i_S, \qquad (g \circ f) \circ h = i_U$$

are satisfied with $h = f^{-1} \circ g^{-1}$ (a mapping from U to S).

By 25.1, the stated result follows immediately.

An even easier application of 25.1 yields:

25.3 If f is a bijection, then f^{-1} is also a bijection and $(f^{-1})^{-1} = f$.

Particular cases of 25.2 and 25.3 give us:

25.4 (i) If f, g are permutations of the set S, then so is $g \circ f$.
(ii) If f is a permutation of the set S, then so is f^{-1}.

Worked examples

1. Let $f : S \to T$, $g : T \to U$, $h : T \to U$ be mappings such that (i) $g \circ f = h \circ f$ and (ii) f is surjective. Prove that $g = h$.

Solution. Let t be an arbitrary element of T. Since f is surjective, $t = f(x)$ for some $x \in S$. Hence we have:

$$g(t) = g(f(x))$$
$$= (g \circ f)(x)$$
$$= (h \circ f)(x) \quad \text{(by (i))}$$
$$= h(f(x))$$
$$= h(t)$$

Since g and h have the same domain and codomain, it follows that $g = h$.

2. Let $f : S \to T, g : T \to U$ be mappings such that $g \circ f$ is injective. Prove that f is injective.

Solution. For $a, b \in S$,

$$
\begin{aligned}
f(a) = f(b) &\Rightarrow g(f(a)) = g(f(b)) \\
&\Rightarrow (g \circ f)(a) = (g \circ f)(b) \\
&\Rightarrow a = b \qquad \text{(since } g \circ f \text{ is injective).}
\end{aligned}
$$

Hence f is injective.

EXERCISES ON CHAPTER FOUR

1. Each of the following attempts to specify a mapping contains a howler. Spot the howler (and resolve never to imitate it).

(i) A mapping $f : \mathbb{N} \to \mathbb{N}$ is specified by $f(x) = x - 7$ $(x \in \mathbb{N})$.

(ii) A mapping $g : \mathbb{R} \to \mathbb{R}$ is specified by $g(x) = 1/x$ $(x \in \mathbb{R})$.

(iii) A mapping $h : \mathbb{R} \to \mathbb{R}$ is specified by $h(x) = \begin{cases} x+1 & \text{if } x \geq 0, \\ 0 & \text{if } x \leq 0. \end{cases}$

(iv) A mapping $\theta : \mathbb{Q} \to \mathbb{Z}$ is specified by $\theta(m/n) = m + n$ $(m, n \in \mathbb{Z}, n \neq 0)$.

2. Look carefully at the definition of equality of mappings (§20, remark (j)) and decide whether, in the context of mappings from the coordinate plane Γ to itself:

(i) rotation anticlockwise about O through $360° = i_\Gamma$;

(ii) rotation anticlockwise about O through $180° =$ rotation clockwise about O through $180°$.

3. The mapping $f : \mathbb{R} \to \mathbb{R}$ is given by $f(x) = x^4 - 1$ $(x \in \mathbb{R})$.

(i) What is the image of 2 under f?

(ii) What is the image under f of the set $[-3, 2]$?

(iii) What is im f? Is f surjective?

4. Let S, T be finite non-empty sets with $|S| = m$, $|T| = n$.

(i) How many different mappings are there from S to T?

(ii) How many different injections are there from S to T (a) if $m > n$, (b) if $m \leq n$?

(iii) Show that if $m = n + 1$, the number of different surjections from S to T is $\frac{1}{2}n\{(n+1)!\}$. Find a corresponding expression for the number of surjections from S to T in the case $m = n + 2$.

5. Let f, g be the mappings from $\mathbb{N}$ to $\mathbb{N}$ given by

$$f(x) = 2x, \quad g(x) = [\tfrac{1}{2}(x+1)] \quad (x \in \mathbb{N}).$$

Prove that f is injective but not surjective, and that g is surjective but not injective.

Give examples of (a) a mapping from $\mathbb{N}$ to $\mathbb{N}$ that has infinite image set but is neither surjective nor injective, (b) a bijection from $\mathbb{N}$ to $\mathbb{Z}$.

6. Suppose that $f : S \to T$ is a surjection. Prove that, for all $A \subseteq S$, $T - f(A) \subseteq f(S - A)$.

7. Suppose that $f : S \to T$ is injective. Prove that, for A, $B \subseteq S$,

$$f(A \cap B) = f(A) \cap f(B).$$

8. Let $f : S \to T$ be a mapping. Prove that

$$f \text{ is injective} \Leftrightarrow \forall A \subseteq S, f(A) \cap f(S - A) = \emptyset.$$

9. Let $f : S \to T$ be a mapping, let A be a subset of S, and let

$$B = \{x \in S : f(x) \in f(A)\}.$$

(i) Show that $A \subseteq B$.

(ii) Prove that $A = B$ if f is injective.

(iii) Determine B in the particular case where $S = T = \mathbb{R}$, $A = [-1, 3]$, and f is given by $x \mapsto x^2$.

10. The relation ρ is defined on $\mathbb{N} \times \mathbb{N}$ by

$$(a, b)\rho(c, d) \Leftrightarrow a + d = b + c$$

Prove that ρ is an equivalence relation on $\mathbb{N} \times \mathbb{N}$.

For $(a, b) \in \mathbb{N} \times \mathbb{N}$, let $C_{(a,b)}$ denote the ρ-class of (a, b); and let S denote the set of all

ρ-classes. Show that the formula

$$\theta(C_{(a,b)}) = b - a \qquad (a, b \in \mathbb{N})$$

consistently defines a mapping from S to $\mathbb{Z}$ and that this mapping θ is bijective.

11. Let $f : S \to T$ be a mapping. Let the relation R be defined on S by

$$xRy \Leftrightarrow f(x) = f(y) \qquad (x, y \in S).$$

Show that R is an equivalence relation on S.

(i) If S, T are finite, what can be said about the number of different R-classes (a) if f is injective, (b) if f is surjective?

(ii) In the general case, let U be the set of all R-classes. Prove that f can be expressed in the form $\alpha \circ \beta$, where α is an injection from U to T and β is a surjection from S to U.

12. Let S be an arbitrary non-empty set. Prove that there does not exist a bijection from S to $\mathscr{P}(S)$. [*Hint*: Suppose $f : S \to \mathscr{P}(S)$ is a bijection and consider $T = \{x \in S : x \notin f(x)\}$.]

13. Let $f : \mathbb{R} \to \mathbb{R}$ be defined by $f(x) = x^2 - 4x + 5$ $(x \in \mathbb{R})$.
(i) Draw the curve $y = f(x)$.
(ii) What is the image set B of f?
(iii) What number c is such that f has bijective restrictions $f_1 : (-\infty, c] \to B$, $f_2 : [c, \infty) \to B$?
(iv) Find formulae for $f_1^{-1}(y)$ and $f_2^{-1}(y)$ $(y \in B)$.

14. Suppose that $f : S \to T$ and $g : T \to U$ are mappings such that $g \circ f$ is surjective. Prove that g is surjective.

15. Suppose that $f : S \to T$, $g : S \to T$, $h : T \to U$ are mappings such that $h \circ f = h \circ g$ and h is injective. Prove that $f = g$.

16. Mappings f, g, h from the set S to itself are such that $f \circ g = h \circ f = i_S$. Prove that f is bijective and that $f^{-1} = g = h$.

17. Suppose that the mapping $f : S \to T$ is not surjective. Let $U = \{0, 1\}$. Prove that there exist two unequal mappings g_1, g_2 from T to U such that $g_1 \circ f = g_2 \circ f$.

18. Suppose that $f : S \to T$ is injective (but not necessarily surjective). Prove the existence of a mapping $g : T \to S$ such that $g \circ f = i_S$.

CHAPTER FIVE

SEMIGROUPS

26. Introduction

At this point in the book we launch into the mainstream of abstract algebra, i.e. into the general study of algebraic systems.

When we refer to an *algebraic system*, we always mean a non-empty set whose elements can be combined by addition, or by multiplication, or by both, or by some other such binary operation(s). In this chapter we begin with a formal look at binary operations, and then we set the scene for abstract algebra by discussing the very simple kind of algebraic system known as a *semigroup*. This discussion enables us to introduce (in a context not cluttered by extraneous details) basic concepts such as *identity elements, inverses, subsystems*—concepts that are relevant to all the species of algebraic system studied in this book. In chapters 6 to 8 we shall consider in detail the important special kind of semigroup that we call a *group*. In a semigroup (in particular in a group) there is just one binary operation. That is why it seems sensible to study semigroups and groups before rings and fields, systems with two binary operations (addition and multiplication) which we explore in the final chapter of the book.

27. Binary operations

Throughout this section S will denote a set.

We say that $\circ$ is a **binary operation** on S iff $x \circ y$ has a meaning for every x, $y \in S$. In this context, we think of $\circ$ not merely as the symbol that is put on paper between x and y to form $x \circ y$ but as the process or rule (call it what you will) by which $x \circ y$ is formed from x and y (see also remark (e) below).

In the case $S = \mathbb{Z}$, familiar examples of binary operations include $+$ (addition), $-$ (subtraction), . (multiplication).

Remarks. (a) We use symbols like $\circ$, $*$ as general symbols for binary operations. Often, however, we deal with binary operations with special names (e.g. addition, subtraction, multiplication) and then we follow familiar conventions in notation for these binary operations (e.g. $+$, $-$, . or $\times$ for addition, subtraction, multiplication, respectively). Moreover, when we deal with a binary operation called multiplication, the simple notation xy (instead of $x.y$ or $x \times y$) is usually acceptable for the "product" of two elements x, y.

(*b*) If $\circ$ is a binary operation on S, it must not be presupposed that $x \circ y$ and $y \circ x$ are equal for all x, $y \in S$.

(*c*) Suppose that $\circ$ is a binary operation on S (so that $x \circ y$ has a meaning for all x, $y \in S$). Then, for any subset T of S, it is obviously true that $x \circ y$ has a meaning for all x, $y \in T$. Thus it is legitimate to claim that $\circ$ is a binary operation on each subset T of S.

(*d*) Let $\circ$ be a binary operation on S. This we have defined to mean simply that $x \circ y$ has a meaning for all x, $y \in S$. Unlike some authors, we do *not* insist that in all cases the object $x \circ y$ must be a member of S. Our point of view has the advantage of giving to the following definition the prominence it merits.

Definition. $\circ$ being a binary operation on S, we say that S is **closed** under $\circ$ iff $x \circ y$ is a member of S for all x, $y \in S$.

In the light of remark (*c*), this means that, more generally, a subset T of S is closed under $\circ$ iff $x \circ y \in T$ for all x, $y \in T$ (e.g. $\mathbb{Z}$ is closed under $-$ because, for every x, $y \in \mathbb{Z}$, $x - y$ is again an integer. But the subset $\mathbb{N}$ is not closed under $-$ since, for example, although $2, 4 \in \mathbb{N}$, $2 - 4 \notin \mathbb{N}$.)

(*e*) Some textbooks define a binary operation to be a mapping. The point is that if $\circ$ is a binary operation (in the sense detailed above) and if, for all x, $y \in S$, $x \circ y$ is an element of the set A, then one may regard $\circ$ as a mapping from $S \times S$ to A, viz. the mapping given by

$$(x, y) \mapsto x \circ y$$

This point of view will not be adopted here, because persistence with it makes certain considerations of subsets of S unnecessarily difficult.

28. Associativity and commutativity

We now think about two basic properties that a binary operation may possess.

Definition 1. Suppose that $\circ$ is a binary operation on the set S and that S is closed under $\circ$. We say that $\circ$ is **associative** (on S) iff

$$\forall x, y, z \in S, \quad (x \circ y) \circ z = x \circ (y \circ z)$$

Most binary operations encountered in elementary mathematics are associative. An obvious exception among binary operations on $\mathbb{R}$ is $-$ (subtraction): although $\mathbb{R}$ is closed under $-$, $-$ is non-associative, since, for example,

$$(1 - 2) - 3 = -4, \text{ while } 1 - (2 - 3) = 2.$$

Definition 2. If $\circ$ is a binary operation on S, we say that $\circ$ is **commutative** (on S) iff $\forall x, y \in S$, $x \circ y = y \circ x$.

For example, addition and multiplication are commutative on $\mathbb{R}$, but subtraction is not commutative on $\mathbb{R}$ since, for example, $1 - 2 \neq 2 - 1$.

More generally, if $\circ$ is a binary operation on S and $x, y \in S$, we say that x, y **commute** with each other (under $\circ$) iff $x \circ y = y \circ x$. So $\circ$ is commutative on S iff every pair of elements of S commute with each other (under $\circ$).

For the rest of this section, suppose that $\circ$ is an associative binary operation on the set S (S being closed under $\circ$).

Because $\circ$ is associative, for any $x, y, z \in S$, we can write

$$x \circ y \circ z \qquad \text{(without brackets!)}$$

without risk of ambiguity: for, by the definition of associativity, the two possible interpretations of $x \circ y \circ z$ (viz. $(x \circ y) \circ z$ and $x \circ (y \circ z)$) are the same.

Likewise, if w, x, y, z all belong to S, then, by repeated application of the definition of associativity, we can show that the five ways of inserting brackets into "$w \circ x \circ y \circ z$" yield identical results:

$$
\begin{aligned}
\{w \circ (x \circ y)\} \circ z &= \{(w \circ x) \circ y\} \circ z \\
&= (w \circ x) \circ (y \circ z) \qquad \text{(since } (a \circ y) \circ z = a \circ (y \circ z)) \\
&= w \circ \{x \circ (y \circ z)\} \qquad \text{(since } (w \circ x) \circ b = w \circ (x \circ b)) \\
&= w \circ \{(x \circ y) \circ z\}
\end{aligned}
$$

More generally ($\circ$ being associative on S), every expression

$$x_1 \circ x_2 \circ x_3 \circ \ldots \circ x_n \qquad \text{(where } x_1, x_2, x_3, \ldots, x_n \in S),$$

of whatever length, is unambiguous in the sense that all ways of bracketing it yield the same answer. This can be proved formally by induction on n.

If $\circ$ is commutative as well as associative on S, it is not difficult to see that, in the expression

$$x_1 \circ x_2 \circ x_3 \circ \ldots \circ x_n \qquad \text{(where each } x_i \in S)$$

one can make any prescribed change in the order of the x_i's without altering the meaning of the expression.

29. Semigroups: definition and examples

Definition. Let S be a non-empty set on which is defined a binary operation $*$. Then the system $(S, *)$ [i.e. the algebraic system consisting of the set S and the binary operation $*$ on S] is called a **semigroup** iff

(1) S is closed under $*$ and (2) $*$ is associative on S.

For example, clearly $(\mathbb{Z}, +)$ is a semigroup. Other examples will be given shortly.

The definition of a semigroup, as given above, has been worded so as to stress, for the beginner's benefit, that a semigroup is not merely a set of objects but a set endowed with a binary operation (satisfying certain conditions). Nevertheless it is customary to use phraseology that hides this point, e.g. as an alternative to the sentence "$(S, *)$ is a semigroup", one frequently says "S is a semigroup with respect to $*$" (or even simply "S is a semigroup", when the binary operation can be surmised from the context); and it is commonplace to speak of "the semigroup S" instead of "the semigroup $(S, *)$". A corresponding remark applies to every kind of algebraic system that we study.

Notice that, in the definition of a semigroup, we do not insist that the binary operation is commutative. A semigroup $(S, *)$ is called a **commutative semigroup** iff $*$ is commutative on S.

A *multiplicative* semigroup means a semigroup in which the binary operation is multiplication. Similarly, an *additive* semigroup means one in which the binary operation is addition.

Examples of semigroups

(1) $\mathbb{Z}$ is a semigroup with respect to multiplication. So is $\mathbb{N}$.

(2) Let Ω be a non-empty set, and let S be the set of all mappings from Ω to Ω. Then $(S, \circ)$ is a semigroup, where $\circ$ denotes mapping composition.

(3) Let $S = \mathscr{P}(\Omega)$, where Ω is a given non-empty set. Then $(S, \cap)$ and $(S, \cup)$ are commutative semigroups. [Note that $(S, \cap)$ and $(S, \cup)$ are different semigroups: for, although they involve the same set S, they involve different binary operations.]

(4) The set of all $n \times n$ real matrices (n being a given positive integer) is a multiplicative semigroup.

Proof of the above assertions is a trivial matter. In each case, we need only remark that the set is non-empty and closed under the binary operation, and that the binary operation is associative.

30. Powers of an element in a semigroup

Let $(S, *)$ be a semigroup, and let $x \in S$. Because of the associativity of $*$, we can unambiguously define (for each $n \in \mathbb{N}$) the nth **power** of x under $*$ by

$$n\text{th power of } x = x * x * x * \ldots * x \qquad (n \text{ factors}).$$

If $*$ is multiplication, the notation x^n is appropriate. If $*$ is addition, the notation nx is appropriate. [nx, meaning

$$x + x + x + \ldots + x \qquad (n \text{ terms}),$$

is usually called the nth *multiple* of x.]

The following *index laws* can be proved for every semigroup.

30.1 For $m, n \in \mathbb{N}$ and x an element of the semigroup $(S, *)$,

   (i) (mth power of x) $*$ (nth power of x) = $(m+n)$th power of x,

   (ii) nth power of (mth power of x) = (mn)th power of x.

It is instructive to write down the forms taken by these laws (a) in a multiplicative semigroup, (b) in an additive semigroup. These are

   (a) (i) $x^m . x^n = x^{m+n}$,   (ii) $(x^m)^n = x^{mn}$;

   (b) (i) $mx + nx = (m+n)x$,   (ii) $n(mx) = (mn)x$.

Proof of 30.1(i). Left-hand side = $\underbrace{(x * x * \ldots * x)}_{m \text{ factors}} * \underbrace{(x * x * \ldots * x)}_{n \text{ factors}}$

$$= \underbrace{x * x * x * x * \ldots * x}_{(m+n) \text{ factors}}$$

$$= (m+n)\text{th power of } x.$$

Part (ii) may be proved similarly.

31. Identity elements and inverses

Definition. In a semigroup $(S, *)$ an element e belonging to S is called an **identity** iff $e * x = x * e = x$ for every $x \in S$.

It should be appreciated that a given semigroup may or may not have an identity, e.g. in the semigroup $(\mathbb{Z}, .)$, the element 1 is an identity; but the semigroup $(E, .)$, where E is the set of all even integers, has no identity element.

31.1 A semigroup cannot contain more than one identity element.

Proof. Suppose that $(S, *)$ is a semigroup and that the elements e, f of S are both identities. [We shall prove that $e = f$.] We have:

$$e = e * f \qquad \text{(since } f \text{ is an identity)}$$
$$= f \qquad \text{(since } e \text{ is an identity).}$$

The result follows.

Notice that the proof of 31.1 does not use associativity.

When we know that a semigroup S has an identity e, we are, by 31.1, justified in calling e *the* identity of S.

In an additive semigroup S, it is usual to call the identity (if there is one) *zero* and denote it by 0: it satisfies

$$x + 0 = 0 + x = x \text{ for all } x \in S.$$

Likewise, in a multiplicative semigroup S, the identity element (if there is one) is often denoted by 1: it satisfies

$$1x = x1 = x \text{ for all } x \in S.$$

We turn attention now to inverses.

Definition. Let $(S, *)$ be a semigroup with an identity element e, and let $x \in S$. An element y belonging to S is called an **inverse** of x (in the semigroup) iff $x * y = y * x = e$.

If $(S, *)$ is a semigroup with identity e, then (since $e * e = e$) the element e has itself as an inverse, but it would be wrong to think that every other element of S must necessarily have an inverse: each may or may not. For example, in the semigroup $(\mathbb{Z}, .)$, which has identity 1, the elements 1 and -1 have themselves as inverses, but no other element has an inverse in the semigroup. [Of course one cannot, for example, claim $\frac{1}{2}$ as an inverse of 2 in this system: for an inverse of 2 must be an element of the semigroup, i.e. of $\mathbb{Z}$.]

We now establish that inverses, when they exist, are unique.

31.2 Let $(S, *)$ be a semigroup with an identity element, and let x be an element of S. Then x cannot have more than one inverse in S.

Proof. Suppose that y, z are both inverses of x in S. Then

$$
\begin{aligned}
y &= y * e &&\text{(where } e \text{ is the identity of the semigroup)} \\
&= y * x * z &&\text{(since } z \text{ is an inverse of } x \text{ [note also that associativity is} \\
&&&\text{tacitly used here])} \\
&= e * z &&\text{(since } y \text{ is an inverse of } x) \\
&= z &&\text{(} e \text{ being the identity).}
\end{aligned}
$$

This proves the result.

Because of 31.2, where a semigroup element x has an inverse, it is justifiable to refer to that inverse as *the* inverse of x.

In a multiplicative semigroup with identity 1, the natural notation for the inverse of element x (if it exists) is x^{-1}: it satisfies $xx^{-1} = x^{-1}x = 1$.

In an additive semigroup with an identity (i.e. a zero, 0), the inverse of element x (if it exists) is naturally called the *negative* of x and denoted by $-x$: it satisfies $x + (-x) = (-x) + x = 0$.

32. Subsemigroups

Suppose that $(S, *)$ is a semigroup and that T is a subset of S. (Then $*$ is a binary operation on T, as explained in §27, remark (c).) We say that T is a **subsemigroup of** S iff the system $(T, *)$ is a semigroup.

For example, in the semigroup $(\mathbb{Z}, .)$, $\mathbb{N}$ is a subsemigroup (since $\mathbb{N}$ is a subset of $\mathbb{Z}$ and $(\mathbb{N}, .)$ is a semigroup).

Clearly, for any semigroup S, S is a subsemigroup of itself.

The following theorem shows how simply one can test whether a given non-empty subset of a semigroup is a subsemigroup.

32.1 Suppose that T is a non-empty subset of S, where $(S, *)$ is a semigroup. Then T is a subsemigroup of S iff T is closed under $*$.

Proof. Obviously, if T is a subsemigroup of S, so that $(T, *)$ is a semigroup, then T is closed under $*$.

Conversely, suppose that T is closed under $*$. Then, since the condition

$$(x * y) * z = x * (y * z)$$

holds for all values of x, y, z in S, it holds, in particular, for all values of x, y, z in T, and thus $*$ is associative on T. From the definition of a semigroup, it now follows that $(T, *)$ is a semigroup, i.e. that T is a subsemigroup of S.

The stated result is now fully proved.

Worked example. Let S be the multiplicative semigroup of all 2×2 real matrices, and let T be the subset of S consisting of the matrices of the form $\begin{bmatrix} x & 0 \\ 0 & 0 \end{bmatrix}$ $(x \in \mathbb{R})$. Prove that T is a subsemigroup of S.

Solution. Clearly $T \neq \emptyset$.

Consider two arbitrary elements $A = \begin{bmatrix} x & 0 \\ 0 & 0 \end{bmatrix}$, $B = \begin{bmatrix} y & 0 \\ 0 & 0 \end{bmatrix}$ of T $(x, y \in \mathbb{R})$.

The product $AB = \begin{bmatrix} xy & 0 \\ 0 & 0 \end{bmatrix}$, which also belongs to T.

Hence T is closed under multiplication.

It follows (by 32.1) that T is a subsemigroup of S.

Post-script to this example. The semigroup S has an identity, viz. the matrix $I = \begin{bmatrix} 1 & 0 \\ 0 & 1 \end{bmatrix}$. And, although $I \notin T$, the subsemigroup T has an identity, viz. the matrix $\begin{bmatrix} 1 & 0 \\ 0 & 0 \end{bmatrix}$. This last assertion is clearly true since

$$\begin{bmatrix} 1 & 0 \\ 0 & 0 \end{bmatrix} \begin{bmatrix} x & 0 \\ 0 & 0 \end{bmatrix} = \begin{bmatrix} x & 0 \\ 0 & 0 \end{bmatrix} \begin{bmatrix} 1 & 0 \\ 0 & 0 \end{bmatrix} = \begin{bmatrix} x & 0 \\ 0 & 0 \end{bmatrix} \quad \text{for all } x \in \mathbb{R}.$$

The point is one which may surprise the uninitiated:

32.2 In a semigroup S with an identity, the identity (if it exists) of a subsemigroup is not necessarily the same as that of S.

EXERCISES ON CHAPTER FIVE

1. A binary operation $\circ$ is defined on $\mathbb{Z}$ by

$$x \circ y = x + y - xy \qquad (x, y \in \mathbb{Z}).$$

Prove that $(\mathbb{Z}, \circ)$ is a semigroup. Let $T = \{x \in \mathbb{Z} : x \leqslant 1\}$. Prove that T is a subsemigroup of the semigroup $(\mathbb{Z}, \circ)$.

2. Let $S = \mathscr{P}(\Omega)$, where Ω is a given set. Do the semigroups $(S, \cap)$ and $(S, \cup)$ have identities? If so, what are they?

3. In a semigroup $(S, \ast)$ an element $e \in S$ is called a *left identity* iff $e \ast x = x$ for all $x \in S$; an element $f \in S$ is called a *right identity* iff $x \ast f = x$ for all $x \in S$.

(i) Let S be a non-empty set, and let an operation $\circ$ be defined on S by $x \circ y = y$ $(x, y \in S)$. Prove that $(S, \circ)$ is a semigroup in which every element is a left identity.

(ii) Prove that, if in a semigroup there is more than one left identity, then there is no right identity.

4. In a semigroup $(S, \ast)$ an element x is called *idempotent* iff $x \ast x = x$.

Let S be a finite multiplicative semigroup, and let $a \in S$. Prove the existence of positive integers k, l such that $a^k = a^{k+l}$, and prove further that a^{kl} is idempotent. (This shows that, for every element a in a finite semigroup, there is a power of a which is idempotent.)

5. Suppose that a multiplicative semigroup S contains a left identity e (see question 3) and that correspondingly every element of S has a left inverse (i.e. for each $x \in S$, there is an element $x' \in S$ satisfying $x'x = e$ but not necessarily $xx' = e$). Prove that, for every $x \in S$, xx' is idempotent (see question 4). Deduce that e is an identity (i.e. a right identity as well as a left identity) and that, for each element x, its left inverse x' is in fact an inverse.

CHAPTER SIX

AN INTRODUCTION TO GROUPS

33. The definition of a group

A **group** may be concisely defined as a semigroup with an identity, in which every element has an inverse.

By 31.1 and 31.2, we can say straight away that:

33.1 In any group, there is just one identity element, and each element of the group has a unique inverse.

A very familiar example of a group is provided by the set of nonzero real numbers (i.e. $\mathbb{R} - \{0\}$): this set, it is easily seen, is a group with respect to multiplication.

Further examples of groups are given in §34. Meanwhile, in the remaining paragraphs of this section, there is a short sequence of remarks whose natural place is immediately beside the definition of a group. The most basic properties of groups will be developed in §35, and later in the chapter attention is given to subgroups (§§36, 37), to the very important concept of the period of a group element (§38), and to cyclic groups (§39), these being groups of a particularly simple kind.

Remarks. (*a*) The binary operation in a group need not be commutative. In the special case when the binary operation is commutative, the group is called an **abelian group**. (The adjective derives from the name of the nineteenth-century mathematician N. H. Abel.)

(*b*) For convenience, we shall concentrate in our general discussion of groups almost exclusively on multiplicative groups, i.e. groups in which the binary operation is multiplication. So, except where otherwise stated, the sentence "G is a group" should be taken to mean that G is a multiplicative group, i.e. that $(G, .)$ is a group. Correspondingly, in general discussion, the identity element of a group will be denoted by 1, and the inverse of an element x by x^{-1}. It is assumed that, if and when the need arises, the student will be capable of translating definitions and results (given here for a multiplicative

69

group) into the form appropriate for an additive group (or, more generally, a group in which the binary operation is $*$). Indeed it is recommended that, in the early stages of his study of groups, the student should systematically practise this capability. As an example, consider the "reversal rule for inverses" (to be proved in §35), which, for a multiplicative group G, states:

$$\forall x, y \in G, \quad (xy)^{-1} = y^{-1}x^{-1}$$

Fairly obviously, the corresponding result for an additive group G is:

$$\forall x, y \in G, \quad -(x+y) = (-y)+(-x)$$

which can be re-stated

$$\forall x, y \in G, \quad -(x+y) = -y-x$$

since in an additive group one defines subtraction by $u-v = u+(-v)$.

(c) Besides the concise definition of a group with which we opened this section, the following fuller (but obviously equivalent) definition of a multiplicative group should be learned.

A set $G(\neq \emptyset)$ on which multiplication is defined is a (multiplicative) group iff all the following postulates are satisfied:

G0: G is closed under multiplication;
G1: multiplication is associative on G;
G2: there is an element $1 \in G$ such that $1x = x1 = x$ for all $x = G$;
G3: corresponding to each element x of G, there is an element $x^{-1} \in G$ such that $x^{-1}x = xx^{-1} = 1$.

The postulates G0, G1, G2, G3 are called the *group postulates* (or **group axioms**). When given a set G on which multiplication is defined, we can prove that G is a group by showing that G0, G1, G2, G3 all hold for the given set G. (Notice that it is not necessary to check separately that $G \neq \emptyset$, since $G \neq \emptyset$ is implied by G2.)

34. Examples of groups

The following examples illustrate the very widespread occurrence of groups in mathematics. Proofs that the systems mentioned are groups are in most cases omitted.

(1) *Examples of groups from the familiar number systems*
$\mathbb{Z}, \mathbb{Q}, \mathbb{R}, \mathbb{C}$ are all abelian groups with respect to addition.
$\mathbb{Q}-\{0\}, \mathbb{R}-\{0\}, \mathbb{C}-\{0\}$ are all abelian groups with respect to multiplication.

(2) *From matrix algebra*
(i) For given $m, n \in \mathbb{N}$, the set of all $m \times n$ matrices with entries in $\mathbb{R}$ is an abelian group with respect to matrix addition. The same is true with $\mathbb{R}$ replaced by $\mathbb{C}$.

(ii) For given $n \in \mathbb{N}$, the set of all $n \times n$ nonsingular matrices with entries in $\mathbb{R}$ is a group with respect to matrix multiplication. (I_n is the identity; each matrix X in the set has an inverse (the matrix X^{-1}) also in the set.) This group is non-abelian if $n \geqslant 2$. It is often denoted by $GL(n, \mathbb{R})$, GL standing for "general linear". If instead we had considered the set of all $n \times n$ matrices with entries in $\mathbb{C}$, we would have described the group $GL(n, \mathbb{C})$.

(3) *From number theory*
For each integer $m \geqslant 2$, $\mathbb{Z}_m$ (see §§18, 19) is a group with respect to addition of congruence classes.

For each prime number p, $\mathbb{Z}_p - \{C_0\}$, i.e. the set of nonzero congruence classes modulo p, is a group with respect to multiplication of congruence classes.

(4) *Groups of permutations*
Let Ω be a non-empty set, and let G be the set of all permutations of Ω (i.e. bijections from Ω to itself). Then G is a group with respect to mapping composition. We call $(G, \circ)$ the **symmetric group** on Ω.

[*Proof* that $(G, \circ)$ is a group. We show that the group axioms all hold for G (taking mapping composition as the "multiplication").
G0: Holds, by 25.4(i).
G1: Holds, since mapping composition is associative (24.1).
G2: Holds since the mapping i_Ω belongs to G (by 24.2) and satisfies (by 24.3) $f \circ i_\Omega = i_\Omega \circ f = f$ for all $f \in G$.
G3: Holds because, for each $f \in G$, the inverse mapping f^{-1} exists, belongs to G (by 25.4(ii)), and (by 24.4) satisfies $f \circ f^{-1} = f^{-1} \circ f = i_\Omega$.]

A number of other matters deserve mention under the current heading of "groups of permutations".

(*a*) The phrase "a group of permutations of the set Ω" means a group whose elements are some (not necessarily all) of the permutations of Ω, the binary operation being mapping composition. Those who pursue their study of group theory beyond the level of this textbook will discover that groups of permutations play a most interesting key role in the development of the general theory of groups.

(*b*) In the context of group theory, when discussing a group G of permutations (of some set Ω), we often abbreviate $f \circ g$ to fg ($f, g \in G$).

(*c*) *The symmetric group of degree n* (where $n \in \mathbb{N}$) means the group of all $n!$ permutations of a set of n objects, which, for convenience, we label $1, 2, 3, \ldots, n$. This group is denoted by S_n.

An element of S_n is often denoted by a $2 \times n$ matrix in which the top row has entries $1, 2, 3, \ldots, n$ (in order) and the second row gives (in order) the images of $1, 2, 3, \ldots, n$, respectively, under the element in question; e.g. the permutation in S_4 that maps 1 to 3, 2 to 2, 3 to 4, and 4 to 1 may be denoted by

$$\begin{pmatrix} 1 & 2 & 3 & 4 \\ 3 & 2 & 4 & 1 \end{pmatrix}$$

(5) *Groups of symmetries of a geometrical figure*
Suppose that Π is a geometrical figure in a euclidean space S.   (S might be a plane or ordinary 3-dimensional space.)   We think of S as a set of points.

An **isometry** of S means a bijective mapping from S to S that preserves distance; i.e. two points at distance d from each other must have images under the mapping which have distance d from each other. Examples include translations, rotations, reflections.

An isometry of S is called a **symmetry** of Π iff it leaves Π as a whole unchanged while perhaps permuting its component elements (points, lines, etc.); e.g. if S is a plane and Π is a square in the plane with centre O, rotation of the plane about O through a right angle (in either direction) is a symmetry of Π.

It can be proved that the set of symmetries of a given figure Π is, in all cases, a group with respect to mapping composition. We call this group the **symmetry group** of Π. In it the identity is the mapping i_S, which fixes (i.e. maps to itself) every point of the space S.

As a specific example, consider the symmetry group of the equilateral triangle Δ with centroid at O and altitudes along the lines l, m, n (see diagram).

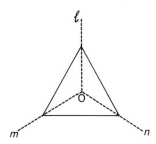

Clearly Δ has the following symmetries:
   $1 =$ identity mapping from the plane to itself,
   $p =$ rotation anticlockwise about O through $120°$,
   $q =$ rotation clockwise about O through $120°$,
   $a =$ reflection in l,
   $b =$ reflection in m,
   $c =$ reflection in n.

In fact, these six symmetries form the complete group G of symmetries of the plane figure Δ.

A multiplication table for this group G is exhibited below. In the table, the entry in the row headed x and the column headed y gives the product xy. (As in the discussion of a group of permutations, so here, we use xy as an abbreviation for $x \circ y$.)

	1	p	q	a	b	c
1	1	p	q	a	b	c
p	p	q	1	c	a	b
q	q	1	p	b	c	a
a	a	b	c	1	p	q
b	b	c	a	q	1	p
c	c	a	b	p	q	1

To illustrate how the information in the table can be obtained, let us consider pa (which means the composition of a followed by p). By means of the diagram below we can follow how the vertices of the triangle are moved when we apply first a and then p, and we observe that the net result is the same as the effect of c: hence we deduce that $pa = c$.

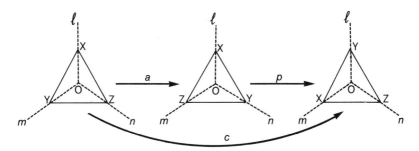

Remarks. (*a*) Since (as the table reveals) $ap \neq pa$, this group G is non-abelian. Later we shall prove that there is no smaller non-abelian group, i.e. that every group with fewer than 6 elements is abelian.

(*b*) The above table illustrates a fact that we shall prove in the next section—that, in each row or column of the multiplication table of a finite group, each element of the group appears precisely once.

(*c*) This subsection has brought to light a point of some practical significance—that wherever there is geometrical symmetry (e.g. in the structure of a molecule or in the configuration of atoms in a crystal), there will be a group to discuss. Often details of the symmetry and details of the group are inter-related in an intelligible way.

(6) More finite groups

If G is a finite group, $|G|$ (i.e. the number of elements in the set G) is always called the **order** of the group G; e.g. the order of S_n is $n!$ (see subsection (4) above), and the order of the group of symmetries of an equilateral triangle is 6.

Here now are details of two further interesting finite groups.

(i) The set of complex numbers $\{1, -1, i, -i\}$ is, with respect to multiplication, a group of order 4. The student will have no difficulty in constructing the multiplication table of this abelian group.

(ii) The set of matrices $Q = \{I, -I, J, -J, K, -K, L, -L\}$, where

$$I = \begin{bmatrix} 1 & 0 \\ 0 & 1 \end{bmatrix}, \quad J = \begin{bmatrix} i & 0 \\ 0 & -i \end{bmatrix}, \quad K = \begin{bmatrix} 0 & 1 \\ -1 & 0 \end{bmatrix}, \quad L = \begin{bmatrix} 0 & i \\ i & 0 \end{bmatrix}$$

is a group with respect to matrix multiplication, I being the identity. This noteworthy non-abelian group of order 8 is called the **quaternion group**. After verifying that

$$JK = -KJ = L, \quad KL = -LK = J, \quad LJ = -JL = K, \quad J^2 = K^2 = L^2 = -I.$$

one can quickly construct a multiplication table for Q. It will be worth while for the student to do so and store the result for future reference.

Trivial groups

A group of order 1 is described as a **trivial** group. In such a group the sole element is the identity, 1, and the full story of the multiplication is told in the equation $1.1 = 1$.

35. Elementary consequences of the group axioms

Throughout this section G will denote an arbitrary multiplicative group.

35.1 (The *cancellation laws* for group G) For $a, x, y \in G$,

  (i) $ax = ay \Rightarrow x = y$,
  (ii) $xa = ya \Rightarrow x = y$.

Proof of (i). Suppose that $ax = ay \, (a, x, y \in G)$. By G3, a has an inverse a^{-1} in G, and hence we have

$$a^{-1}ax = a^{-1}ay \quad \text{(associativity (G1) being tacitly used here)},$$

i.e. $1x = 1y,$

i.e. $x = y.$

This proves (i). Part (ii) may be proved similarly.

It should be obvious why 35.1 is entitled "cancellation laws". The next result is sometimes referred to as "the *division laws* (for group *G*)".

35.2 For $a, b, x \in G$,

   (i) $ax = b \Leftrightarrow x = a^{-1}b$,

   (ii) $xa = b \Leftrightarrow x = ba^{-1}$.

   Proof of (i). $ax = b \Rightarrow a^{-1}ax = a^{-1}b$ (using G3 and G1)

$$\Rightarrow 1x = a^{-1}b$$
$$\Rightarrow x = a^{-1}b$$

and, conversely,

$$x = a^{-1}b \Rightarrow ax = aa^{-1}b \Rightarrow ax = 1b \Rightarrow ax = b.$$

Part (ii) is proved similarly.

These "division laws" prompt the remark that, in a non-abelian group *G*, it would be a great mistake to employ the notation $\dfrac{b}{a}$ for the solution of either of the equations $ax = b$, $xa = b$: for the notation $\dfrac{b}{a}$ is ambiguous—one does not know whether it means ba^{-1} or $a^{-1}b$.

Hitherto, for y, u belonging to the group *G*, in order to be able to draw the conclusion "*u* is the inverse of *y*", we have had to establish *both* that $uy = 1$ *and* that $yu = 1$. However, as a particular case of 35.2, we now have:

35.3 For u, y belonging to the group *G*, if we know *either* that $uy = 1$ *or* that $yu = 1$, then we may conclude that $u = y^{-1}$.

   Proof. By 35.2(i), $yu = 1 \Rightarrow u = y^{-1}1 = y^{-1}$;
and, by 35.2(ii), $uy = 1 \Rightarrow u = 1y^{-1} = y^{-1}$.

This last result enables all of the following to be rapidly proved.

35.4 In the group *G*, the inverse of the identity is itself (i.e. $1^{-1} = 1$).

35.5 For every x in the group *G*, $(x^{-1})^{-1} = x$.

35.6 For all x, y in the group *G*, $(xy)^{-1} = y^{-1}x^{-1}$.

The result 35.6 is known as the *reversal rule* for group inverses. It extends to longer products, i.e.

35.7 For elements $x_1, x_2, \ldots, x_n$ belonging to the group *G*,

$$(x_1 x_2 \ldots x_{n-1} x_n)^{-1} = x_n^{-1} x_{n-1}^{-1} \ldots x_2^{-1} x_1^{-1}.$$

To illustrate the usefulness of 35.3, here is a proof of 35.6.

   Proof. Let $x, y \in G$. Then $y^{-1}x^{-1}$ is an element of *G*, and we have

$$(xy)(y^{-1}x^{-1}) = xyy^{-1}x^{-1} = x1x^{-1} = 1$$

By 35.3, it follows that $(xy)^{-1} = y^{-1}x^{-1}$.

As an important postscript to the above results, we introduce, for each element a of the group G, two mappings from G to G which we shall always denote by λ_a and ρ_a and which we call "left multiplication by a" and "right multiplication by a", respectively. These are specified by

$$\lambda_a(x) = ax, \quad \rho_a(x) = xa \qquad (x \in G).$$

It is useful to consider these mappings, largely because:

35.8 For each a belonging to the group G, λ_a and ρ_a are permutations of G, i.e. bijections from G to G.

Proof. Consider λ_a where a is an arbitrary element of G.
λ_a is injective because (cf. 22.1)

$$\lambda_a(x) = \lambda_a(y) \Rightarrow ax = ay \Rightarrow x = y \qquad \text{(by 35.1)}.$$

And λ_a is surjective because (cf. 21.2) the arbitrary element y of G is the image under λ_a of $a^{-1}y$, since $\lambda_a(a^{-1}y) = aa^{-1}y = 1y = y$.
It follows that, for each $a \in G$, λ_a is bijective.
Similarly one can prove that ρ_a is bijective for each $a \in G$.

We can now prove a fact mentioned in passing in §34, viz:

35.9 If G is a finite group, then in every row or column of the multiplication table of G each element of G appears exactly once.

Proof. Let G be a finite group of order n, and let $x_1, x_2, \ldots, x_n$ be a list of the elements of G in the order in which they appear in the headings of the multiplication table. Consider an arbitrary row of the table—say the row headed a ($a \in G$). The entries in this row of the table are, respectively,

$$ax_1, ax_2, \ldots, ax_n$$

i.e.

$$\lambda_a(x_1), \lambda_a(x_2), \ldots, \lambda_a(x_n).$$

So, since λ_a is a permutation of $G = \{x_1, x_2, \ldots, x_n\}$, it is clear that each element of G appears exactly once in the row. This proves the result for rows; and a similar discussion proves it for columns.

Powers of a group element
Let x be an element of the group G. By virtue of our discussion of positive integral powers of an element of a semigroup (in §30), x^n already has a meaning for every $n \in \mathbb{N}$. We can also define non-positive integral powers of the group element x as follows:
zeroth power: $x^0 = 1$;

negative powers: for $n \in \mathbb{N}$, $x^{-n} = (x^n)^{-1}$, which (by 35.7) is the same as $(x^{-1})^n$.

The following *index laws* are universally true in group theory.

35.10 For a group element x,

$$\text{(i) } x^m . x^n = x^{m+n} \quad \text{(ii) } (x^m)^n = x^{mn} \qquad (m, n \in \mathbb{Z})$$

To prove these, one needs to separate out various cases and patiently establish the laws in each case, making use of 30.1. By way of illustration, here is a proof of (i) for the case where $m > 0$, $n < 0$ and $|m| > |n|$.

Proof. Write $n = -l$, so that $l \in \mathbb{N}$ and $l < m$. Then

$$
\begin{aligned}
x^m . x^n &= x^m . x^{-l} \\
&= x^{m-l} . x^l . x^{-l} && \text{(using 30.1 and the fact that } m - l > 0) \\
&= x^{m-l} . x^l . (x^l)^{-1} && \text{(by definition of a negative power)} \\
&= x^{m-l} . 1 && \text{(since } yy^{-1} = 1 \text{ for every } y \text{ in the group)} \\
&= x^{m-l} \\
&= x^{m+n}
\end{aligned}
$$

36. Subgroups

Suppose that G is a (multiplicative) group, i.e. that $(G, .)$ is a group, and that H is a subset of G. (As explained in §27, remark (c), multiplication is defined on H.)   We say that H is a **subgroup** of G iff $(H, .)$ is a group.

Simple illustrations:

(1) $\mathbb{Q} - \{0\}$ is a subgroup of the multiplicative group $\mathbb{R} - \{0\}$.

(2) $\{1, -1, i, -i\}$ is a subgroup of the multiplicative group $\mathbb{C} - \{0\}$.

(3) If H is a group of permutations of the set Ω and G is the symmetric group on Ω (see §34(4)), then H is a subgroup of G.

It is clear that:

36.1 For every group G, $\{1\}$ and G are subgroups of G.

We use the following terminology for subgroups of a group G:

(*a*) $\{1\}$ is the **trivial subgroup** of G;

(*b*) a **proper subgroup** of G is a subgroup different from G;

(*c*) a *non-trivial proper subgroup*, therefore, is a subgroup equal to neither $\{1\}$ nor G.

Two elementary facts about a subgroup are presented in the next theorem. Anyone tempted to suggest that it is unnecessary to remark on these points should scrutinize his opinions in the light of 32.2.

36.2 Suppose that H is a subgroup of the group G. Then:

(i) the identity 1_H of the group H is the same as the identity 1 of the group G;

(ii) for each $h \in H$, the inverse of h in the group H is the same as the inverse of h in the group G.

Proof. (i) Since 1_H is the identity of H, $1_H 1_H = 1_H$, and, since 1 is the identity of G, $11_H = 1_H$. Hence

$$1_H 1_H = 11_H$$

and hence, by cancellation in the group G (35.1(ii)), $1_H = 1$.

(ii) Let h be an arbitrary element of H. Let h' denote its inverse in the group H, and let h^{-1} denote its inverse in the group G. In view of part (i), $hh' = 1$. By 35.3, (since $h' \in G$) it follows that $h' = h^{-1}$. This proves the result.

Tests for a subgroup

If H is a subset of the group G, one can test whether H is a subgroup of G by checking whether the group axioms G0, G1, G2, G3 hold for H. The following results, however, give quicker and more efficient tests.

36.3 Let H be a subset of the group G. Then H is a subgroup of G iff the following conditions all hold:

(1) $H \neq \emptyset$,

(2) H is closed under multiplication,

(3) $x \in H \Rightarrow x^{-1} \in H$.

[*Note*: Condition (3) is sometimes expressed as "H is closed under the taking of inverses".]

Proof. (i) Suppose that H is a subgroup. Then (1) and (2) hold by virtue of the group axioms G2, G0 as they apply to H. Moreover (3) holds because, if $x \in H$, x has an inverse in H (since G3 holds for H) and (by 36.2(ii)) this inverse equals x^{-1}, so that $x^{-1} \in H$.

(ii) Conversely, suppose that the conditions (1), (2), (3) all hold. We prove that the group axioms G0, G1, G2, G3 hold for H.

G0 holds for H, by condition (2).

G1 clearly holds for H, since the multiplication on G is associative.

To prove G2, introduce into the discussion an element h of H (a legitimate step, by condition (1)). By (3), $h^{-1} \in H$; and hence, by (2), $hh^{-1} \in H$, i.e. $1 \in H$. Hence G2 holds for H.

G3 holds for H, by condition (3).

It follows that H is a subgroup.

Proof of the stated result is now complete.

An even simpler but not always more useful test is:

36.4 Let H be a subset of the group G. Then H is a subgroup of G iff (1) $H \neq \emptyset$ and (2) x and y belong to $H \Rightarrow x^{-1}y \in H$.

Proof. (i) If H is a subgroup, then certainly $H \neq \emptyset$ and

$$x \text{ and } y \in H \Rightarrow x^{-1} \text{ and } y \in H \qquad \text{(by G3 for } H\text{)}$$
$$\Rightarrow x^{-1}y \in H \qquad \text{(by G0 for } H\text{)}.$$

(ii) Conversely, suppose that the conditions (1), (2) hold.

By (1), we may introduce an element h of H. By (2), $h^{-1}h \in H$, i.e. $1 \in H$. Therefore G2 holds for H.

Next, knowing that $1 \in H$, we deduce from (2) that

$$x \in H \Rightarrow x^{-1}1 \in H,$$

i.e.

$$x \in H \Rightarrow x^{-1} \in H.$$

Therefore G3 holds for H.

We now observe that

$$x \text{ and } y \in H \Rightarrow x^{-1} \text{ and } y \in H \qquad \text{(by the previous paragraph)}$$
$$\Rightarrow (x^{-1})^{-1}y \in H \qquad \text{(by (2))}$$
$$\Rightarrow xy \in H.$$

Therefore G0 holds for H.

Finally, G1 certainly holds for H, since the multiplication on G is associative.

It follows that H is a subgroup.

Proof of the stated result is now complete.

The final variation on this theme relates to the case H finite:

36.5 Suppose that H is a *finite* subset of the group G. Then H is a subgroup of G iff (1) $H \neq \emptyset$ and (2) H is closed under multiplication.

Proof. (i) If H is a subgroup of G, then clearly the conditions (1) and (2) hold.

(ii) Conversely, suppose that the conditions (1), (2) hold.

Suppose that x is an element of H. Since H is closed under multiplication, all elements in the list $x, x^2, x^3, x^4, \ldots$ belong to H. But H is finite. Therefore the list cannot be repetition-free, i.e. we must have

$$x^r = x^s \qquad (\alpha)$$

for some $r, s \in \mathbb{N}$ with $r < s$. The equation (α) can be re-written $x^r . 1 = x^r . x^{s-r}$; and hence, by cancellation in the group G, we deduce that

$$x^{s-r} = 1 \qquad (\beta)$$

Since $s - r \in \mathbb{N}$, it follows that $1 \in H$. Further, (β) can be re-written $x . x^{s-r-1} = 1$, showing (by 35.3) that $x^{-1} = x^{s-r-1}$. Since $s - r - 1$ is either a positive integer or zero, i.e. x^{s-r-1} is either a positive power of x or 1, it follows that $x^{-1} \in H$.

This shows that $x \in H \Rightarrow x^{-1} \in H$. And from this and the conditions (1) and (2), it follows (by 36.3) that H is a subgroup.

Proof of the stated result is now complete.

Worked example. Let $G = GL(n, \mathbb{R})$ [see §34(2)], and let H be the subset of G consisting of the matrices in G with determinant 1. Show that H is a subgroup of G.

[*Note.* The following facts about determinants will be assumed:
  (i) for $n \times n$ matrices A, B, $\det(AB) = \det A \times \det B$; and (consequently)
  (ii) for a nonsingular matrix A, $\det(A^{-1}) = 1/\det A$.]

Solution. We use 36.4.

(1) $H \neq \emptyset$, since the identity matrix I_n belongs to H.

(2) Let X, Y be arbitrary members of H, so that $\det X = \det Y = 1$. Then, using the facts about determinants noted above, we have

$$\det(X^{-1}Y) = \det(X^{-1}) \times \det Y = (1/\det X) \times \det Y = (1/1) \times 1 = 1;$$

and so $X^{-1}Y \in H$.

This proves that X, $Y \in H \Rightarrow X^{-1}Y \in H$.

(By 36.4) it follows that H is a subgroup of G.

Further illustrations of the use of 36.3 and 36.4 are provided by the proofs of the following theorem and of theorems in §37.

36.6 Let H and K be subgroups of a group G. Then $H \cap K$ is also a subgroup of G.

Proof. (1) Since H, K are subgroups of G, $1 \in H$ and $1 \in K$. Hence $1 \in H \cap K$. In particular, $H \cap K \neq \emptyset$.

(2) Let x, y be arbitrary elements of $H \cap K$. Then x, y both belong to H, and so, since H is a subgroup, $x^{-1}y \in H$. Similarly, $x^{-1}y \in K$. Hence $x^{-1}y \in H \cap K$.

Thus x, $y \in H \cap K \Rightarrow x^{-1}y \in H \cap K$.

By 36.4, it follows that $H \cap K$ is a subgroup of G.

More generally, one can prove that:

36.7 If G is a group, then the intersection of any finite or infinite family of subgroups of G is again a subgroup of G.

37. Some important general examples of subgroups

(1) *Centralizers*

Let G be a given group and let a be an element of G (a to be fixed throughout the following discussion). Let

$$C(a) = \{x \in G : xa = ax\}$$

i.e. the set of all members of G that commute with the particular element a. We call $C(a)$ the **centralizer** of a (in G).

37.1 $C(a)$ is a subgroup of G.

Proof. We use 36.3.
(1) Since $1a = a1$, $1 \in C(a)$. So $C(a) \neq \emptyset$.
(2) Let $x, y \in C(a)$. Then $xa = ax$ and $ya = ay$; and hence

$$(xy)a = x(ya) = x(ay) = (xa)y = (ax)y = a(xy),$$

which shows that $xy \in C(a)$.

Thus $C(a)$ is closed under multiplication.
(3) Let $x \in C(a)$. Then $ax = xa$. Hence $x^{-1}(ax)x^{-1} = x^{-1}(xa)x^{-1}$, i.e. $x^{-1}a = ax^{-1}$, i.e. $x^{-1} \in C(a)$.

Thus $x \in C(a) \Rightarrow x^{-1} \in C(a)$.

It follows (by 36.3) that $C(a)$ is a subgroup.

(2) *The centre of a group*
Definition. The **centre** of a group G is the subset

$$\{x \in G : \forall g \in G,\ xg = gx\}$$

i.e. it is the subset of G comprising those elements that commute with every element of G. The centre of G is usually denoted by $Z(G)$.

On comparing the definition of $Z(G)$ with that of the centralizer of an element of G, we see that, for $x \in G$, x belongs to $Z(G)$ iff x belongs to the centralizer of every individual element of G. Therefore $Z(G)$ equals the intersection of all the centralizers of individual elements of G. By 36.7 and 37.1, it follows that:

37.2 For every group G, $Z(G)$ is a subgroup of G.

The result 37.2 may also be proved directly by using 36.3 instead of proceeding via 36.7 and 37.1.

In chapter 8, an illustration will be given of how useful consideration of the centre of a group can be. In the meantime, note that:

37.3 For a group G, $Z(G)$ is the whole of G iff G is abelian.

(3) *The subgroup generated by an individual element*
Let G be a group, and let a be an element of G (a to be fixed throughout the following discussion). Noting that (as a simple consequence of the group axioms) a^n is an element of G for every $n \in \mathbb{Z}$, consider

$$H = \{a^n : n \in \mathbb{Z}\}$$

i.e. the subset of G comprising all elements expressible as powers of a.

37.4 $H (=\{a^n : n \in \mathbb{Z}\})$ is a subgroup of G.

Proof. (1) Clearly $H \neq \emptyset$: e.g. $a \in H$.
(2) Let $x, y \in H$. Then $x = a^m$, $y = a^n$ for some $m, n \in \mathbb{Z}$. Hence (by 35.10)
$$x^{-1}y = (a^m)^{-1}a^n = a^{-m}a^n = a^{n-m}$$
which shows that $x^{-1}y \in H$.
Thus $x, y \in H \Rightarrow x^{-1}y \in H$.
It follows (by 36.4) that H is a subgroup.

We call the subgroup $H (=\{a^n : n \in \mathbb{Z}\})$ the **subgroup of G generated by** a, and we denote it by $\langle a \rangle$. Note that:

37.5 If G is a group and $a \in G$, then $\langle a \rangle$ is the smallest subgroup of G containing the element a, i.e. if L is any subgroup of G such that $a \in L$, then $\langle a \rangle \subseteq L$.

Proof. Let L be a subgroup of the group G such that $a \in L$. By the group axioms as they apply to L, it is clear that every integral power of an element of L must be an element of L. In particular, therefore, $a^n \in L$ for every $n \in \mathbb{Z}$, and thus $\langle a \rangle \subset L$. This proves the stated result.

38. Period of an element

Let a be a given element of the group G. We are concerned throughout this section with the powers of a, i.e. the elements of the subgroup $\langle a \rangle$. We begin with a twofold definition that draws attention to an important distinction between two possible cases.

Definition. (1) If there is no positive power of a that equals 1, we say that a has **infinite period**.
(2) If there is a positive power of a that equals 1 and the kth power of a is the lowest such positive power (i.e. $a^k = 1$ but $a^n \neq 1$ if $1 \leqslant n \leqslant k-1$), we say that a has **period** k.

Remarks. (*a*) In definition (2), we know that there is such a thing as the "lowest such positive power" by virtue of the well-ordering principle (10.1).
(*b*) The word "order" is often used instead of "period".

Illustrations. (i) In the group $\mathbb{R} - \{0\}$, 2 has infinite period (because, $\forall n \in \mathbb{N}$, $2^n \neq 1$).
(ii) In the group $\mathbb{C} - \{0\}$, i has period 4 (because $i^4 = 1$ while i^1, i^2, i^3 are all different from 1).
Clearly:

38.1 In every group, the identity (and only the identity) has period 1.

There follows a detailed analysis of each of the two cases distinguished in the above definitions.

(1) *Infinite period case*

38.2 Suppose that the group element a has infinite period. Then, for $m, n \in \mathbb{Z}$, $a^m \neq a^n$ if $m \neq n$ (and consequently $\langle a \rangle$ is infinite).

Proof. For $m, n \in \mathbb{Z}$,

$$a^m = a^n \Rightarrow a^m (a^n)^{-1} = 1 \text{ and } a^n (a^m)^{-1} = 1$$
$$\Rightarrow a^{m-n} = a^{n-m} = 1$$
$$\Rightarrow a^{|m-n|} = 1$$
$$\Rightarrow |m-n| = 0 \qquad \text{(since no positive power of } a \text{ equals 1)}$$
$$\Rightarrow m = n$$

The result follows (by the law of contraposition).

An immediate corollary of 38.2 is:

38.3 In a finite group G, no element can have infinite period.

(2) *Finite period case*

38.4 Suppose that the group element a has finite period k ($k \in \mathbb{N}$). Then:
   (i) if the integer n has principal remainder r on division by k, then $a^n = a^r$;
   (ii) for $n \in \mathbb{Z}$, $a^n = 1 \Leftrightarrow k|n$;
   (iii) $1, a, a^2, \ldots, a^{k-1}$ is a complete repetition-free list of the elements of $\langle a \rangle$;
   (iv) $|\langle a \rangle| = k$, i.e. the order of the subgroup generated by a equals the period of a.

Proof. (i) Let $n \in \mathbb{Z}$ and let n have principal remainder r on division by k. Then $n = qk + r$ for some integer q. Hence, using the index laws and the fact that $a^k = 1$, we have

$$a^n = a^{qk+r} = (a^k)^q a^r = 1^q a^r = a^r$$

as required.
   (ii) Let n denote an integer.
   First suppose that $a^n = 1$. Let r be the principal remainder on division of n by k. Then, by part (i), $a^r = a^n = 1$. But $0 \leqslant r \leqslant k-1$ and, since k is the period of a, $a^s \neq 1$ if $1 \leqslant s \leqslant k-1$. It follows that $r = 0$, i.e. that $k|n$.
   Thus $a^n = 1 \Rightarrow k|n$.
   Part (ii) follows on observing that, conversely:

$$k|n \Rightarrow n = lk \text{ for some integer } l$$
$$\Rightarrow a^n = a^{lk} = (a^k)^l = 1^l = 1$$

   (iii) By (i), every power of a is equal to one appearing in the list

$1, a, a^2, \ldots, a^{k-1}$. Moreover, this list is repetition-free: for if it contained a repetition, we would have $a^m = a^n$ with $0 \leqslant m < n \leqslant k-1$ and hence $a^{n-m} = 1$ with $1 \leqslant n-m \leqslant k-1$—a contradiction since the period of a is k. These observations establish part (iii).

(iv) It follows immediately from (iii) that $|\langle a \rangle| = k$.

Among all the important detail of the various parts of 38.4, the student should not overlook the simple fact that, when the group element a has period k, the sequence of successive powers of a (i.e. the sequence $\ldots, a^{-2}, a^{-1}, a^0, a, a^2, a^3, a^4, \ldots$) repeats itself with periodicity k. The case $a = i$ in the group $\mathbb{C} - \{0\}$ provides a helpful illustration of this and of all the points brought out in 38.4.

For further illustrations of group elements with finite period, let us turn to the group of all isometries of the coordinate plane. In this group, let r be rotation anticlockwise about the origin through $2\pi/n$ (i.e. through one nth of a revolution), n being an arbitrary positive integer. Clearly r has period n, and so $\langle r \rangle$ is (by 38.4(iv)) a group of order n. This illustration shows that:

38.5 For every $n \in \mathbb{N}$, there exists a group of order n.

39. Cyclic groups

Definition. A group (or subgroup) G is called **cyclic** iff there is an element a ($\in G$) such that $\langle a \rangle$ is the whole of G.

So in any group G, the subgroup generated by any element a (i.e. $\langle a \rangle$) is a cyclic subgroup of G. Thus it is easy to provide illustrations of cyclic groups, e.g.:

(1) the subgroup $\langle 2 \rangle = \{2^n : n \in \mathbb{Z}\}$ of $\mathbb{R} - \{0\}$ is an infinite cyclic group;
(2) the subgroup $\langle i \rangle = \{1, i, i^2, i^3\} = \{1, i, -1, -i\}$ of $\mathbb{C} - \{0\}$ is a cyclic group of order 4 (cf. 38.4, (iii) and (iv)).

It is helpful to note at the outset the following very simple point.

39.1 If G is a group such that (1) $x \in G$ and (2) every element of G is expressible as a power of x, then $G = \langle x \rangle$ (and thus G is cyclic).

Proof. Suppose that the conditions (1) and (2) hold for the group G. Because of (1), $\langle x \rangle$ is a subset of the group G. And, because of (2), every element of G belongs to the subset $\langle x \rangle$. Therefore $\langle x \rangle = G$.

If G is a cyclic group, an element $x \in G$ such that $\langle x \rangle$ is the whole of G is called a **generator** of G. (This fits in with the description of $\langle a \rangle$, where a is a group element, as the subgroup generated by a.)

By the meaning of the word "cyclic", for any cyclic group G there must be at least one generator of G; there may be more than one such generator, though it should not be imagined that every element of G must be a generator of G. For example, consider the case $G = \{2^n : n \in \mathbb{Z}\}$ $(= \langle 2 \rangle)$. Obviously 2 is a generator. So is $\frac{1}{2}$, since $\langle \frac{1}{2} \rangle$ is the whole of G. [This is clear from 39.1, since (1) $\frac{1}{2} \in G$ and (2) the arbitrary element 2^n of G is expressible as a power of $\frac{1}{2}$, viz. $(\frac{1}{2})^{-n}$.] On the other hand, the element 4 $(= 2^2)$ of G is not a generator of G, because the subgroup $\langle 4 \rangle$ is not the whole of G. [For example, $2 \notin \langle 4 \rangle$, since 2 cannot be expressed as an integral power of 4.]

If G is a finite cyclic group of order k and x is a generator of G, so that $G = \langle x \rangle$, then (making use of 38.4(iv)), we have

$$\text{period of } x = |\langle x \rangle| = |G| = k$$

Thus:

39.2 Every generator of a finite cyclic group of order k has period k.

Properties of cyclic groups
39.3 Every cyclic group is abelian.

Proof. Let G be a cyclic group, so that $G = \langle a \rangle$ for some element $a \in G$.

Let x, y be arbitrary elements of G. Since $G = \langle a \rangle$, $x = a^m$ and $y = a^n$ for some $m, n \in \mathbb{Z}$. Hence

$$xy = a^m . a^n = a^{m+n} = a^{n+m} = a^n . a^m = yx$$

Since x, y were arbitrary, this proves that G is abelian.

39.4 Every subgroup of a cyclic group is also cyclic.

Proof. Let $G = \langle a \rangle$ be an arbitrary cyclic group, and let H be a subgroup of G.

If H is the trivial subgroup (i.e. $\{1\}$), then $H = \langle 1 \rangle$, which is cyclic.

So we suppose henceforth that H is non-trivial. Consequently H contains a nonzero power of a, and (since $x \in H \Rightarrow x^{-1} \in H$) it follows that H contains at least one positive power of a. By the well-ordering principle (10.1), we may legitimately consider the lowest positive power of a that belongs to H: let this be the dth power of a. [We shall show that $H = \langle a^d \rangle$.]

Let x be an arbitrary element of H. Since $H \subseteq G = \langle a \rangle$, $x = a^n$ for some $n \in \mathbb{Z}$. By the division algorithm (11.1), we can write $n = qd + r$, for some $q, r \in \mathbb{Z}$ with $0 \leqslant r \leqslant d - 1$; and, since a^n $(= x)$ and a^d belong to the subgroup H, we deduce that

$$a^n (a^d)^{-q} \in H, \quad \text{i.e. } a^{n-qd} \in H, \quad \text{i.e. } a^r \in H.$$

But $0 \leqslant r \leqslant d - 1$, and a^d is the lowest positive power of a that belongs to H. Hence $r = 0$, and so $x = a^n = (a^d)^q$.

Thus every element of H is expressible as a power of a^d, which belongs to H. It follows, by 39.1, that $H = \langle a^d \rangle$.

The stated result now follows.

Worked problem. Let G be a finite cyclic group of order n, and let d be a divisor of n. Show that G has exactly one subgroup of order d.

Solution. Let x be a generator of G, so that, by 39.2, x has period n. Clearly $x^{n/d}$ has period d, and so G has a subgroup of order d, viz. the subgroup $H = \langle x^{n/d} \rangle$.

Let K be an arbitrary subgroup of G of order d. [We aim to prove that $K = H$.] By 39.4, K is cyclic: say $K = \langle x^t \rangle$ ($t \in \mathbb{Z}$). [Of course, since $K \subseteq G$, any generator of K belongs to G and is therefore a power of x.] By 39.2, x^t has period d and therefore $(x^t)^d = 1$, i.e. $x^{td} = 1$. Hence, by 38.4(ii), since x has period n, $n|dt$, i.e. $dt = qn$ for some $q \in \mathbb{Z}$. So $t = (n/d) \times q$, and hence $x^t = (x^{n/d})^q$. It follows that x^t belongs to $\langle x^{n/d} \rangle$, i.e. $x^t \in H$. Therefore, since H is a group, $\langle x^t \rangle \subseteq H$, i.e. $K \subseteq H$. As $|K| = |H|$, it follows that $K = H$.

Hence H is the one and only subgroup of order d in G.

EXERCISES ON CHAPTER SIX

N.B. The symbol G, wherever it occurs in the following questions, should be taken to denote a multiplicative group (arbitrary apart from restrictions imposed in the question).

1. Let G_1, G_2 be groups, and let multiplication be defined on $G_1 \times G_2$ by
$$(x_1, y_1)(x_2, y_2) = (x_1 x_2, y_1 y_2) \qquad (x_1, x_2 \in G_1, y_1, y_2 \in G_2).$$
Prove that (with respect to the multiplication just defined) $G_1 \times G_2$ is a group. (This group, to which we refer in some future exercises, is called the *direct product* of G_1, G_2.) Prove that the direct product $G_1 \times G_2$ is abelian iff both G_1 and G_2 are abelian.

2. Prove that, $\forall x, y \in G, \lambda_x \circ \lambda_y = \lambda_{xy}, \rho_x \circ \rho_y = \rho_{yx}, \lambda_x \circ \rho_y = \rho_y \circ \lambda_x$. (See §35 for the meanings of $\lambda_a, \rho_a \ (a \in G)$.)

3. Let S be a finite multiplicative semigroup in which the cancellation laws hold (i.e. for $a, x, y \in S, ax = ay \Rightarrow x = y$ and $xa = ya \Rightarrow x = y$). Prove that S is a group.

Give an example of an infinite semigroup T such that the cancellation laws hold in T but T is not a group.

4. Suppose that $x, y \in G$ and that $(xy)^{-2} = y^{-2}x^{-2}$. Show that x and y commute.

5. In $GL(2, \mathbb{R})$, let H be the subset comprising all the matrices of the form $\begin{bmatrix} x & y \\ 0 & z \end{bmatrix}$

$(x, y, z \in \mathbb{R}, xz \neq 0)$, and let K be the subset comprising all the matrices $\begin{bmatrix} 1 & t \\ 0 & 1 \end{bmatrix} (t \in \mathbb{R})$.

Prove that H is a subgroup of $GL(2, \mathbb{R})$ and that K is a subgroup of H.

6. Let G be a group of permutations of the non-empty set Ω and, for each $t \in \Omega$, let $G_t = \{ f \in G : f(t) = t \}$. Prove that, for each $t \in \Omega$, G_t is a subgroup of G. (G_t is called the *stabilizer* of t in G.)

7. (i) For $x, y \in G$, prove that x, y commute iff $xyx^{-1} = y$.
(ii) Prove that, for $n \geq 3$, $Z(S_n)$ is trivial.

8. Let $x, y \in G$. Prove that if $C(x) \subseteq C(y)$, then $xy = yx$. Prove further that $C(x) \subseteq C(y) \Leftrightarrow y \in Z(C(x))$.

9. Let H, K be subgroups of G such that neither $H \subseteq K$ nor $K \subseteq H$. Prove that $H \cup K$ is not a subgroup of G.

10. Give an example where H is a non-empty subset of G and H is closed under multiplication, but H is not a subgroup of G.

11. In $GL(2, \mathbb{R})$, let A be the matrix $\begin{bmatrix} 1 & -1 \\ 1 & 0 \end{bmatrix}$. By calculating the first few powers of A, determine the period of A.

12. Prove that, for $x \in G, x^2 = 1 \Leftrightarrow x = x^{-1}$.
Hence or otherwise prove that if every element of $G - \{1\}$ has period 2, then G is abelian.

13. Prove that $\forall x \in G$, x and x^{-1} have the same period. Deduce that if G is finite and has even order, then G contains an odd number of elements of period 2 (and, in particular, G contains at least one element of period 2).

14. If $x \in G - \{1\}$ and $x^{12} = x^2$, what periods are possible for x?

15. Let a, $x \in G$. Prove that, $\forall n \in \mathbb{N}$, $(xax^{-1})^n = xa^n x^{-1}$. Deduce that xax^{-1} has the same period as a.

Deduce further that (i) $\forall b$, $c \in G$, bc and cb have the same period, (ii) G cannot contain exactly two elements of period 2.

16. If x is a group element of period 20, what are the periods of x^4, x^{10}, x^{12}?

17. Draw up the multiplication table of the quaternion group. Show that it contains 6 elements of period 4 and just 1 element of period 2.

18. Let x be an element with odd period in G. Prove that $\exists y \in G$ s.t. $y^2 = x$.

19. Prove that the group formed by the nonzero congruence classes modulo 7 is cyclic.

20. Prove that the multiplicative group of positive rational numbers is non-cyclic.

21. Let $G_1 = \langle x \rangle$, $G_2 = \langle y \rangle$ be finite cyclic groups of orders m, n, respectively. Prove that, in $G_1 \times G_2$ (see question 1) the period of (x, y) is the l.c.m. of m and n (i.e. the least positive integer which both m and n divide). Deduce that the direct product of two cyclic groups of relatively prime orders is also cyclic.

22. Let $G = \langle x \rangle$, where x has period n, and let r denote a positive integer. Prove that x^r generates G iff hcf $(r, n) = 1$.

How many different generators has a cyclic group of order $2^k (k \in \mathbb{N})$?

How many different generators has an infinite cyclic group?

23. The non-trivial group G has no non-trivial proper subgroup. Prove that G is finite cyclic of prime order.

CHAPTER SEVEN

COSETS AND LAGRANGE'S THEOREM ON FINITE GROUPS

40. Introduction

The highlight of this chapter is the theorem of Lagrange which states that the order of a subgroup of a finite group G is a factor of $|G|$. This simple fact is the foundation stone of the theory of finite groups. After proving the theorem in §43, we shall develop some straightforward applications of it in §44, e.g. we prove that every group of prime order is cyclic and that every group of order less than 6 is abelian. The ease with which we can prove these results in §44 is a measure of how significant an addition to our understanding of finite groups Lagrange's theorem is.

   The proof of Lagrange's theorem itself comes easily enough, once we show (see diagram) that, given a finite group G with subgroup H, we can partition G into a number of equal-sized subsets, one of which is H. We arrive at the

relevant partition of G by considering one or other of two equivalence relations called "left congruence modulo H" and "right congruence modulo H". These equivalence relations are defined and discussed in §42. The equivalence classes they determine turn out to be what we term *left* and *right cosets* of H. Cosets are introduced early in the chapter—in §41, where we prove the fact (needed for the proof of Lagrange's theorem) that each is the same size as the subgroup H.

41. Multiplication of subsets of a group

Throughout this section, G will denote a group.

We begin by giving a meaning to the product XY, where X, Y are subsets of G.

Definition. Let X, Y be subsets of the group G. Then XY means the subset

$$\{xy : x \in X, y \in Y\}$$

of G, i.e. the set of all elements of G that can be expressed as

(an element of X).(an element of Y)　　　[in that order].

XY is $\emptyset$ if X or Y is $\emptyset$.

Illustration. In the group of symmetries of the equilateral triangle (whose multiplication table was given in §34(5)), let $X = \{1, p\}$ and $Y = \{a, c\}$. Then

$$XY = \{1a, 1c, pa, pc\}$$

Now $1a, 1c, pa, pc$ are respectively equal to a, c, c, b. Hence in this case XY is the 3-element subset $\{a, b, c\}$.

It is worth while noting that:

41.1 With respect to multiplication of subsets of G, $\mathscr{P}(G)$ is a semigroup.

Proof. It is clear that $\mathscr{P}(G)$ is non-empty and is closed under multiplication (of subsets of G).

Let X, Y, Z be three arbitrary subsets of G. For $g \in G$, we have:

$$
\begin{aligned}
g \in (XY)Z &\Leftrightarrow g = tz \text{ for some } t \in XY, z \in Z \\
&\Leftrightarrow g = (xy)z \text{ for some } x \in X, y \in Y, z \in Z \\
&\Leftrightarrow g = x(yz) \text{ for some } x \in X, y \in Y, z \in Z \quad \text{(by the associativity} \\
&\qquad\qquad\qquad\qquad\qquad\qquad\qquad\qquad\qquad \text{of the multiplication defined on } G) \\
&\Leftrightarrow g = xu \text{ for some } x \in X, u \in YZ \\
&\Leftrightarrow g \in X(YZ).
\end{aligned}
$$

Hence $X(YZ) = (XY)Z$.

This proves that the multiplication defined on $\mathscr{P}(G)$ is associative; and it follows that $(\mathscr{P}(G), .)$ is a semigroup.

As an important special case, let us consider the product XY (where X, $Y \subseteq G$) when either X or Y is a singleton (i.e. a one-membered set).

If $X = \{a\}$ (a being an element of G), the product of the subsets is $\{a\}Y$, which we normally write simply as aY. Thus

$$
\begin{aligned}
aY &= \{ay : y \in Y\} \\
&= \text{set of elements of } G \text{ obtainable from elements of } Y \text{ by left multipli-} \\
&\qquad \text{cation by } a \\
&= \lambda_a(Y).
\end{aligned}
$$

[See §35 for the meaning of λ_a.]

Similarly, if $Y = \{b\}$ (b being an element of G), the product XY is $X\{b\}$, which we normally write simply as Xb. Thus

$Xb = \{xb : x \in X\}$
   = set of elements of G obtainable from elements of X by right multiplication by b
   $= \rho_b(X)$.

Since (see 35.8) both λ_a and ρ_b are injective (for all $a, b \in G$), we deduce by 22.2 the following result about the sizes of the subsets aY, Xb:

41.2 (i) If Y is a finite subset of G and $a \in G$, then $|aY| = |Y|$.
(ii) If X is a finite subset of G and $b \in G$, then $|Xb| = |X|$.

Cosets

We are going to be much interested in the subsets xH and Hx, where x is an element of the group G and H is a subgroup of G. Such subsets are termed *cosets* of H. More precisely:

Definition. If H is a subgroup of the group G, then (1) for each $x \in G$, the subset xH of G is called the **left coset** of H determined by x, and (2) for each $x \in G$, the subset Hx of G is called the **right coset** of H determined by x.

As a particular case of 41.2, note that:

41.3 If H is a finite subgroup of the group G, then each left or right coset of H has the same order as H.

42. Another approach to cosets

Throughout this section, G will denote a group and H will denote a subgroup of G.

We are about to define two equivalence relations on G. It will turn out that the equivalence classes determined by these relations are in one case the left cosets of H and in the other case the right cosets of H.

Definitions. (1) A relation called **left congruence modulo H** is defined on G by

$$x \equiv^l y \,(\text{mod } H) \Leftrightarrow x^{-1}y \in H \qquad (x, y \in G)$$

["$x \equiv^l y \,(\text{mod } H)$" is read "$x$ is left congruent to y modulo H".]

(2) A relation called **right congruence modulo H** is defined on G by

$$x \equiv^r y \,(\text{mod } H) \Leftrightarrow xy^{-1} \in H \qquad (x, y \in G)$$

["$x \equiv^r y \,(\text{mod } H)$" is read "$x$ is right congruent to y modulo H".]

42.1 Left and right congruence modulo H are equivalence relations on G.

Proof. We give the proof for left congruence, the proof for right congruence being similar.

(1) For every $x \in G$, $x \equiv^l x \pmod{H}$ because $x^{-1}x = 1$ and $1 \in H$ (since H is a subgroup). Hence left congruence (modulo H) is reflexive.

(2) For $x,\ y \in G$,

$$x \equiv^l y \pmod{H} \Rightarrow x^{-1}y \in H$$
$$\Rightarrow (x^{-1}y)^{-1} \in H \qquad \text{(since } H \text{ is a subgroup)}$$
$$\Rightarrow y^{-1}x \in H \qquad \text{(by reversal rule, 35.6)}$$
$$\Rightarrow y \equiv^l x \pmod{H}.$$

Hence left congruence (modulo H) is symmetric.

(3) For $x,\ y,\ z \in G$,

$$x \equiv^l y \pmod{H} \text{ and } y \equiv^l z \pmod{H} \Rightarrow x^{-1}y \in H \text{ and } y^{-1}z \in H$$
$$\Rightarrow (x^{-1}y)(y^{-1}z) \in H$$
$$\text{(since } H \text{ is a subgroup)}$$
$$\Rightarrow x^{-1}z \in H$$
$$\Rightarrow x \equiv^l z \pmod{H}.$$

Hence left congruence (modulo H) is transitive.

It follows that left congruence (modulo H) is an equivalence relation.

With reference to the definitions of left and right congruence (modulo H) note that $x^{-1}y$, xy^{-1} are, so to speak, measures of how x, y differ multiplicatively, i.e. they are the answers to the questions

$$y = x.(\text{WHAT?}), \qquad x = (\text{WHAT?}).y$$

respectively. So, when we assert that elements x, y are left or right congruent to each other modulo H, we are saying that they differ from each other multiplicatively by an element of H (on the right or left, respectively). This can be indicated more formally by giving various obviously equivalent versions of the meanings of $x \equiv^l y \pmod{H}$ and $x \equiv^r y \pmod{H}$, as follows.

42.2 For $x,\ y \in G$,

(i) $x \equiv^l y \pmod{H} \Leftrightarrow x^{-1}y \in H$
$$\Leftrightarrow x^{-1}y = h \text{ for some } h \in H$$
$$\Leftrightarrow y = xh \text{ for some } h \in H;$$

(ii) $x \equiv^r y \pmod{H} \Leftrightarrow xy^{-1} \in H$
$$\Leftrightarrow xy^{-1} = h \text{ for some } h \in H$$
$$\Leftrightarrow x = hy \text{ for some } h \in H$$

In an abelian group, the distinction between left and right congruence modulo H disappears, and we can refer to both relations simply as "congruence modulo H".

In an additive abelian group G, the condition for elements x, y to be congruent to each other modulo the subgroup H is

$$x - y \in H$$

i.e. x, y differ (additively) by an element of H.

At this point, the student is likely to realize that he has already met a particular case of congruence modulo a subgroup. In discussing integers, when we say "$x \equiv y \pmod{m}$" (x, y, m being integers with m positive), we are stating that x and y are congruent to each other modulo the subgroup M of $(\mathbb{Z}, +)$ comprising all the integral multiples of m (i.e. that x, y differ additively by an element of that subgroup M).

Returning to the general case of a multiplicative group G with subgroup H, we come to the identification of the equivalence classes determined by left and right congruence modulo H.

42.3 For each $x \in G$,
(i) the left congruence class of x (mod H) is the left coset xH;
(ii) the right congruence class of x (mod H) is the right coset Hx.

Proof. Let x be an arbitrary element of G, and let C_x be the left congruence class of x (mod H). Then, for $t \in G$,

$t \in C_x \Leftrightarrow x \equiv^l t \pmod{H}$ (by the definition of "equivalence class")
        $\Leftrightarrow t = xh$ for some $h \in H$ (see 42.2)
        $\Leftrightarrow t \in xH$ (by meaning of xH).

Hence $C_x = xH$.

This proves part (i). The proof of part (ii) is similar.

It should be noted that 42.3 tells us both that each congruence class is a coset and that each coset is a congruence class! Knowing this, we can deduce important results about cosets from general theorems about equivalence relations, e.g. from 17.5 we deduce:

42.4 The set of left cosets of H is a partition of G; and so is the set of right cosets of H.

And from 17.6 it follows that:

42.5 For each $x \in G$,
(i) the one and only left coset (of H) to which x belongs is xH;
(ii) the one and only right coset (of H) to which x belongs is Hx.

The remark following 17.3 and the clarificatory note following 17.5 are also highly relevant to cosets. In particular, it must be borne in mind that each coset of H has more than one label of the form xH (or Hx) provided it contains more than one element; e.g. if a left coset of H contains the elements $x_1, x_2, x_3, \ldots$, then that left coset is equal to $x_1 H$ and to $x_2 H$ and to $x_3 H \ldots$. Correspondingly, due care must be exercised in applying 42.5, e.g. if we are given that $x \in yH$ (where $x, y \in G$), then 42.5 entitles us to say that $yH = xH$, but it does not entitle us to infer that $x = y$.

The diagram below illustrates the partitions described in 42.4. It also indicates certain simple facts about cosets, e.g. (a) in each partition H itself is one of the cosets (since $1H = H$ and $H1 = H$), and (b) every coset has the same size (see 41.3).

(i) partition of G into left cosets: shaded area represents xH.

(ii) partition of G into right cosets: shaded area represents Hx.

A re-reading of §40 will reveal that the road to proof of Lagrange's theorem is now open. Before taking that road, however, let us pause to complete the job of recording consequences of 42.3. The point to be made is that 42.3 enables us to translate a condition involving cosets of H into an equivalent condition involving elements of G (but not mentioning cosets), and vice versa. Consider, for instance, the condition $x \in yH$ $(x, y \in G)$. Because yH is the left congruence class of $y \pmod{H}$, the condition $x \in yH$ is equivalent to $x \equiv^l y \pmod{H}$, i.e. to $x^{-1} y \in H$. By similar thinking, the student should be able to prove the following list of equivalences.

42.6 For $x, y \in G$,

(1L) $x \in yH \Leftrightarrow x^{-1} y \in H$ (1R) $x \in Hy \Leftrightarrow xy^{-1} \in H$

(2L) $xH = yH \Leftrightarrow x^{-1} y \in H$ (2R) $Hx = Hy \Leftrightarrow xy^{-1} \in H$

(3L) $xH = H \Leftrightarrow x \in H$ (3R) $Hx = H \Leftrightarrow x \in H$

(4L) x, y belong to the same left coset of $H \Leftrightarrow x^{-1} y \in H$

(4R) x, y belong to the same right coset of $H \Leftrightarrow xy^{-1} \in H$

As the worked examples below indicate, it is important to have all the information in 42.6 at one's fingertips. Some students may find it helpful to note that (2L), (2R) can be reduced to the easily remembered (3L), (3R) through the observation that (for $x, y \in G$)

$$xH = yH \Leftrightarrow x^{-1}(xH) = x^{-1}(yH) \Leftrightarrow H = (x^{-1}y)H$$

and

$$Hx = Hy \Leftrightarrow (Hx)y^{-1} = (Hy)y^{-1} \Leftrightarrow H(xy^{-1}) = H$$

Worked examples

1. Take G to be the group $\{1, p, q, a, b, c\}$ of symmetries of the equilateral triangle (as in §34(5)), and take H to be the subgroup $\langle a \rangle = \{1, a\}$. Determine explicitly how G partitions into left and right cosets of H in this case.

Solution. The left cosets of H are as follows:

$$H \text{ itself} = \{1, a\} = 1H = aH$$
$$bH = \{b1, ba\} = \{b, q\} = qH$$
$$cH = \{c1, ca\} = \{c, p\} = pH$$

[The left-hand diagram below illustrates the partition of G into these left cosets. In the above analysis, notice how, as soon as we know that a subset X is a left coset and that it contains the element t, we can at once conclude (by 42.5) that $X = tH$. E.g. having found above that $bH = \{b, q\}$, we know that $\{b, q\}$ is a left coset containing q and can infer straight away that $\{b, q\}$ equals qH.]

The right cosets of H are as follows:

$$H \text{ itself} = \{1, a\} = H1 = Ha$$
$$Hb = \{1b, ab\} = \{b, p\} = Hp$$
$$Hc = \{1c, ac\} = \{c, q\} = Hq$$

The diagram on the right illustrates the partition of G into these right cosets.

 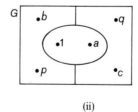

(i)   (ii)

[*Note*: as this case illustrates, the partitions of G into left and right cosets may be different.]

2. Let a be a fixed element of the group G and let $C(a)$ be the centralizer of a in G (see §37(1)). Prove that, for $g, h \in G$, $gag^{-1} = hah^{-1} \Leftrightarrow g, h$ belong to the same left coset of $C(a)$.

Solution. For $g, h \in G$,

$$gag^{-1} = hah^{-1} \Leftrightarrow g^{-1}(gag^{-1})h = g^{-1}(hah^{-1})h \qquad \text{(right to left impli-}$$
$$\text{cation by cancellation laws)}$$
$$\Leftrightarrow ag^{-1}h = g^{-1}ha$$
$$\Leftrightarrow g^{-1}h \in C(a) \qquad \text{(by definition of } C(a))$$
$$\Leftrightarrow g, h \text{ belong to same left coset of } C(a) \qquad \text{(see 42.6).}$$

3. Let $G = \langle a \rangle$ be an infinite cyclic group, and let H be the subgroup $\langle a^s \rangle$, where s is a given positive integer. Show that

$$H, aH, a^2H, \ldots, a^{s-1}H \qquad \qquad (*)$$

is a complete repetition-free list of the cosets of H in G.

Solution. The problem subdivides into proving (i) that the list $(*)$ is complete, i.e. that every element of G lies in one of the cosets listed, and (ii) that the list $(*)$ is repetition-free.

(i) Let x be an arbitrary element of G. Then $x = a^n$ for some $n \in \mathbb{Z}$. By the division algorithm, $n = qs + r$ for some $q, r \in \mathbb{Z}$ with $0 \leqslant r \leqslant s - 1$. Hence we have

$$x = a^{qs+r} = a^r(a^s)^q = a^r . \text{(an element of } H)$$

and thus $x \in a^rH$, which (since $0 \leqslant r \leqslant s - 1$) is one of the cosets in the list $(*)$.

It follows that every element of G belongs to (at least) one of the cosets in the list $(*)$.

(ii) Suppose (with a view to obtaining a contradiction) that

$$a^iH = a^jH \text{ with } 0 \leqslant i < j \leqslant s - 1$$

Then (see 42.6) $(a^i)^{-1}a^j \in H$, i.e. $a^{j-i} \in H$, and so $a^{j-i} = a^{ks}$ for some $k \in \mathbb{Z}$. Since a has infinite period, we deduce (by 38.2) that $j - i = ks$—a contradiction since $0 < j - i < s$.

From this contradiction it follows that the list $(*)$ is repetition-free.

The desired result is now established.

43. Lagrange's theorem

One further definition is necessary before we come to the full detail of Lagrange's theorem.

Definition. If H is a subgroup of the group G, the (left) **index** of H in G is the number of different left cosets of H in G.

Whenever this number is finite, we denote it by $|G:H|$.

43.1 (Lagrange's theorem) Let G be a finite group and let H be a subgroup of

G. Then $|G| = |G:H| \times |H|$. In particular, therefore, both $|H|$ and $|G:H|$ are factors of $|G|$.

Proof. Let $|H| = m$ and $|G:H| = k$. Consider the partition of G into left cosets of H (see diagram).

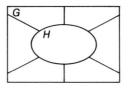

Since $|G:H| = k$, there are in the partition k different cosets and, by 41.3, each contains m elements. Hence altogether G contains km elements,

i.e. $$|G| = km = |G:H| \times |H|$$

which is the result.

Worked examples

1. In a group G, H and K are different subgroups, both of order p, where p is prime. Show that $H \cap K = \{1\}$.

Solution. $H \cap K$ is a subgroup of G (by 36.6) and indeed, since $H \cap K \subseteq H$, $H \cap K$ is a subgroup of H. So, by Lagrange's theorem, $|H \cap K|\,|\,|H|$, i.e. $|H \cap K|\,|\,p$. Therefore, since p is prime, $|H \cap K| = 1$ or p.

Since H, K are unequal sets, each of order p, it is clear that H is not contained in K and therefore (cf. 7.3(ii)) $H \cap K \neq H$. Hence $H \cap K \subset H$ and so $|H \cap K| < |H| = p$.

It follows that $|H \cap K| = 1$. Hence, since $H \cap K$, being a subgroup, must contain the element 1, we deduce that $H \cap K = \{1\}$.

2. Let G be a finite group, and let H, K be subgroups of G such that $K \subseteq H$. Show that $|G:H|$ is a factor of $|G:K|$.

Solution. By Lagrange's theorem, $|G:H| = |G|/|H|$ and $|G:K| = |G|/|K|$. Hence

$$|G:K| = (|H|/|K|) \times |G:H|$$

But, since K is a subgroup of H, it follows from Lagrange's theorem that $|H|/|K|$ is an integer. So the last equation gives the result.

The rest of this section is devoted to a point arising from Lagrange's theorem and the way in which we proved it. Consider a group G with subgroup

H. In 43.1 we showed, by considering the partition of G into left cosets of H, that, if G is finite,

$$\text{number of different left cosets of } H \text{ in } G = |G|/|H|.$$

Clearly, if instead we had considered the partition of G into right cosets of H, the same argument would have led us to the conclusion that, if G is finite,

$$\text{number of different right cosets of } H \text{ in } G = |G|/|H|.$$

Therefore, if G is finite, the numbers of different left and right cosets of the subgroup H in G are equal. We now use a different method to prove that this fact remains true whenever there are finitely many different left cosets of H in G, even if G is infinite.

43.2 Suppose that H is a subgroup of the group G and that there are exactly r different left cosets of H in G; and let $a_1 H, a_2 H, \ldots, a_r H$ be a complete repetition-free list of them ($a_1, a_2, \ldots, a_r \in G$). Then there are exactly r different right cosets of H in G, and $Ha_1^{-1}, Ha_2^{-1}, \ldots, Ha_r^{-1}$ is a complete repetition-free list of them.

Proof. By hypothesis, each element of G belongs to precisely one of the left cosets $a_1 H, a_2 H, \ldots, a_r H$.

Let x be an arbitrary element of G. We note that, for $1 \leqslant i \leqslant r$,

$$x \in Ha_i^{-1} \Leftrightarrow x(a_i^{-1})^{-1} \in H \qquad \text{(see 42.6)}$$
$$\Leftrightarrow xa_i \in H$$
$$\Leftrightarrow (x^{-1})^{-1}a_i \in H$$
$$\Leftrightarrow x^{-1} \in a_i H \qquad \text{(42.6 again)}.$$

Since (by the opening sentence of the proof) the condition $x^{-1} \in a_i H$ is true for precisely one value of i, it follows that the equivalent condition $x \in Ha_i^{-1}$ is true for precisely that same one value of i.

This proves that each element of G belongs to precisely one of the right cosets $Ha_1^{-1}, Ha_2^{-1}, \ldots, Ha_r^{-1}$, and the stated result follows.

44. Some consequences of Lagrange's theorem

The consequences we discuss in this section have largely to do with periods of elements and stem from the following corollary of Lagrange's theorem.

44.1 In a finite group G, the period of each element divides $|G|$.

Proof. Let x be an arbitrary element of the finite group G. By 38.4(iv), the period of x equals the order of the subgroup $\langle x \rangle$ and is, therefore, by 43.1, a factor of $|G|$. This proves the result.

We use 44.1 to prove the next result, which is a simple example of something of great interest in finite group theory—namely the extent to which the properties of a finite group are determined by the prime decomposition of its order. (Another example, which we shall establish in chapter 8, is that every group whose order is the square of a prime is abelian.)

44.2 Every group of prime order is cyclic. Moreover, in a group of prime order p, every non-identity element (i.e. every element other than 1) has period p.

Proof. Let G be a group of order p, where p is prime.

Consider an arbitrary element x of the non-empty set $G - \{1\}$. By 44.1, the period of x is a factor of the prime p; but (cf. 38.1) the period of x is not 1, since $x \neq 1$. Therefore the period of x is p. Hence the subgroup $\langle x \rangle$ has order p and is therefore the whole of G.

Thus G is cyclic, every non-identity element of it having period p.

We are now in a position to prove:

44.3 Every group of order less than 6 is abelian.

For this purpose, we re-state a result set as an exercise on chapter 6:

44.4 A group in which every non-identity element has period 2 is abelian.

Proof of 44.3. A group of order 1 (i.e. a trivial group) is clearly abelian. And, since 2, 3, 5 are primes, it follows from 44.2 that every group of order 2, 3 or 5 is cyclic and hence (by 39.3) abelian. So it remains to prove that every group of order 4 is abelian.

Let then G be a group of order 4; and suppose, with a view to obtaining a contradiction, that G is non-abelian. No element of G can have period 4: for if $x (\in G)$ had period 4, $\langle x \rangle$ would be the whole of G and so G would be cyclic and therefore abelian. But, by 44.1, every element of G has period dividing 4, i.e. has period 1, 2 or 4. Hence every non-identity element of G has period 2, and so, by 44.4, G is abelian—a contradiction.

From this contradiction the stated result follows.

Four-groups

As the proof of 44.3 might lead one to suspect, there do exist groups of order 4 in which every non-identity element has period 2. Any group of order 4 with this property is called a **Klein four-group** (or simply a **four-group**).

There is an important sense in which all Klein four-groups deserve to be described as "the same" or "the same in structure". In particular, apart from variations in labelling the elements, there is only one possible multiplication table for a four-group.

44.5 For a four-group whose non-identity elements are denoted by a, b, c, the multiplication table can only be:

	1	a	b	c
1	1	a	b	c
a	a	1	c	b
b	b	c	1	a
c	c	b	a	1

Outline of proof. In constructing the multiplication table, one can begin by filling in the row and column headed 1 and then entering the information $a^2 = b^2 = c^2 = 1$ (this being the defining feature of a four-group). One then finds that, by virtue of 35.9, there is only one way to complete the table.

So far no justification of the assertion that four-groups exist has been given. That omission is fully repaired by the following two illustrations.

Illustration 1. Let G consist of the following symmetries of the rectangle drawn below:

$$1 = \text{identity mapping from the plane to itself,}$$

$$a = \text{reflection in } x \text{ axis,} \qquad b = \text{reflection in } y \text{ axis,}$$

$$c = \text{half-turn (i.e. rotation through } 180°\text{) about } O.$$

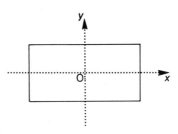

It is easily verified that:

$$a^2 = b^2 = c^2 = 1; \quad ab = ba = c; \quad bc = cb = a; \quad ca = ac = b$$

Hence it is clear that G, which is a finite non-empty subset of the group of all isometries of the plane, is closed under multiplication. Therefore, by 36.5, G is a group. And, since $a^2 = b^2 = c^2 = 1$, G is a four-group.

Illustration 2. Let G_1 consist of the following elements of S_4:

$$1_1 = \text{identity permutation of } \{1, 2, 3, 4\},$$

$$a_1 = \begin{pmatrix} 1 & 2 & 3 & 4 \\ 2 & 1 & 4 & 3 \end{pmatrix}, \quad b_1 = \begin{pmatrix} 1 & 2 & 3 & 4 \\ 3 & 4 & 1 & 2 \end{pmatrix}, \quad c_1 = \begin{pmatrix} 1 & 2 & 3 & 4 \\ 4 & 3 & 2 & 1 \end{pmatrix}$$

Then (imitating the method employed in illustration 1) one can show that

G_1 is a four-group. Its multiplication table is exactly the same as the table in 44.5, except that all the elements have subscript 1s in their labels.

Worked examples

1. Let G be a non-abelian group of order 10. Prove that G contains at least one element of period 5.

Solution. By 44.1, the period of each element of G is a factor of 10, i.e. is 1 or 2 or 5 or 10. However, no element of G has period 10 (since G is non-abelian and therefore non-cyclic); and only the identity has period 1. Hence each non-identity element has period 2 or 5. But, if all the non-identity elements had period 2, G would be abelian (cf. 44.4)—which is not the case. Therefore, at least one element of G has period 5.

2. Let G be a group of order 27. Prove that G contains a subgroup of order 3.

Solution. Choose any element $x \in G - \{1\}$. Since $x \neq 1$ and since the period of x divides $|G|$ ($=27$), it follows that the period of x is 3 or 9 or 27. Correspondingly, x or x^3 or x^9 has period 3, and $\langle x \rangle$ or $\langle x^3 \rangle$ or $\langle x^9 \rangle$ is a subgroup of G of order 3.

Remark. It must not be supposed that there is a general converse of Lagrange's theorem that says that if d is a factor of the order of a finite group G, then G must have a subgroup of order d. For example, the group S_5 has order $5!$, i.e. 120 which has 15 as a factor, but there is no subgroup of order 15 in S_5—a fact whose proof is one of the matters to be dealt with in one of the exercises following chapter 9. However, it is true (though proof is beyond the scope of this book) that if d is a prime power (i.e. an integer of the form p^r, where p is a prime and $r \in \mathbb{N}$) and d divides the order of a finite group G, then G has at least one subgroup of order d. This is part of the information provided by a set of beautiful and remarkable theorems proved in the 1870s by the Norwegian mathematician Sylow, of which students will hear a lot more if they pursue the study of finite group theory beyond the level of this textbook.

EXERCISES ON CHAPTER SEVEN

N.B. As in the exercises on chapter 6, the symbol G, wherever it occurs in the following questions, should be taken to denote a multiplicative group (arbitrary apart from restrictions imposed in the question).

1. Suppose that $a \in G$ and that $X = \{1, a^2\}$, $Y = \{1, a, a^3\}$. How many different elements has the subset XY (i) if a has period 4, (ii) if a has period 6?

2. Prove that if H is a subgroup of G, then $HH = H$.

3. Suppose that G is finite and that X is a subset of G such that $|X| > \frac{1}{2}|G|$. Prove that $XX = G$.

4. Let X, Y, Z be subsets of G. (i) Show that

$$X(Y \cap Z) \subseteq XY \cap XZ. \tag{*}$$

(ii) By considering the case $X = \{a, a^{-1}\}$, $Y = \{a\}$, $Z = \{a^{-1}\}$, where a is an element of G such that $a^2 \neq 1$, show that it is possible to have $X(Y \cap Z) \neq XY \cap XZ$. (iii) Show that equality holds in (*) when $|X| = 1$, i.e. that, for Y, $Z \subseteq G$ and $x \in G$, $x(Y \cap Z) = xY \cap xZ$.

5. Let H be a subgroup of G, and let $g \in G$. Prove that gHg^{-1} is also a subgroup of G and that, if H is finite, $|gHg^{-1}| = |H|$.

6. For $X \subseteq G$, one defines X^{-1} to mean $\{x^{-1} : x \in X\}$. Show that (i) for all $X, Y \subseteq G$, $(XY)^{-1} = Y^{-1}X^{-1}$, (ii) if L is a subgroup of G, then $L^{-1} = L$, (iii) if X is a non-empty subset of G, then X is a subgroup iff $X^{-1}X \subseteq X$.

Deduce that, if H, K are subgroups of G, then HK is a subgroup iff $HK = KH$.

7. Translate each of the following into a statement not involving cosets. (Here H denotes a subgroup of G, and x, y, a, b, c, s, t denote elements of G.) (i) $x \in yH$; (ii) a, b belong to the same right coset of H; (iii) $Hc = H$; (iv) $sH = tH$.

8. Suppose that H, L are subgroups of G with $H \subseteq L$. Show that, for each $x \in G$, either $xH \subseteq L$ or $xH \cap L = \emptyset$.

9. Suppose that $aH \cap bK \neq \emptyset$, where H, K are subgroups of G and a, $b \in G$. Prove that $aH \cap bK$ is a left coset of the subgroup $H \cap K$. (*Hint*: see last part of question 4.)

10. Suppose that H and K are subgroups of G and that $xH \subseteq yK$ for certain elements $x, y \in G$. Prove that $H \subseteq K$.

11. Let H, K be subgroups of G. Show that if a_1, $a_2, \ldots, a_n \in H$ and if the cosets $a_1(H \cap K)$, $a_2(H \cap K), \ldots, a_n(H \cap K)$ of $H \cap K$ in H are all different, then the cosets a_1K, $a_2K, \ldots, a_nK$ of K in G are all different. Deduce that if $|G:K|$ is finite, $|H : H \cap K| \leqslant |G : K|$.

Suppose now that $|G:H| = 8$, $|G:K| = 9$, and $|H \cap K| = 5$. Show that G is finite, and determine its order. (Note that $|G : H \cap K| = |G : H| \times |H : H \cap K|$—cf. worked example 2 in §43.)

12. Let H, K be finite subgroups of G. Prove that, for $h_1, h_2 \in H$, $h_1K = h_2K \Leftrightarrow h_1, h_2$ belong to the same left coset of $H \cap K$ in H. Hence or otherwise prove that (whether or not HK is a subgroup)

$$|HK| = |H||K|/|H \cap K|.$$

Deduce that, for all $g \in G$, $|HgK| = |H||K|/|H \cap gKg^{-1}|$.

13. In a finite group G there are subgroups H, K of orders 25, 36, respectively. Prove that (i) $H \cap K = \{1\}$, (ii) $900 \mid |G|$.

14. Suppose that H, K are unequal subgroups of G, each of order 16. Prove that $24 \leqslant |H \cup K| \leqslant 31$.

15. Suppose that G is finite and that H, K are subgroups of G such that $K \subset H$ and $|G:K|$ is prime. Prove that $H = G$.

16. Let G be finite. Prove that, for each $x \in G$, the period of x divides $|C(x)|$.

17. Let G be a group of permutations of the non-empty set Ω, let $t \in \Omega$, and let G_t be the stabilizer of t in G (see exercise 6 on chapter 6). Show that each left coset of G_t in G comprises all the elements of G that map t to some particular element of Ω. Deduce that, if G is finite, $|G:G_t|$ equals the number of different images that t has under elements of G, and conclude that this number divides $|G|$.

18. (i) Suppose that G is finite. Prove that, $\forall x \in G$, $x^{|G|} = 1$.

(ii) Let p be a prime number. Assuming that the nonzero congruence classes modulo p form a group with respect to multiplication, use part (i) to prove Fermat's theorem (see §19) that if $n \in \mathbb{Z}$ and $p \nmid n$, then $n^{p-1} \equiv 1 \pmod{p}$.

19. Suppose that the elements x, y of G have relatively prime (finite) periods m, n, respectively. Show that $\langle x \rangle \cap \langle y \rangle = \{1\}$. Deduce that, if x, y commute with each other, then the period of xy is mn.

20. Prove that a non-abelian group of order 8 contains at least one element of period 4.

21. Suppose that G has exactly m subgroups of order p (p being prime). Show that the total number of elements of period p in G is $m(p-1)$.

Deduce the following results. (i) A non-cyclic group of order 55 has at least one element of period 5 and at least one element of period 11. (ii) A non-cyclic group of order p^2 (p prime) has altogether $p+3$ subgroups.

22. Suppose that G is finite and abelian and that G contains more than 2 elements of period 3. By considering an appropriate subgroup, show that $9 \mid |G|$.

State a generalization with 3 replaced by an arbitrary prime p.

Deduce that if G is abelian of order pq, where p, q are distinct primes, then G is cyclic.

CHAPTER EIGHT

HOMOMORPHISMS, NORMAL SUBGROUPS, AND QUOTIENT GROUPS

45. Introduction

In the context of group theory, the word *homomorphism* means a mapping from one group to another which, so to speak, respects the operations of multiplication defined on the two groups. To be more explicit, a homomorphism from a group G to a group G_1 is a mapping $\theta : G \to G_1$ which satisfies

$$\theta(xy) = \theta(x)\theta(y) \quad \text{for all } x, y \in G$$

As students work through this chapter, they will come to see that it is such mappings that enable us to express and explore relationships between groups. For example, as will be seen in §46, the existence of a bijective homomorphism from a group G to a group G_1 tells us that the groups G, G_1 are the same in structure, the same in all respects that abstract algebra is concerned with (like the two illustrations of four-groups given in §44).

A discussion of homomorphisms in general (not just bijective ones) and of their basic properties is given in §47.

Sections 48 to 52 are devoted to different but related topics. In §49 we define what we mean by a *normal* subgroup of a given group G, and in §50 we show that if K is such a subgroup of G, we can make the set of left cosets of K in G into a group—a procedure foreshadowed in §18 when we made $\mathbb{Z}_m$ into an algebraic system. One of the first-fruits of this idea is a proof of the fact that a group whose order is the square of a prime must be abelian (§51).

46. Isomorphic groups

In geometry the recognition of congruent triangles is important because two congruent triangles, though in general distinct objects with different positions in space, have identical internal properties (sides of the same length, same area, etc.). Analogously, we saw in §44 that there is a sense in which all four-groups (whether their elements are matrices or mappings or anything else) are the same: they have, we could say, the same internal properties, the same algebraic structure. This was most clearly seen in the result 44.5.

Mathematicians describe two algebraic systems with the same algebraic structure as "isomorphic to each other". By derivation *isomorphic* means *of equal form*. This idea is of great importance in the study of algebraic systems of all kinds. So far, our formulation of it has been rather vague, and our immediate aim must be to move to a precise definition of what we mean by *isomorphic groups*.

Let us begin by considering two finite groups G, G_1. In the case of a finite group, all the information about its "algebraic structure" is contained in the multiplication table. Accordingly, what we require is a concise criterion for G, G_1 to have (once elements have been re-labelled if necessary) the same multiplication table. When that is the case, there must be a bijection $\theta : G \to G_1$, associating each element of G with the element of G_1 that plays the same role in the multiplication table of G_1; and, because of the similarity of the multiplication tables, we must have

$$xy = z \text{ in } G \Rightarrow \theta(x)\theta(y) = \theta(z) \text{ in } G_1$$

so that

$$\forall x, y \in G, \quad \theta(xy) = \theta(x)\theta(y) \qquad (*)$$

It can also be seen that, conversely, if a bijection $\theta : G \to G_1$ satisfying (*) is given, then, by re-labelling $\theta(x)$ as x for each $x \in G$, we can arrive at the situation where G, G_1 have identical multiplication tables.

Since these considerations extend to infinite groups if we interpret "multiplication table" to mean the totality of information about multiplication in the group, it becomes apparent that it makes sense to give the following formal definition of *isomorphic groups*.

Definition. Let G, G_1 be groups. Then G is said to be **isomorphic** to G_1 iff there exists a bijection θ from G to G_1 such that

$$\forall x, y \in G, \quad \theta(xy) = \theta(x)\theta(y) \qquad (*)$$

Moreover, a mapping θ from G to G_1 that is bijective and satisfies (*) is called an **isomorphism** from G to G_1. The standard notation for "G is isomorphic to G_1" is

$$G \cong G_1$$

46.1 For any set S of groups, $\cong$ is an equivalence relation on S.

Sketch of proof. (1) For every $G \in S$, $G \cong G$ because i_G is an isomorphism from G to G.

(2) For $G_1, G_2 \in S$,

$$G_1 \cong G_2 \Rightarrow \text{there is an isomorphism } \theta : G_1 \to G_2$$
$$\Rightarrow \text{there is an isomorphism (viz. } \theta^{-1}) \text{ from } G_2 \text{ to } G_1$$
$$\Rightarrow G_2 \cong G_1.$$

(3) For $G_1, G_2, G_3 \in S$,

$$G_1 \cong G_2 \text{ and } G_2 \cong G_3 \Rightarrow \text{there are isomorphisms } \theta : G_1 \to G_2, \phi : G_2 \to G_3$$
$$\Rightarrow \text{there is an isomorphism (viz. } \phi \circ \theta) \text{ from } G_1 \text{ to } G_3$$
$$\Rightarrow G_1 \cong G_3.$$

Remark. Clearly (by 22.4) isomorphic finite groups must have the same order. But the converse is not true, e.g. a four-group and a cyclic group of order 4 are non-isomorphic groups of order 4.

Worked example. Prove that two finite cyclic groups of the same order are isomorphic to each other.

Solution. Let G, G_1 be finite cyclic groups, both of order k.

Let a be a generator of G and b a generator of G_1. Then (by 39.2) a and b both have period k. Moreover (by 38.4) $1, a, a^2, \ldots, a^{k-1}$ is a complete repetition-free list of the elements of G, and $1, b, b^2, \ldots, b^{k-1}$ is a complete repetition-free list of the elements of G_1.

An obvious bijection from G to G_1 is the mapping θ specified by

$$\theta(a^n) = b^n \qquad (n = 0, 1, 2, \ldots, k-1)$$

We shall prove that θ is an isomorphism.

As a preliminary, we prove that the equation $\theta(a^n) = b^n$ holds for all integral values of n (not just those in the range 0 to $k-1$). For this purpose, let n be an arbitrary integer, and let r be its principal remainder on division by k. Then, by 38.4(i), $a^n = a^r$ and $b^n = b^r$. Hence

$$\theta(a^n) = \theta(a^r)$$
$$= b^r \qquad \text{(by definition of } \theta, \text{ since } 0 \leqslant r \leqslant k-1)$$
$$= b^n$$

Now let x, y be arbitrary elements of G. Since $G = \langle a \rangle$, $x = a^s$ and $y = a^t$ for some integers s, t, and hence

$$\theta(xy) = \theta(a^s \cdot a^t) = \theta(a^{s+t}) = b^{s+t} = b^s \cdot b^t = \theta(a^s)\theta(a^t) = \theta(x)\theta(y).$$

Since x, y are arbitrary and θ is bijective, it follows that θ is an isomorphism from G to G_1.

Hence $G \cong G_1$, which proves the result.

47. Homomorphisms and their elementary properties

Definition. Let G, G_1 be groups. A mapping $\theta : G \to G_1$ (whether bijective or not) is called a **homomorphism** from G to G_1 iff $\theta(xy) = \theta(x)\theta(y)$ for all $x, y \in G$.

Remarks. (*a*) Obviously "isomorphism" is synonymous with "bijective homomorphism". An illustration of a non-bijective homomorphism is the mapping from $GL(n, \mathbb{R})$ to $\mathbb{R} - \{0\}$ given by $X \mapsto \det X$, where $n \geqslant 2$. This mapping det is a homomorphism because $\det (XY) = \det X \times \det Y$ for all X, $Y \in GL(n, \mathbb{R})$; and it is not injective (and therefore not bijective) because, for example, there is certainly more than one member of $GL(n, \mathbb{R})$ with determinant equal to 1.

(*b*) If θ is a homomorphism from the group G to the group G_1, then, for any elements $x_1, x_2, \ldots, x_n \in G$,

$$\theta(x_1 x_2 \ldots x_n) = \theta(x_1)\theta(x_2)\ldots\theta(x_n);$$

and, in particular, for every $x \in G$ and every $n \in \mathbb{N}$,

$$\theta(x^n) = \{\theta(x)\}^n$$

These assertions are easily proved by repeated application of the definition of a homomorphism. E.g. in the case $n = 3$ we have

$$\theta(x_1 x_2 x_3) = \theta(x_1 x_2)\theta(x_3) = \theta(x_1)\theta(x_2)\theta(x_3).$$

In the following discussion and theorems up to 47.8 (inclusive), θ will denote an arbitrary homomorphism from G to G_1 (G, G_1 being groups). For clarity, we shall denote the identities of G, G_1 by 1, 1_1, respectively.

47.1 $\theta(1) = 1_1$.

Proof. $\theta(1) = \theta(1.1)$
$\qquad\qquad = \theta(1)\theta(1)$ $\qquad$ (since θ is a homomorphism);
i.e. $\quad \theta(1).1_1 = \theta(1)\theta(1)$.
Hence, by cancellation in G_1, $1_1 = \theta(1)$.

47.2 $\forall x \in G, \quad \theta(x^{-1}) = \{\theta(x)\}^{-1}$.

Proof. Let x be an arbitrary element of G. We have

$$\theta(x)\theta(x^{-1}) = \theta(xx^{-1}) \qquad \text{(since } \theta \text{ is a homomorphism)}$$
$$= \theta(1)$$
$$= 1_1 \qquad \text{(by 47.1)}$$

This shows (by 35.3) that $\theta(x^{-1})$ is the inverse of $\theta(x)$ in the group G_1, i.e. $\theta(x^{-1}) = \{\theta(x)\}^{-1}$.

47.3 If H is a subgroup of G, then $\theta(H)$ is a subgroup of G_1.

Proof. Let H be a subgroup of G.

(1) Since $H \neq \emptyset$, $\theta(H) \neq \emptyset$.

(2) Let x, y be arbitrary elements of $\theta(H)$. Then $x = \theta(h_1)$ and $y = \theta(h_2)$ for some $h_1, h_2 \in H$. Hence, using 47.2, we have

$$x^{-1}y = \{\theta(h_1)\}^{-1}\theta(h_2) = \theta(h_1^{-1})\theta(h_2) = \theta(h_1^{-1}h_2).$$

Since H is a subgroup, $h_1^{-1}h_2 \in H$, and hence $x^{-1}y \in \theta(H)$.

This proves that x, $y \in \theta(H) \Rightarrow x^{-1}y \in \theta(H)$.

It follows (by 36.4) that $\theta(H)$ is a subgroup of G_1.

An important special case of 47.3 (the case $H = G$) is:

47.4 $\operatorname{im} \theta$ is a subgroup of G_1.

The above theorems, it should be noted, are all variations on the theme "homomorphisms respect features of groups". By combining 47.2 with the remark (*b*) on the definition of a homomorphism, it is easy to prove further that:

47.5 For every $x \in G$ and for every integer n, $\theta(x^n) = \{\theta(x)\}^n$.

We use words for special kinds of homomorphisms as follows.

(*a*) an injective homomorphism is called a **monomorphism**;

(*b*) a surjective homomorphism is called an **epimorphism**;

(*c*) (as already noted) a bijective homomorphism is called an **isomorphism**.

If θ is a monomorphism, then the restriction of θ obtained by reducing the codomain to $\operatorname{im} \theta$ (a group, by 47.4) is a bijection (and still a homomorphism), i.e. it is an isomorphism from G to $\operatorname{im} \theta$. Therefore:

47.6 If θ is a monomorphism with domain G, then $\operatorname{im} \theta \cong G$.

Kernel of a homomorphism

As before, θ denotes an arbitrary homomorphism from G to G_1.

Definition. The **kernel** of θ (denoted by $\ker \theta$) is the subset

$$\{x \in G : \theta(x) = 1_1\}$$

of the domain G of θ.

Straight away we show that:

47.7 $\ker \theta$ is a subgroup of G.

Proof. (1) Since (by 47.1) $\theta(1) = 1_1$, $1 \in \ker \theta$. So $\ker \theta \neq \emptyset$.

(2) Let x, $y \in \ker \theta$, so that $\theta(x) = \theta(y) = 1_1$. Then

$$\begin{aligned}
\theta(x^{-1}y) &= \theta(x^{-1})\theta(y) \qquad \text{(since } \theta \text{ is a homomorphism)} \\
&= \{\theta(x)\}^{-1}\theta(y) \qquad \text{(by 47.2)} \\
&= 1_1^{-1}1_1 \\
&= 1_1
\end{aligned}$$

and so $x^{-1}y \in \ker \theta$.

This shows that $x, y \in \ker \theta \Rightarrow x^{-1}y \in \ker \theta$.

It follows (by 36.4) that $\ker \theta$ is a subgroup of G.

Illustration. Consider, as before, the homomorphism $\det : GL(n, \mathbb{R}) \to \mathbb{R} - \{0\}$. The kernel of this homomorphism is the subgroup of $GL(n, \mathbb{R})$ consisting of all the matrices in $GL(n, \mathbb{R})$ with determinant 1.

The kernel of a homomorphism turns out to be a very useful object to consider. For example, as the next theorem shows, a knowledge of the kernel of a homomorphism tells us immediately whether the homomorphism is injective.

47.8 Let $\theta : G \to G_1$ be a homomorphism. Then θ is a monomorphism if and only if $\ker \theta$ is trivial.

Proof. (i) Suppose that θ is a monomorphism. By 47.1, $1 \in \ker \theta$. If $\ker \theta$ contained a second element $t \neq 1$, we would have $\theta(t) = \theta(1)$ $(= 1_1)$—a contradiction since θ is injective. Therefore $\ker \theta$ must consist of 1 alone.

Thus θ is a monomorphism $\Rightarrow \ker \theta = \{1\}$.

(ii) Conversely, suppose that $\ker \theta = \{1\}$. Then, for $x, y \in G$,

$$\theta(x) = \theta(y) \Rightarrow \theta(x)\{\theta(y)\}^{-1} = \theta(y)\{\theta(y)\}^{-1}$$
$$\Rightarrow \theta(xy^{-1}) = 1_1 \qquad \text{(by 47.2)}$$
$$\Rightarrow xy^{-1} \in \ker \theta \qquad \text{(by definition of } \ker \theta)$$
$$\Rightarrow xy^{-1} = 1 \qquad \text{(since } \ker \theta = \{1\})$$
$$\Rightarrow x = y$$

so that θ is injective.

Thus $\ker \theta = \{1\} \Rightarrow \theta$ is a monomorphism.

Homomorphisms where the group operations need not be multiplication

Suppose that $(G, \circ)$ and $(G_1, *)$ are groups, with binary operations $\circ, *$, respectively. Then a homomorphism from G to G_1 means a mapping $\theta : G \to G_1$ such that

$$\forall x, y \in G, \ \theta(x \circ y) = \theta(x) * \theta(y)$$

All our results on homomorphisms extend to this more general situation with suitable adjustments in notation.

Illustration. Let G be the multiplicative group of all positive real numbers, and let G_1 be the additive group of all real numbers. Then the logarithm function, $\log : G \to G_1$, given by $x \mapsto \log x$, is (for any definite choice of base of logarithms) a homomorphism from G to G_1, since

$$\log(xy) = \log x + \log y \quad \text{for all} \quad x, y \in G.$$

Indeed log is actually an isomorphism from G to G_1.

Similarly the exponential function is (in the real or complex context) a homomorphism, since

$$\exp(x+y) = (\exp x).(\exp y) \quad \text{for all relevant } x, y.$$

The fact that the logarithm function is an isomorphism from the multiplicative group of positive numbers (a system in which calculation is relatively difficult) to the additive group of real numbers (a system in which calculation is relatively easy) explains in a sophisticated way the arithmetical usefulness of logarithms.

Cayley's theorem

The final result in this section (a theorem associated with the nineteenth-century British mathematician Cayley) gives a striking hint of the value, for group theory as a whole, of the study of groups of permutations.

47.9 Every group is isomorphic to a group of permutations.

Proof. Let G be an arbitrary group.

Let S be the symmetric group on the set G, i.e. the group of all permutations of G. [We shall show that G is isomorphic to a subgroup of S.] By 35.8, we know that $\lambda_x \in S$ for each $x \in G$ (where λ_x denotes left multiplication by x). So we can define a mapping $\theta : G \to S$ by $\theta(x) = \lambda_x$.

We now note that, for all $x, y, g \in G$,

$$(\lambda_x \circ \lambda_y)(g) = \lambda_x(\lambda_y(g)) = \lambda_x(yg) = x(yg) = (xy)g = \lambda_{xy}(g)$$

This proves that, for all $x, y \in G$,

$$\lambda_x \circ \lambda_y = \lambda_{xy}, \quad \text{i.e.} \quad \theta(x) \circ \theta(y) = \theta(xy)$$

and so θ is a homomorphism.

Consider now $\ker \theta$. For $x \in G$,

$$x \in \ker \theta \Rightarrow \theta(x) = \text{identity of the group } S$$
$$\Rightarrow \lambda_x = i_G$$
$$\Rightarrow \lambda_x(1) = i_G(1)$$
$$\Rightarrow x = 1 \quad \text{(by the definitions of } \lambda_x \text{ and } i_G).$$

Therefore $\ker \theta$ can contain no element other than 1, and so, since certainly $1 \in \ker \theta$, $\ker \theta = \{1\}$.

By 47.8, it follows that θ is a monomorphism. Hence, by 47.6, $G \cong \text{im } \theta$. Since $\text{im } \theta$ is a subgroup of S and, therefore, $\text{im } \theta$ is a group of permutations, the result follows.

48. Conjugacy

Throughout this section, G will denote a group.

For $a, x \in G$, the **conjugate of** a **by** x means the element xax^{-1} of G. And, for given $a \in G$, each element of G expressible as xax^{-1} for some $x \in G$ is called a **conjugate** of a (in G).

Conjugacy (in G) is the relation $\sim$ on G defined by
$$a \sim b \Leftrightarrow b \text{ is a conjugate of } a,$$
i.e.
$$a \sim b \Leftrightarrow \exists x \in G \text{ s.t. } b = xax^{-1}$$

Leaving the proof as an exercise, we record that:

48.1 Conjugacy is an equivalence relation on G.

Because of this fact, G can be partitioned into $\sim$-classes, which we call **conjugacy classes**. Thus the conjugacy class of a (where $a \in G$) means the set of all conjugates of a in G: we denote this subset of G by $\mathrm{cl}(a)$.

To obtain information about $|\mathrm{cl}(a)|$ when G is finite, we must consider the question: when are two conjugates gag^{-1}, hah^{-1} of a equal and when are they different $(g, h \in G)$? In fact, this question has already been answered—in worked example 2 in §42: from the result proved there, we know that the conjugates of a by two elements of G are
$$\left\{ \begin{array}{l} \text{the same if the elements belong to the same left coset of } C(a), \\ \text{different if the elements belong to different left cosets of } C(a). \end{array} \right\}$$

It follows that:

48.2 In a finite group G, for each $a \in G$, the number of different conjugates of a in G equals the number of different left cosets of $C(a)$ in G, i.e.
$$|\mathrm{cl}(a)| = |G : C(a)|$$
and consequently $|\mathrm{cl}(a)|$ is a divisor of $|G|$.

(The final assertion follows by Lagrange's theorem (43.1).)

Note that $\mathrm{cl}(a)$ may consist of a alone $(a \in G)$. In fact:

$$\mathrm{cl}(a) = \{a\} \Leftrightarrow \text{every conjugate of } a \text{ is equal to } a$$
$$\Leftrightarrow \forall x \in G, \; xax^{-1} = a$$
$$\Leftrightarrow \forall x \in G, \; xa = ax$$
$$\Leftrightarrow a \in Z(G), \text{ the centre of } G \text{ (see §37(2))}.$$

Therefore:

48.3 If Z is the centre of G, the elements of Z form single-membered conjugacy classes, and the elements of $G - Z$ belong to multi-membered conjugacy classes.

This situation is illustrated in the diagram below, which portrays the partition of G into conjugacy classes: the centre Z divides into single-membered classes (one of which is $\{1\}$), while $G - Z$ divides into multi-membered classes.

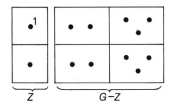

We end this section with a lemma which is of great significance because it can be used to open up an important seam of results about finite groups whose order is a power of a prime, e.g. later in the chapter, we shall use this lemma in our proof that a group whose order is the square of a prime must be abelian.

48.4 Let G be a group of order p^r, where p is a prime and $r \in \mathbb{N}$. Then the centre of G is non-trivial.

Proof. The result is certainly true if G is abelian: for then (cf. 37.3) the centre of G is the whole of G.

So henceforth we consider the case G non-abelian. Let Z be the centre of G, and let $C_1, C_2, \ldots, C_t$ be the multi-membered conjugacy classes of G, i.e. the conjugacy classes into which the non-empty set $G - Z$ partitions (see diagram). Clearly, then, $|G - Z| = \sum\limits_{j=1}^{t} |C_j|$. As $|Z| = |G| - |G - Z|$, we deduce that

$$|Z| = |G| - \sum_{j=1}^{t} |C_j| \qquad (\alpha)$$

Now each $|C_j|$ is, by 48.2, a factor of $|G|$, i.e. of p^r, and is greater than 1. So each $|C_j|$ is p or p^2 or p^3 or $\ldots$, and in particular each $|C_j|$ is divisible by p. Therefore, in the equation (α) every term of the right-hand side is divisible by p, and hence $|Z|$ is divisible by p. Hence $|Z| \neq 1$.

The result now follows.

49. Normal subgroups

Throughout this section, G will denote a group and H a subgroup of G.

As we saw in earlier discussion of cosets, for $x \in G$, the left coset xH need not equal the right coset Hx. This is the background for the definition of a normal subgroup.

Definition. Given that H is a subgroup of G, we say that H is a **normal** subgroup of G (or that H is normal in G) iff $xH = Hx$ for all $x \in G$. The standard notation for "H is a normal subgroup of G" is

$$H \triangleleft G$$

When H is a normal subgroup of G, it is clear from the definition that we may forget the distinction between left and right cosets of H and speak simply of "cosets of H". In §50 it will be shown that the cosets of a normal subgroup of G form a group with respect to multiplication of subsets of G—a fact without any parallel in the case of a non-normal subgroup. Meanwhile, we consider theorems that provide tests for normality and show that subgroups of certain kinds are normal.

49.1 In every group G, the subgroups G and $\{1\}$ are normal.

Proof. Clearly, $\forall x \in G$, $xG = Gx = G$ and $x\{1\} = \{1\}x = \{x\}$.

49.2 A subgroup of index 2 is always normal.

Proof. Let H be a subgroup with index 2 in the group G.

Then there are just two left cosets of H in G. One of these is H, and, since the left cosets form a partition of G, the other must be $G - H$. But also, by 43.2, there are just 2 right cosets of H in G: again one of them is H and hence the other is $G - H$. It follows, by 42.5, that, for every $x \in G$,

$$xH = Hx = \begin{cases} H & \text{if } x \in H \\ G - H & \text{if } x \notin H \end{cases}$$

Therefore $H \triangleleft G$, which proves the result.

The next theorem gives a very useful test for normality.

49.3 Let H be a subgroup of the group G. Then $H \triangleleft G$ iff the condition

$$h \in H \text{ and } x \in G \Rightarrow xhx^{-1} \in H$$

holds (i.e. iff every conjugate of an element of H by an element of G belongs to the subgroup H).

Proof. (i) Suppose that $H \triangleleft G$. Then

$$h \in H \text{ and } x \in G \Rightarrow xh \in xH$$
$$\Rightarrow xh \in Hx \quad \text{(by the meaning of } H \triangleleft G\text{)}$$
$$\Rightarrow xhx^{-1} \in H \quad \text{(cf. 42.6)}.$$

Thus $H \triangleleft G$ implies that the condition holds.
(ii) Conversely, suppose that the condition holds.

Let x be an arbitrary element of G. For $t \in G$, we have

$$t \in xH \Rightarrow t^{-1}x \in H \qquad \text{(cf. 42.6)}$$
$$\Rightarrow t(t^{-1}x)t^{-1} \in H \qquad \text{(by the condition)}$$
$$\Rightarrow xt^{-1} \in H$$
$$\Rightarrow t \in Hx \qquad \text{(cf. 42.6)}.$$

and conversely

$$t \in Hx \Rightarrow xt^{-1} \in H$$
$$\Rightarrow t^{-1}(xt^{-1})(t^{-1})^{-1} \in H \qquad \text{(by the condition)}$$
$$\Rightarrow t^{-1}x \in H$$
$$\Rightarrow t \in xH.$$

Hence $t \in xH \Leftrightarrow t \in Hx \ (t \in G)$; and so $xH = Hx$.
This shows that $\forall x \in G$, $xH = Hx$, i.e. that $H \lhd G$.
Thus the condition implies $H \lhd G$.
Proof of the stated result is now complete.

We apply 49.3 in proving the next two results.

49.4 The kernel of a homomorphism with G as domain is a normal subgroup of G.

Proof. Let $\theta : G \to G_1$ be a homomorphism (G, G_1 being groups). By 47.7, $\ker \theta$ is a subgroup of G.
Suppose that $k \in \ker \theta$ and $x \in G$. Then

$$\theta(xkx^{-1}) = \theta(x)\theta(k)\{\theta(x)\}^{-1}$$
$$= \theta(x)1_1\{\theta(x)\}^{-1}, \quad \text{where } 1_1 \text{ is identity of } G_1 \text{ (since } k \in \ker \theta)$$
$$= \theta(x)\{\theta(x)\}^{-1}$$
$$= 1_1$$

so that $xkx^{-1} \in \ker \theta$.
Thus $k \in \ker \theta$ and $x \in G \Rightarrow xkx^{-1} \in \ker \theta$.
So, by 49.3, $\ker \theta \lhd G$, which proves the result.

49.5 Every subgroup of G that is contained in the centre Z of G (including Z itself) is normal in G.

Proof. Let H be a subgroup of G such that $H \subseteq Z$, where Z is the centre of G. Then, for every $h \in H$ and every $x \in G$,

$$xhx^{-1} = hxx^{-1} \qquad (h \text{ commutes with } x \text{ since } h \in H \text{ and } H \subseteq Z)$$
$$= h$$

so that $xhx^{-1} \in H$.
Thus $h \in H$ and $x \in G \Rightarrow xhx^{-1} \in H$.

So, by 49.3, $H \lhd G$, which proves the result.

A particular case of 49.5 is the following result, which is, in fact, immediately obvious from the original definition of normality.

49.6 In an abelian group, every subgroup is a normal subgroup.

50. Quotient groups

Once again, G will, throughout the section, denote a group.

We now set off to show that if $K \lhd G$, the cosets of K in G form a group with respect to multiplication of subsets of G. We shall require the result of one of the exercises on chapter 7, viz:

50.1 For every subgroup H of G, $HH = H$.

Henceforth suppose that $K \lhd G$. For $x, y \in G$, we have:

$$(xK)(yK) = xKyK \qquad \text{(brackets can be dropped because subset multiplication is associative (41.1))}$$
$$= xyKK \qquad (Ky = yK \text{ because } K \lhd G)$$
$$= xyK \qquad \text{(by 50.1)}.$$

Hence:

50.2 When $K \lhd G$, the set of cosets of K in G is closed under multiplication (of subsets of G); and we have the simple rule

$$(xK)(yK) = xyK \qquad (x, y \in G).$$

Note: the "simple rule" says that, for $x, y \in G$, $C_x C_y = C_{xy}$, where C_t denotes the coset of K to which t belongs.

We now build on 50.2 to produce the main result.

50.3 When $K \lhd G$, the set of cosets of K in G is a group (with respect to multiplication of subsets of G).

Proof. Let S be the set of cosets of K in G ($K \lhd G$). We check the group axioms in turn for S.

G0: Holds for S by 50.2.

G1: Holds for S since the multiplication of subsets of G is associative.

G2: The coset K itself acts as an identity in S: for, $\forall x \in G$,

$$K(xK) = (1K)(xK) = (1x)K = xK, \text{ and } (xK)K = (xK)(1K) = (x1)K = xK.$$

Thus G2 holds for S.

G3: Each coset xK ($x \in G$) has an inverse in S (viz. $x^{-1}K$), since

$$(xK)(x^{-1}K) = (xx^{-1})K = 1K = K, \text{ and } (x^{-1}K)(xK) = (x^{-1}x)K = 1K = K$$

and since K has been recognized as identity element. Thus G3 holds for S.
It now follows that S is a group, as asserted.

Nomenclature. When $K \lhd G$, the group formed by the cosets of K in G is called the **quotient group** (or **factor group**) of G modulo K. It is denoted by G/K. Note explicitly:

50.4 When $K \lhd G$,
    (i) the identity of the group G/K is K,
    (ii) the inverse of the element xK in G/K is $x^{-1}K$ $(x \in G)$,
    (iii) if $|G:K|$ is finite, then $|G/K| = |G:K|$.
    ((i) and (ii) emerged in the course of the proof of 50.3; and (iii) is obvious).

It should also be noted that the simple rule for multiplying elements of G/K can, through repeated application in conjunction with 50.4, be extended to longer products, powers, etc. For example,

$$(x_1 K)(x_2 K)\dots(x_n K) = x_1 x_2 \dots x_n K \qquad (x_1, x_2, \dots, x_n \in G)$$

$$(xK)(yK)^{-1}(zK) = xy^{-1}zK \qquad (x, y, z \in G)$$

$$(xK)^n = x^n K \qquad (x \in G, n \in \mathbb{Z}).$$

Such observations are associated with the fact that:

50.5 When $K \lhd G$, the mapping $v: G \to G/K$ given by $x \mapsto xK$ is an epimorphism with kernel K.

Proof. When $K \lhd G$, we have, for every $x, y \in G$,

$$v(xy) = xyK \qquad \text{(by definition of } v)$$
$$= (xK)(yK) \qquad \text{(by rule of multiplication in } G/K)$$
$$= v(x)v(y)$$

Therefore v is a homomorphism. Moreover, v is surjective, since the arbitrary element xK of G/K $(x \in G)$ is equal to $v(x)$. So v is an epimorphism. Finally, for $x \in G$,

$$x \in \ker v \Leftrightarrow v(x) = \text{identity of } G/K$$
$$\Leftrightarrow xK = K$$
$$\Leftrightarrow x \in K \qquad \text{(by 42.6)},$$

so that $\ker v = K$.

Remarks. (a) The epimorphism v introduced in 50.5 is called the **natural epimorphism** from G onto G/K.

(b) In 49.4 it was proved that the kernel of a homomorphism with the group G as domain is a normal subgroup of G. From 50.5 it emerges that, conversely, every normal subgroup of the group G is the kernel of at least one homomorphism with G as domain.

(c) We have shown that, when $K \lhd G$, the set of (left) cosets of K in G can be made into a group with the rule of multiplication

$$(\text{coset of } x) \cdot (\text{coset of } y) = \text{coset of } xy \qquad (x, y \in G)$$

It is natural to ask: is it ever possible, when H is a non-normal subgroup of G, to make the set of left cosets of H in G into a group with the rule of multiplication

$$(\text{left coset of } x) \cdot (\text{left coset of } y) = \text{left coset of } xy \qquad (x, y \in G)?$$

The answer is no: for, if it were possible, the mapping from G to the group formed by the cosets given by $x \mapsto (\text{left coset of } x)$ would, fairly obviously, be a homomorphism with H as kernel—an impossibility unless $H \lhd G$.

51. The quotient group G/Z

Let G be a group, and let Z be its centre. By 49.5, $Z \lhd G$, and so the quotient group G/Z exists. An interesting and useful lemma about this quotient group is:

51.1 The group G/Z, where Z is the centre of the group G, cannot be non-trivial cyclic.

Proof. Suppose that G/Z is cyclic.

Then $G/Z = \langle \tau \rangle$ for some $\tau \in G/Z$, i.e. $G/Z = \langle tZ \rangle$ for some $t \in G$. So each element of G/Z, i.e. each coset of Z in G, is equal to $(tZ)^i$, i.e. to $t^i Z$, for some integer i.

Let x, y be arbitrary elements of G. Suppose that the cosets of Z to which they belong are $t^m Z$, $t^n Z$, respectively ($m, n \in \mathbb{Z}$). Then $x = t^m z_1$ and $y = t^n z_2$ for some $z_1, z_2 \in Z$. Hence

$$\begin{aligned} xy &= t^m z_1 t^n z_2 \\ &= t^m t^n z_1 z_2 \qquad (z_1 \text{ commutes with } t^n \text{ since } z_1 \in Z) \\ &= t^{m+n} z_1 z_2 \end{aligned}$$

and

$$\begin{aligned} yx &= t^n z_2 t^m z_1 \\ &= t^n t^m z_1 z_2 \qquad (z_2 \text{ commutes with } t^m z_1 \text{ since } z_2 \in Z) \\ &= t^{m+n} z_1 z_2 \end{aligned}$$

and hence $xy = yx$.

Since x, y are arbitrary, this proves that G is abelian, i.e. that Z is the whole of G, i.e. that G/Z is trivial.

It follows that G/Z cannot be cyclic and non-trivial.

As a corollary, we now prove:

51.2 If G is a group of order p^2, where p is prime, then G is abelian.

Proof. Let G be a group of order p^2, where p is prime.

Let Z be the centre of G. By Lagrange's theorem, $|Z|\,|\,|G|$ and so $|Z|$ is 1 or p or p^2.

But $|Z| \neq 1$, by 48.4; and $|Z| \neq p$ since

$$|Z| = p \Rightarrow |G/Z| = |G|/|Z| = p^2/p = p$$
$$\Rightarrow G/Z \text{ is non-trivial and cyclic (by 44.2)}$$

—an impossibility, by 51.1.

Therefore $|Z| = p^2$, i.e. Z is the whole of G, i.e. G is abelian. This proves the result.

52. The first isomorphism theorem

We saw in §50 that, when K is a normal subgroup of the group G, there is a homomorphism v such that

$$\operatorname{im} v = G/K = G/\ker v.$$

The following result, known as the first isomorphism theorem, may be seen as a generalization of what we have just mentioned.

52.1 Let $\theta : G \to G_1$ be a homomorphism, G, G_1 being groups. Then

$$\operatorname{im} \theta \cong G/\ker \theta.$$

Proof. Let $K = \ker \theta$. (The group G/K exists, by virtue of 49.4.)

The proof hinges on the possibility of defining a mapping $\phi : G/K \to G_1$ consistently by

$$\phi(xK) = \theta(x) \qquad (x \in G) \tag{ω}$$

The problem of consistency (encountered before in §18) arises here because each member of G/K is likely to have several labels of the form tK $(t \in G)$, and so one must investigate the question: if the element ξ of G/K is equal both to xK and to yK $(x, y \in G)$, do we necessarily come, in using the equation (ω) to determine $\phi(\xi)$, to the same conclusion when using the label xK for ξ as when using the label yK, i.e. is $\theta(x)$ necessarily the same as $\theta(y)$? The answer is yes, because, for $x, y \in G$,

$$\begin{aligned}
xK = yK &\Leftrightarrow x^{-1}y \in K \ (= \ker \theta) && \text{(by 42.6)} \\
&\Leftrightarrow \theta(x^{-1}y) = 1_1 && (1_1 \text{ being the identity of } G_1) \\
&\Leftrightarrow \{\theta(x)\}^{-1}\theta(y) = 1_1 && (\text{since } \theta \text{ is a homomorphism}) \\
&\Leftrightarrow \theta(x) = \theta(y)
\end{aligned}$$

So we may proceed knowing that the equation (ω) consistently defines a mapping ϕ from G/K to G_1. Moreover, since the above analysis shows that, for

$x, y \in G$,

$$\theta(x) = \theta(y) \Rightarrow xK = yK$$

i.e.

$$\phi(xK) = \phi(yK) \Rightarrow xK = yK$$

the mapping ϕ is injective. We also observe that

$$\operatorname{im} \phi = \{\phi(xK) : x \in G\} = \{\theta(x) : x \in G\} = \operatorname{im} \theta.$$

We note finally that ϕ is a homomorphism, since, for all $x, y \in G$,

$\phi(xK \cdot yK) = \phi(xyK)$	(by rule of multiplication in G/K)
$= \theta(xy)$	(by definition of ϕ)
$= \theta(x)\theta(y)$	(since θ is a homomorphism)
$= \phi(xK)\phi(yK)$	(by definition of ϕ)

To sum up, we have proved that ϕ is a monomorphism whose image set equals $\operatorname{im} \theta$. By 47.6, it follows that

$$\operatorname{im} \theta \cong \text{domain of } \phi, \quad \text{i.e.} \quad \operatorname{im} \theta \cong G/K, \quad \text{i.e.} \quad \operatorname{im} \theta \cong G/\ker \theta.$$

The first isomorphism theorem has many uses in the more advanced development of the theory of groups. Here we can illustrate its usefulness in telling us the structure of certain quotient groups in terms of something familiar.

For example, in $GL(n, \mathbb{R})$, the subset H comprising all the matrices with determinant 1 is, we saw in an earlier section, a subgroup. And $H \lhd GL(n, \mathbb{R})$, because H is the kernel of the homomorphism $\det : GL(n, \mathbb{R}) \to \mathbb{R} - \{0\}$ given by $X \mapsto \det X$. What, we might ask, is the structure of the factor group $GL(n, \mathbb{R})/H$? Since H is the kernel of the homomorphism det and since the image set of this homomorphism is clearly the whole of $\mathbb{R} - \{0\}$, 52.1 answers our question, telling us that $GL(n, \mathbb{R})/H \cong \mathbb{R} - \{0\}$.

Another illustration arises from the homomorphism θ from the additive group $(\mathbb{R}, +)$ to the multiplicative group $\mathbb{C} - \{0\}$ given by

$$\theta(x) = \cos(2\pi x) + i \sin(2\pi x) \qquad (x \in \mathbb{R}).$$

Here the kernel is the additive group $(\mathbb{Z}, +)$ and the image set is the multiplicative group T of all complex numbers with modulus 1. Hence the additive quotient group $\mathbb{R}/\mathbb{Z}$ is isomorphic to the multiplicative group T.

Worked problem. Let θ be a homomorphism from G to G_1, G and G_1 being finite groups. Show that $|\operatorname{im} \theta|$ divides $\operatorname{hcf}(|G|, |G_1|)$.

Solution. By the first isomorphism theorem, $\operatorname{im} \theta \cong G/K$, where $K = \ker \theta$, and hence

$$|\operatorname{im} \theta| = |G/K| = |G : K|,$$

from which it follows, by Lagrange's theorem that $|\operatorname{im}\theta| \, | \, |G|$. But since $\operatorname{im}\theta$ is a subgroup of G_1 (cf. 47.4), it follows also from Lagrange's theorem that $|\operatorname{im}\theta| \, | \, |G_1|$. Hence (by 12.4) $|\operatorname{im}\theta| \, | \, \operatorname{hcf}(|G|, |G_1|)$.

EXERCISES ON CHAPTER EIGHT

1. Let X, Y be finite sets with $|X| = |Y|$, and let S_X, S_Y be the symmetric groups on X, Y, respectively. Prove that $S_X \cong S_Y$, by introducing a bijection $\alpha : X \to Y$ and considering the mapping $\theta : S_X \to S_Y$ given by $\theta(f) = \alpha \circ f \circ \alpha^{-1}$ $(f \in S_X)$.

2. Let K be the multiplicative group of all matrices of the form $\begin{bmatrix} 1 & t \\ 0 & 1 \end{bmatrix}$ $(t \in \mathbb{R})$ (cf. exercise 5 on chapter 6). Prove that $(K, .) \cong (\mathbb{R}, +)$.

3. (i) Let G_1, G_2, G_3 be groups, and let $\theta : G_1 \to G_2$, $\phi : G_2 \to G_3$ be homomorphisms. Prove that $\phi \circ \theta$ is a homomorphism.
(ii) Suppose that $\theta : G \to G_1$ is an isomorphism (G, G_1 being groups). Prove that θ^{-1} is an isomorphism from G_1 to G.
[*Note*: This exercise fills in some of the detail omitted from the proof of 46.1.]

4. Let $\theta : G \to G_1$ be a homomorphism (G, G_1 being groups), and let x be an element of G.
(i) Prove that $\theta(\langle x \rangle) = \langle \theta(x) \rangle$.
(ii) Suppose that x has finite period k. Prove that $\theta(x)$ has finite period dividing k and that, if θ is injective, then the period of $\theta(x)$ equals k.

5. Let G_1, G_2 be groups, and let θ be the mapping from the direct product $G_1 \times G_2$ to G_1 given by $(x, y) \mapsto x$ $(x \in G_1, y \in G_2)$. Prove that θ is an epimorphism whose kernel is isomorphic to G_2.

6. Let $\theta : G \to G_1$ be an epimorphism (G, G_1 being groups), and let Z, Z_1 be the centres of G, G_1, respectively. Prove that $\theta(Z) \subseteq Z_1$.

7. Prove that the intersection of two normal subgroups of G is also a normal subgroup of G.

8. (i) Let H be a subgroup of G. Show that $H \lhd G$ iff $xHx^{-1} = H$ for every $x \in G$. Deduce (with the aid of exercise 5 on chapter 7) that if G contains exactly one subgroup of order m ($m \in \mathbb{N}$), then that subgroup is normal in G.
(ii) Prove that the quaternion group is an example of a non-abelian group in which every subgroup is normal.

9. In the group G, let H, K be subgroups such that $K \lhd G$. Prove that the subset KH is a subgroup. Show also that if $H \lhd G$, then $KH \lhd G$.

10. Let G be the group of order 8 consisting of the matrices $\pm I$, $\pm A$, $\pm B$, $\pm C$, where $A = \begin{bmatrix} 1 & 0 \\ 0 & -1 \end{bmatrix}$, $B = \begin{bmatrix} 0 & -1 \\ 1 & 0 \end{bmatrix}$, $C = \begin{bmatrix} 0 & 1 \\ 1 & 0 \end{bmatrix}$, let $H = \{I, -I, A, -A\}$, and let $K = \{I, A\}$ $(= \langle A \rangle)$. Leaving aside verification that G is a group, show that $H \lhd G$ and $K \lhd H$, while K is not normal in G. [*Moral*: normality is not transitive.]

11. Suppose that in a group G of order 2^n (where n is an integer $\geqslant 2$) the element x has period 2^{n-1}. Show that $x^{2^{n-2}}$ belongs to the centre of G.

12. Let G be a group. For any subset A of G, define $N(A)$ to be the subset $\{x \in G : xA = Ax\}$ of G.
(i) Prove that, for every subset A of G, $N(A)$ is a subgroup of G. Prove further that if A is a subgroup of G, then $A \lhd N(A)$ and that $N(A)$ is the largest subgroup of G in which A is normal (in the sense that, for a subgroup B, $A \lhd B \Rightarrow B \subseteq N(A)$).
(ii) Let L, M be abelian subgroups of G, and let H be the smallest subgroup that contains $L \cup M$. (That there is such a subgroup can be seen by considering the

intersection of all subgroups containing $L \cup M$). By considering $N(L \cap M)$, prove that $L \cap M \lhd H$.

13. Let $\theta: G \to G_1$ be a homomorphism, G and G_1 being finite groups, let $a \in G$, and suppose that $\theta(a)$ has period m in G_1. Show that the period of a in G is a divisor of $m|\ker \theta|$.

14. A subgroup H of the group G has the property that $\forall x \in G, \ x^2 \in H$. By considering the identity $xhx^{-1} = (xh)^2 h^{-1}(x^{-1})^2 \ (x \in G, h \in H)$, prove that $H \lhd G$. Prove further that the quotient group G/H is abelian.

15. In an infinite abelian group G, let H be the set of all elements of finite period. Prove that $H \lhd G$ and that in G/H no non-identity element has finite period.

16. For elements x, y in a group G, the *commutator* of x, y (denoted by $[x, y]$) is defined to be the element $x^{-1}y^{-1}xy$. Prove that x, y commute iff $[x, y] = 1$.
Let H, K be normal subgroups of the group G such that $H \cap K = \{1\}$. By considering the commutator $[h, k]$, where h, k are arbitrary elements of H, K, respectively, prove that each element of H commutes with every element of K.

17. For a normal subgroup N of the group G, prove that, for $x, \ y \in G, \ [xN, yN] = [x, y]N$, and deduce that G/N is abelian iff $[x, y] \in N$ for all $x, \ y \in G$.
Deduce that, for normal subgroups H, K of the group G,

$$G/H \text{ and } G/K \text{ both abelian} \Rightarrow G/(H \cap K) \text{ is abelian.}$$

If H, K are normal subgroups of the group G such that both G/H and G/K are cyclic, need it be true that $G/(H \cap K)$ is also cyclic?

18. Let G be a group and Z its centre. Show that if xZ and yZ (where $x, \ y \in G$) are inverses of each other in the group G/Z, then x and y commute with each other in G.

19. Give an example of a pair of groups G_1 and G_2 containing normal subgroups K_1 and K_2, respectively, such that $K_1 \cong K_2$ and $G_1/K_1 \cong G_2/K_2$ but G_1 and G_2 are not isomorphic to each other. (*Moral*: the structures of K and of G/K do not determine the structure of G.)

20. Let G be a finite group with the property that $x^2 = 1$ for all $x \in G$. Prove that $|G|$ must be a power of 2.

21. (i) Let H be a subgroup of index 2 in the group G. Prove that $\forall x \in G, \ x^2 \in H$.
(ii) Is it true that if H is a subgroup of index 3 in G, then x^3 must belong to H for every $x \in G$?

22. (i) Show that if x is an element of the group G and $x \notin Z(G)$, then $Z(G) \subset C(x)$ (strict containment).
(ii) Use part (i) in conjunction with 48.4 to prove (without mentioning G/Z) that a group of order p^2, where p is prime, must be abelian.

23. Let H be a subgroup of the group G. Show that $H \lhd G$ iff H is expressible as the union of some entire conjugacy classes of G.

24. Let G be a group of order p^r, where p is prime and $r \in \mathbb{N}$, and let N be a non-trivial normal subgroup of G. Use exercise 23 to prove that $N \cap Z(G)$ is non-trivial. Deduce that if $|N| = p$, then $N \subseteq Z(G)$.

25. Let G be a non-abelian group of order p^3, where p is prime.
  (i) Show that $|Z(G)| = p$.
  (ii) Using exercise 22(i), deduce that the number of conjugacy classes in G is $p^2 + p - 1$.

(iii) Show also, using exercise 24, that G contains just one normal subgroup of order p.

(iv) Prove finally that $x^p \in Z(G)$ for every $x \in G$.

26. As a generalization of 51.1, show that if H and N are subgroups of a group G such that $N \subseteq H$, $N \subseteq Z(G)$ and H/N is cyclic, then H is abelian.

27. Let G be a non-abelian group of order pq, where p, q are distinct primes. (i) Prove that $Z(G)$ is trivial. (ii) Noting that (by exercise 15 on chapter 6) all elements in a conjugacy class have the same period, and using the result of exercise 16 on chapter 7, prove that the number of elements of period p in G is a multiple of q. (This number is also a multiple of $p-1$, by exercise 21 on chapter 7.)

28. Using the last exercise and exercise 22 on chapter 7, prove that every group of order 15 is cyclic.

29. Let H be a subgroup of the group G. A subgroup C of G is called a *complement* of H in G iff $G = CH$ and $H \cap C = \{1\}$. Prove that if $H \lhd G$, then every such complement of H is isomorphic to G/H.

Show that in the quaternion group there is a normal subgroup not possessing a complement.

30. Let G be a finite group.

(i) Let θ be a homomorphism from G to a group G_1, and let H be a subgroup of G. By an application of the first isomorphism theorem, prove that $|\theta(H)|$ divides $|H|$.

(ii) Let K be a normal subgroup of G, and suppose that the order m of K and the index $|G:K|$ of K in G are relatively prime. By applying part (i) with $\theta = v$, the natural epimorphism from G to G/K, and with H an arbitrary subgroup of order m in G, prove that K is the only subgroup of order m in G.

31. Let G be a non-abelian group of order $2p$, where p is an odd prime. Prove that G contains at least one element of period p. Let a be such an element, and let $K = \langle a \rangle$. Prove that $K \lhd G$. Prove further that every element of $G - K$ has period 2 and that if b is any such element, then $bab = a^{-1}$.

32. This and the remaining problems in this sequence involve *automorphisms* of groups. An automorphism of a group G means an isomorphism from G to G. Prove that the set of all automorphisms of G is a group (with respect to mapping composition). We denote this group by Aut G.

33. Let G be a group. For each $g \in G$, let τ_g be the mapping from G to G defined by $\tau_g(x) = gxg^{-1}$ $(x \in G)$. Prove that, $\forall g \in G$, $\tau_g \in$ Aut G.

Let θ be the mapping from G to Aut G defined by $\theta(g) = \tau_g$ $(g \in G)$. Prove that θ is a homomorphism whose kernel is $Z(G)$.

Let $I = \{\tau_g : g \in G\}$. Deduce from the last part that I is a subgroup of Aut G and that $I \cong G/Z(G)$. Show also that $I \lhd$ Aut G.

34. Suppose that α is an automorphism of the finite group G with the property that $\forall t \in G - \{1\}$, $\alpha(t) \neq t$. Show that, for $x, y \in G$,

$$x^{-1}\alpha(x) = y^{-1}\alpha(y) \Rightarrow x = y$$

and deduce that every element of G is expressible in the form $g^{-1}\alpha(g)$ for some $g \in G$.

If $\alpha^2 = i_G$ (i.e. $\alpha \circ \alpha = i_G$), show that $\alpha(g) = g^{-1}$ for all $g \in G$ and deduce that G is abelian of odd order.

35. Let G be a cyclic group. Show that Aut G is abelian. Show further that:

(i) if G has prime order p, then $|\text{Aut } G| = p - 1$;

(ii) if $|G| = 8$, then Aut G is a four-group;

(iii) if G is infinite, then $|\text{Aut } G| = 2$.

CHAPTER NINE

THE SYMMETRIC GROUP S_n

53. Introduction

In §34 (pp. 71–2) the symmetric group S_n of degree n (n an arbitrary positive integer) was introduced. To recall the most fundamental facts, S_n is the group consisting of all permutations of the set $\{1, 2, \ldots, n\}$, the binary operation being mapping composition; and the order of S_n is $n!$

It should be appreciated that if G is any group of permutations of a set of n objects, then the natural step of labelling these objects $1, 2, \ldots, n$ makes G a group of permutations of $\{1, 2, \ldots, n\}$ and thus a subgroup of the group which we have named S_n. So the study of S_n and its subgroups covers the study of all groups of permutations of finite sets. That such a study is likely to be of central importance in the general development of finite group theory is strongly signalled by Cayley's theorem (47.9): any group is isomorphic to a group of permutations (of a finite set if the group is finite). It is, moreover, true (today as in the past) that the groups encountered in different branches of mathematics and in its applications are very often groups of permutations of finite sets—e.g. of the set of roots of a polynomial equation, or the set of vertices in a network (or "graph"), or the set of hydrogen atoms in a methane molecule, etc. Thus there are plenty of powerful reasons for embarking on a detailed study of symmetric groups, and it is to the most basic aspects of such a study that this chapter is devoted.

In §§54 and 55 we prepare the way for major advances in our insight by focusing attention on *cycles* (permutations of a specially simple kind) and by showing that every element of S_n is expressible in a standard way in terms of cycles. This enables us in §§56 and 57 to discuss periods of elements of S_n and to obtain details of how S_n partitions into conjugacy classes. The concluding sections of the chapter (§§58, 59, 60) reveal matters which might well escape the notice of the uninitiated, but turn out to be of far-reaching importance— in particular, the existence of a natural classification of permutations into two kinds, termed *even* and *odd*, and the fact that (for each $n \geqslant 2$) the even permutations in S_n form an interesting subgroup of index 2 called the *alternating group* of degree n.

Before we proceed to detailed consideration of these topics, let us clarify matters of nomenclature and notation. In discussing S_n we shall refer to members of the set $\{1, 2, \ldots, n\}$ as **objects**. So each member of S_n is a mapping whose action is to permute "the objects" in some particular way. To eliminate any risk of confusion between objects and permutations, we shall throughout the chapter use letters of the English alphabet (e.g. a, b, c) for objects, and letters of the Greek alphabet (e.g. $\alpha, \beta, \sigma, \tau$) for member elements of S_n. In particular, we shall use ι for the identity element of the group S_n (i.e. the identity mapping of the set $\{1, 2, \ldots, n\}$ to itself). Although the binary operation in S_n is mapping composition, we shall use the notation of multiplicative group theory to the extent of writing $\sigma\tau$ for $\sigma \circ \tau$ and describing this as the "product" of σ and τ ($\sigma, \tau \in S_n$).

The specification of an element of S_n by a $2 \times n$ matrix was explained at the top of p. 72, and, if necessary, that should now be revised. The student should also make sure that he can fluently obtain the product (i.e. composition) of any two given members of S_n and the inverse of a member of S_n. E.g., in S_5 let

$$\sigma = \begin{pmatrix} 1 & 2 & 3 & 4 & 5 \\ 5 & 4 & 2 & 1 & 3 \end{pmatrix}, \quad \tau = \begin{pmatrix} 1 & 2 & 3 & 4 & 5 \\ 5 & 1 & 3 & 2 & 4 \end{pmatrix}.$$

Then, as the student should verify in detail,

$$\sigma\tau = \begin{pmatrix} 1 & 2 & 3 & 4 & 5 \\ 3 & 5 & 2 & 4 & 1 \end{pmatrix}, \quad \text{and} \quad \tau^{-1} = \begin{pmatrix} 1 & 2 & 3 & 4 & 5 \\ 2 & 4 & 3 & 5 & 1 \end{pmatrix}.$$

If a is an object and $\rho \in S_n$, we say that ρ **fixes** a iff $\rho(a) = a$; and we say that ρ **moves** a iff $\rho(a) \neq a$. In the above illustration, σ moves all of the objects 1, 2, 3, 4, 5, while τ fixes 3.

54. Cycles

Suppose that $a_1, a_2, \ldots, a_r$ are distinct objects (i.e. members of $\{1, 2, 3, \ldots, n\}$), r being greater than 1. We denote by

$$(a_1\, a_2\, a_3 \ldots a_{r-1}\, a_r)$$

the element of S_n that maps a_1 to a_2, a_2 to $a_3, \ldots, a_{r-1}$ to a_r, and a_r to a_1, and fixes all other objects—i.e. the element of S_n that moves each of $a_1, a_2, \ldots, a_r$ one place round a circle as suggested by the diagram on p. 126, while fixing all other objects.

Such a permutation is called a **cycle**, of **length** r.

In S_6, for example, $(1\, 6\, 2\, 5)$ is a cycle of length 4; it is the permutation that maps 1 to 6, 6 to 2, 2 to 5, and 5 to 1, while fixing 3 and 4.

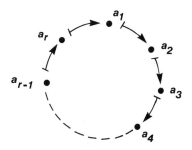

Remarks

(a) It is clear (especially on considering the above diagram) that

$$(a_1 \, a_2 \, a_3 \ldots a_r), (a_2 \, a_3 \, a_4 \ldots a_r \, a_1), (a_3 \, a_4 \, a_5 \ldots a_r \, a_1 \, a_2), \text{ etc.,}$$

all mean the same thing: that is, the choice of starting-point is immaterial when one employs the notation we have introduced for a cycle. Thus, for example, we have four possible ways of writing the cycle (1 6 2 5), namely:

$$(1\,6\,2\,5), \quad (6\,2\,5\,1), \quad (2\,5\,1\,6), \quad \text{and} \quad (5\,1\,6\,2);$$

and more generally there are r possible ways of writing a cycle of length r.

(b) The description of the action of a cycle α in terms of moving certain objects one place round a circle is both a natural and a helpful one. Notice that α^2 (i.e. $\alpha \circ \alpha$) will move the objects in question two places round the circle, α^3 (i.e. $\alpha \circ \alpha \circ \alpha$) will move them three places round the circle, etc. Further, α^{-1} will move all the objects one place round the circle in the reverse direction. Thus, for example, if α is (1 6 2 5), it is readily seen that α^2 is the permutation which interchanges 1 and 2 and interchanges 5 and 6, while $\alpha^{-1} = (1\,5\,2\,6)$.

(c) The above observations about powers of a general cycle $\alpha = (a_1 \, a_2 \ldots a_r)$ make it apparent that the lowest positive power of α that maps the objects $a_1, a_2, \ldots, a_r$ all the way round the circle of our earlier diagram and back to themselves is α^r. That is to say, the lowest positive power of α that equals ι is the rth power. So we can record the fact that:

54.1 The period of a cycle is equal to its length.

(d) A cycle of length 2 is termed a **transposition**. Such a permutation interchanges two objects while fixing all the others. A useful fact about transpositions is:

54.2 For $n \geqslant 2$, every element of S_n can be expressed as a product of transpositions.

Proof. Suppose $n \geqslant 2$, and let σ be an arbitrary member of S_n. The case $\sigma = \iota$ is easy to deal with, since $\iota = (1\,2)(1\,2)$, the product of two (equal) transpositions.

So henceforth we can confine attention to the case $\sigma \neq \iota$. In this case there are objects moved by σ. Let a be any such object. Observe that σ also moves $\sigma(a)$. [For, if $\sigma(\sigma(a))$ were equal to $\sigma(a)$, it would follow by the injectivity of σ that $\sigma(a) = a$, contrary to the fact that σ moves a.] Let τ_1 be the transposition $(a \, \sigma(a))$. Then clearly $\tau_1 \sigma$ fixes all the objects that σ fixes; and we have

$$(\tau_1 \sigma)(a) = \tau_1(\sigma(a)) = a,$$

showing that, in addition, $\tau_1 \sigma$ fixes a. Thus, given $\sigma \in S_n - \{\iota\}$, we have produced a transposition τ_1 such that $\tau_1 \sigma$ fixes more objects than σ.

If $\tau_1 \sigma \neq \iota$, we can repeat the process to produce a transposition τ_2 such that $\tau_2 \tau_1 \sigma$ fixes more objects than $\tau_1 \sigma$; and, if $\tau_2 \tau_1 \sigma \neq \iota$, we can then produce a transposition τ_3 such that $\tau_3 \tau_2 \tau_1 \sigma$ fixes yet more objects, etc. Since the number of objects is finite, it is evident that we must eventually arrive at a sequence of transpositions

$$\tau_1, \tau_2, \ldots, \tau_l$$

such that

$$\tau_l \tau_{l-1} \ldots \tau_2 \tau_1 \sigma = \iota.$$

From this equation it follows that

$$\begin{aligned}
\sigma &= (\tau_l \tau_{l-1} \ldots \tau_2 \tau_1)^{-1} \\
&= \tau_1^{-1} \tau_2^{-1} \ldots \tau_{l-1}^{-1} \tau_l^{-1} \quad\quad \text{(cf. 35.7)} \\
&= \tau_1 \tau_2 \ldots \tau_{l-1} \tau_l,
\end{aligned}$$

since, for each j, $\tau_j^2 = \iota$ (cf. 54.1) and therefore $\tau_j^{-1} = \tau_j$.

Thus σ is indeed equal to a product of transpositions, and the result is proved.

The result 54.2 is often expressed by saying that S_n is *generated* by the transpositions in it.

55. Products of disjoint cycles

We describe two cycles as **disjoint** iff the sets of objects that they move are disjoint sets. E.g., in S_8 the cycles $(1\,3\,7\,6)$ and $(2\,8\,5)$ are disjoint because the sets of objects that they move (namely, $\{1, 3, 6, 7\}$ and $\{2, 5, 8\}$) are disjoint. On the other hand, the cycles $(1\,3\,7\,6)$ and $(2\,7\,5)$ are not disjoint, because they both move the object 7.

More generally, we describe a sequence of cycles $\alpha_1, \alpha_2, \ldots, \alpha_k$ as disjoint iff they are pairwise disjoint, i.e. there is no object moved by two of them.

It turns to be important to explore, and familiarize oneself with, the action on the set of objects of a product of given disjoint cycles.

Let us begin with the case of two disjoint cycles, say $\alpha = (a_1 \, a_2 \ldots a_r)$ and

$\beta = (b_1 b_2 \ldots b_s)$. By appeal merely to the meaning of $\alpha\beta$ (i.e. $\alpha \circ \beta$), one soon sees that $\alpha\beta$ is the permutation which maps a_1 to a_2, a_2 to $a_3, \ldots, a_r$ to a_1, b_1 to b_2, b_2 to $b_3, \ldots, b_s$ to b_1, while fixing all other objects; that is, $\alpha\beta$ has the effect indicated in the following diagram.

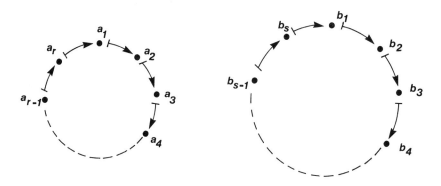

It is immediately clear that $\beta\alpha$ has precisely the same effect on all the objects, and therefore $\alpha\beta = \beta\alpha$. (They are equal mappings.) This reveals the fact that:

55.1 Disjoint cycles commute with each other.

An informal way to describe the action of the product $\alpha\beta$ of two disjoint cycles α, β as above is to say that the effect of $\alpha\beta$ is the "union" of the effects of α and β. This refers to the fact that $\alpha\beta$ has the same effect as α on the objects α moves and the same effect as β on the objects β moves—i.e. that we get a diagram indicating the action of $\alpha\beta$ by putting together a circle showing the action of the cycle α and a circle showing the action of β.

By appeal to nothing more sophisticated than the meaning of a product $\alpha_1 \alpha_2 \ldots \alpha_k$ in S_n (i.e. $\alpha_1 \circ \alpha_2 \circ \ldots \circ \alpha_k$), one can readily see that the remarks just made extend to products of any number of disjoint cycles. Informally one could say that the effect of a product $\sigma = \alpha_1 \alpha_2 \ldots \alpha_k$ of disjoint cycles $\alpha_1, \alpha_2, \ldots, \alpha_k$ is the union of the effects of $\alpha_1, \alpha_2, \ldots$, and α_k—i.e. that one obtains a diagram indicating the action of σ by bringing together in one diagram circles showing the actions of the cycles $\alpha_1, \alpha_2, \ldots, \alpha_k$. In more formal terms, the point which has emerged is this:

55.2 If the permutation σ is the product of the disjoint cycles $\alpha_1, \alpha_2, \ldots, \alpha_k$, then σ has the same effect as α_1 on the objects α_1 moves, the same effect as α_2 on the objects α_2 moves, etc., and σ fixes the objects moved by none of $\alpha_1, \alpha_2, \ldots, \alpha_k$.

The student should realise that if a particular permutation σ is given to us in the form of a product of disjoint cycles, then because of 55.2 we can see *at a glance* what $\sigma(a)$ is for each object a. Consider for example the permutation

$$\sigma = (1\ 4)(2\ 8\ 10\ 7)(3\ 9\ 5) \qquad (\in S_{10}).$$

If we require (for instance) to know what $\sigma(7)$ is, we can see *at once* that the answer is 2 (because (cf. 55.2) σ has the same effect on 7 as the cycle in which 7 appears, namely $(2\ 8\ 10\ 7)$).

Having thought through the details that lead to 55.2, one can readily see that it generalizes to any product of permutations (cycles or other permutations) which move pairwise disjoint sets of objects. The more general result, which we shall make use of in §56, is as follows.

55.3 If the permutation σ is the product $\rho_1 \rho_2 \ldots \rho_k$ of permutations $\rho_1, \rho_2, \ldots, \rho_k$ which move pairwise disjoint sets of objects, then σ has the same effect as ρ_1 on the objects ρ_1 moves, the same effect as ρ_2 on the objects ρ_2 moves, etc., and σ fixes the objects moved by none of $\rho_1, \rho_2, \ldots, \rho_k$.

Our next aim is to demonstrate the immensely useful fact that any given non-identity permutation σ in S_n can be expressed as a product of disjoint cycles (in an essentially unique way). To clarify the explanation of how an arbitrary given σ may be so expressed, we shall illustrate each stage by referring to the particular permutation

$$\sigma^* = \begin{pmatrix} 1 & 2 & 3 & 4 & 5 & 6 & 7 & 8 & 9 & 10 \\ 7 & 4 & 1 & 10 & 9 & 6 & 3 & 2 & 5 & 8 \end{pmatrix} \qquad (\in S_{10}).$$

We begin by choosing an object moved by σ: we find its image under σ, then the image of that image under σ, and so on until we get back to the object we started with.

To illustrate with σ^*, let us start with the object 1. Its image under σ^* is 7; the image of 7 under σ^* is 3; and the image of 3 is 1, which brings us back to the object we started with. Thus we have found that, under σ^*,

$$1 \mapsto 7 \mapsto 3 \mapsto 1.$$

and hence there comes to our notice the cycle $(1\ 7\ 3)$, a cycle with the key property of having the same effect on the objects it moves as has the given permutation.

There is a vital detail in what has been said so far that cries out for justification. Why is it that, in the process just described and illustrated, we get back to the object we started with? Obviously we must (because the set of objects is finite) generate a repetition at some point in our chain

$$1 \mapsto 7 \mapsto \ldots;$$

but why is it that the first sign of a repetition is the reappearance of the starting-object 1? Might we not, in some other example, generate a chain such as

$$1 \mapsto \underline{4} \mapsto 8 \mapsto 3 \mapsto \underline{4} \ldots, \qquad (*)$$

with the first repeated object (4) different from the starting-object? The answer comes from the injectivity of the permutation that we are dealing with. The chain $(*)$ indicates that our permutation maps both 1 and 3 to 4—an impossibility for an injective mapping. In the same way, the first repetition in any chain produced in the way that has been described cannot be other than the reappearance of the starting-object.

So far, one stage of a process has been discussed, and the outcome of that stage is the production of a cycle $((1\,7\,3)$ in our illustration) with the same effect on the objects it moves as has the given permutation. To continue the process, we choose some so far unmentioned object that is moved by σ. In our illustration with σ^* let us take 2. We generate, as before, a chain of successive images under the given permutation, starting with the new object just chosen. In the case of σ^*, the chain we obtain on starting with 2 is

$$2 \mapsto 4 \mapsto 10 \mapsto 8 \mapsto 2,$$

and hence we discover another cycle, $(2\,4\,10\,8)$, which has the same effect as σ^* on certain of the objects. It is no fluke that this second cycle is disjoint from $(1\,7\,3)$, the one found previously: for the occurrence of one of the objects from $(1\,7\,3)$ in the chain starting from 2 would be an obvious contradiction of the injectivity of σ^*. (For example, we could not possibly generate a chain like

$$2 \mapsto 5 \mapsto 8 \mapsto \underline{7} \ldots$$

after finding the cycle $(1\,7\,3)$ as above, since this would mean that the permutation maps both 1 and 8 to 7—a contradiction of its injectivity.)

For any given permutation σ, we continue in this way until we have exhausted all the objects moved by σ. From what has already been said, it is apparent that we shall produce a sequence of disjoint cycles, each of which has the same effect on the objects it moves as has σ. Moreover, since the process has been continued until all the objects moved by σ are exhausted, the objects fixed by σ will be precisely those not moved by any of the cycles we have produced.

In the case of σ^*, we have already found the cycles $(1\,7\,3)$ and $(2\,4\,10\,8)$. Further continuation gives us the cycle $(5\,9)$, at which point all objects moved by σ^* are exhausted. So we end up with three cycles $(1\,7\,3)$, $(2\,4\,10\,8)$, and $(5\,9)$, each of which has the same effect on the objects it moves as has σ^*. The effect of the product of these three cycles is (cf. 55.2 and the preceding informal remarks) the union of the effects of the cycles; and so the product has the same effect as σ^* on all nine objects that σ^* moves, while fixing the object (6) that

σ^* fixes. So it is clear (cf. the meaning of "equal mappings") that this product is equal to σ^*. We have thus succeeded in expressing σ^* as a product of disjoint cycles, namely

$$\sigma^* = (1\ 7\ 3)(2\ 4\ 10\ 8)(5\ 9);$$

and surely enough has been said to make it clear that our method can be applied to achieve the same goal with any given $\sigma \in S_n - \{\iota\}$.

By appeal once again to 55.2 an important supplementary point can quickly be brought to light. This is the fact that there is no possibility of variation in the cycles which appear when any given σ is expressed as a product of disjoint cycles: these cycles are uniquely determined by σ. This can be made clear through reference with 55.2 in mind to our illustration σ^*. Our previous uses of 55.2 have very much emphasised the fact that it tells us that the effect of a product of disjoint cycles can be inferred from the effects of the individual cycles. But of course we can equally well use 55.2 to make a deduction about each cycle in the product from a knowledge of the action of the product: each cycle must, by 55.2, have the same effect, so far as the objects it moves are concerned, as the product. So, for example, if the permutation σ^* is to be expressed as a product of disjoint cycles, it follows by 55.2 that the cycle in which 1 appears *must* map 1 to $\sigma^*(1)$, i.e. 7; equally it *must* map 7 to 3, and it *must* map 3 to 1, so that it must be $(1\ 7\ 3)$. Similarly, the cycles $(2\ 4\ 10\ 8)$ and $(5\ 9)$ would have to appear, and thus a product of disjoint cycles equal to σ^* can only be

$$(1\ 7\ 3)(2\ 4\ 10\ 8)(5\ 9),$$

scope for variation being limited to our choice of starting-point in writing each cycle down and the order in which we put the three cycles. (That order can, of course, be varied at will since disjoint cycles commute (cf. 55.1).) As the argument just deployed could obviously be applied to an arbitrary σ, we have now completed justification of the following major conclusion.

55.4 Any given non-identity permutation in S_n can be expressed as a product of disjoint cycles, in a way that is unique apart from the possibility of varying the order in which the cycles appear and the starting-point in each.

Remarks

(a) We shall call the expression for a given permutation σ as a product of disjoint cycles the **d.c. form** of σ ("d.c." standing for "disjoint cycles").

(b) It has taken many sentences to describe how to obtain the d.c. form of a given permutation and to justify 55.4. However, once the process is understood, one can obtain the d.c. form of any particular permutation σ almost instantaneously! All one has to do is to write down the cycles which appear through generating chains of successive images under σ (e.g. the cycle (1 7 3) in the case of our illustration σ^*).

(c) Sometimes it is convenient, in giving a permutation in d.c. form, to indicate fixed objects as cycles of length 1. Thus, an alternative notation for our permutation σ^* is

$$(1\ 7\ 3)(2\ 4\ 10\ 8)(5\ 9)(6).$$

When this convention is adopted, the identity element of S_n (which has so far been excluded from our considerations) can be expressed in d.c. form as

$$(1)(2)(3)\ldots(n).$$

(d) In several contexts it is useful to have a means of indicating how many cycles there are of each possible length in the d.c. form of a permutation σ. Such a means is provided by what we call the **cycle pattern** of σ: we say that σ has cycle pattern

$$1^{\lambda_1}.2^{\lambda_2}.3^{\lambda_3}\ldots.$$

if the d.c. form of σ contains λ_1 cycles of length 1 (i.e. σ fixes λ_1 objects), λ_2 cycles of length 2, λ_3 cycles of length 3, etc. The "1^{λ_1}" indication of the fixed objects may not be omitted if there are any fixed objects. On the other hand, in the notation $1^{\lambda_1}.2^{\lambda_2}.3^{\lambda_3}\ldots.$ we omit "j^{λ_j}" whenever λ_j is 0 (i.e. there are no cycles of length j); and we usually write "j^{λ_j}" simply as j if $\lambda_j = 1$. So, for example, with these conventions, the permutation

$$(1\ 4)(3\ 11)(5\ 8)(6\ 12\ 7\ 9)(2)(10)$$

in S_{12} has cycle pattern $1^2.2^3.4$. A cycle of length 5 in S_7 is a permutation with cycle pattern $1^2.5$; and a cycle of length 7 in S_7 has cycle pattern simply 7. The identity element of S_n has (for every n) cycle pattern 1^n.

(e) The d.c. form of a permutation being a standard and useful form of that permutation, it is important to acquire confidence in handling permutations in d.c. form (e.g. finding a product of two permutations) without feeling any desire to translate the permutations into any other form. In this connection one should note how extremely easy it is to write down the inverse or any positive power of a permutation σ given in d.c. form. Say $\sigma = \alpha_1 \alpha_2 \ldots \alpha_k$, where $\alpha_1, \alpha_2, \ldots, \alpha_k$ are disjoint cycles. Then, *because disjoint cycles commute*,

$$\sigma^{-1} = \alpha_1^{-1}\alpha_2^{-1}\ldots\alpha_k^{-1}$$

and

$$\sigma^m = \alpha_1^m\alpha_2^m\ldots\alpha_k^m \qquad\qquad \text{(for any } m \in \mathbb{N});$$

and, as was explained in §54, it is a simple matter to write down the inverse or mth power of any of the cycles α_j. So, for example, returning again to our example permutation

$$\sigma* = (1\ 7\ 3)(2\ 4\ 10\ 8)(5\ 9)(6),$$

we find it a very straightforward task to write down the inverse, square and cube of $\sigma*$, namely:

$$(\sigma*)^{-1} = (1\ 3\ 7)(2\ 8\ 10\ 4)(5\ 9)(6);$$
$$(\sigma*)^2 = (1\ 3\ 7)(2\ 10)(4\ 8)(5)(6)(9);$$
$$(\sigma*)^3 = (2\ 8\ 10\ 4)(5\ 9)(1)(3)(6)(7).$$

56. Periods of elements of S_n

It has already been pointed out (see 54.1) that the period of a cycle is equal to its length. The following more general result shows how the period of any non-identity element of S_n can be deduced from its d.c. form (or indeed from its cycle pattern).

56.1 Let $\sigma \in S_n - \{\iota\}$. Suppose that σ is the product $\alpha_1 \alpha_2 \ldots \alpha_k$ of disjoint cycles $\alpha_1, \alpha_2, \ldots, \alpha_k$, of respective lengths $l_1, l_2, \ldots, l_k$, all ≥ 2. Then the period of σ is the l.c.m. of $l_1, l_2, \ldots, l_k$ (i.e. the lowest positive integer divisible by all of $l_1, l_2, \ldots, l_k$).

Proof. Let m denote an arbitrary positive integer. As observed in remark (e) at the end of §55,

$$\sigma^m = \alpha_1^m \alpha_2^m \ldots \alpha_k^m.$$

Since $\alpha_1, \alpha_2, \ldots, \alpha_k$ are disjoint cycles, it is clear that the permutations $\alpha_1^m, \alpha_2^m, \ldots, \alpha_k^m$ move disjoint sets of objects. So if one of them (say α_j^m) moves object a, it follows (by 55.3) that σ^m also moves a. In particular, therefore,

$$\alpha_j^m \neq \iota \text{ for some } j \Rightarrow \sigma^m \neq \iota;$$

or, equivalently (cf. 4.1),

$$\sigma^m = \iota \Rightarrow \text{every } \alpha_j^m = \iota \qquad (j = 1, 2, \ldots, k).$$

The converse of this last statement is obviously true too, and so we have:

$$\sigma^m = \iota \Leftrightarrow \text{every } \alpha_j^m = \iota\ (1 \leq j \leq k)$$
$$\Leftrightarrow l_j | m \text{ for each } j\ (1 \leq j \leq k)$$

(by 38.4(ii), since each α_j has period l_j, by 54.1). Hence the lowest value of m ($\in \mathbb{N}$) for which $\sigma^m = \iota$ is the lowest positive integer divisible by all of $l_1, l_2, \ldots, l_k$, i.e. the l.c.m. of $l_1, l_2, \ldots, l_k$. This proves the result.

Note. In order to exclude unnecessary complications from the proof, mention of cycles of length 1 (i.e. fixed objects) was deliberately omitted in the above statement of 56.1. However, it is easily seen that an obvious alternative statement of 56.1, namely that the period of a permutation given in d.c. form is the l.c.m. of the lengths of the disjoint cycles appearing, is true whether we include or exclude cycles of length 1 in our calculations. Our calculations of periods will be very slightly less cluttered if we exclude such cycles.

As a specific example of the use of 56.1, consider the permutation

$$(1\ 14)(2\ 8)(3\ 15\ 7\ 6)(5\ 12\ 9\ 13\ 11\ 10)(4)$$

in S_{15}. By 56.1, the period of this permutation is l.c.m. $\{2, 2, 4, 6\}$, i.e. 12. And the same is true of every permutation of cycle pattern $1 \cdot 2^2 \cdot 4 \cdot 6$.

It is also easy now to answer a question such as: what is the smallest symmetric group that contains an element of period 10? The point is that a permutation of period 10, when expressed in d.c. form, must (by 56.1) contain at least either a cycle of length 5 and a cycle of length 2 or a cycle of length 10. Clearly this is possible iff the number of objects is at least 7, and so the answer to the question is S_7.

57. Conjugacy in S_n

The discussion of conjugacy in S_n begins with the observation that it is very easy indeed, when given permutations $\sigma, \tau \in S_n$ with σ expressed in d.c. form, to calculate the conjugate of σ by τ (i.e. the permutation $\tau\sigma\tau^{-1}$). The following result gives the details.

57.1 Let

$$\sigma = (a_1\ a_2 \ldots a_r)(b_1\ b_2 \ldots b_s)(c_1\ c_2 \ldots c_t) \ldots$$

be a permutation in S_n expressed in d.c. form, and let τ be an arbitrary element of S_n. Then the conjugate $\tau\sigma\tau^{-1}$ of σ is

$$(\tau(a_1)\ \tau(a_2) \ldots \tau(a_r))(\tau(b_1) \ldots \tau(b_s))(\tau(c_1) \ldots \tau(c_t)) \ldots$$

(obtainable by going through the given expression for σ, replacing each object where it occurs by its image under τ).

The proof is simply a matter of verifying that $\tau\sigma\tau^{-1}$ has the alleged effect on the objects. E.g. it *is* true that $\tau\sigma\tau^{-1}$ maps $\tau(a_1)$ to $\tau(a_2)$, because

$$(\tau\sigma\tau^{-1})(\tau(a_1)) = \tau[\sigma(\tau^{-1}(\tau(a_1)))]$$
$$= \tau[\sigma(a_1)] \quad (\tau^{-1}(\tau(x))\ \text{being always}\ x)$$
$$= \tau(a_2);$$

and similarly $\tau\sigma\tau^{-1}$ maps $\tau(a_2)$ to $\tau(a_3)$, etc., etc.

For a numerical illustration of 57.1, take

$$\sigma = (1\ 4\ 7)(2\ 6\ 3)(5), \quad \tau = (1\ 6)(2\ 7\ 5\ 4)(3) \quad (\text{in}\ S_7).$$

To obtain $\tau\sigma\tau^{-1}$, we simply go through the given expression for σ as indicated above: the 1 gets replaced by $\tau(1)$, i.e. 6; the 4 gets replaced by $\tau(4)$, i.e. 2; etc. Once one is familiar with the procedure, one can immediately write down

$$\tau\sigma\tau^{-1} = (6\ 2\ 5)(7\ 1\ 3)(4),$$

or (same thing in a less arbitrary order)

$$\tau\sigma\tau^{-1} = (1\ 3\ 7)(2\ 5\ 6)(4).$$

A significant corollary of 57.1 is:

57.2 In S_n two permutations are conjugate to each other iff they have the same cycle pattern. Consequently, each conjugacy class in S_n consists of all the permutations with a particular cycle pattern.

Proof. It is clear from 57.1 (and illustrated in the above numerical example) that if two permutations are conjugate to each other in S_n, then they have the same cycle pattern. It remains to demonstrate the truth of the converse—i.e. the fact that permutations with the same cycle pattern are conjugate to each other.

For this purpose consider two permutations in S_n with the same cycle pattern—say

$$\sigma = (a_1\, a_2 \ldots a_r)(b_1\, b_2 \ldots b_s)\ldots \underbrace{(f_1)(f_2)\ldots(f_t)}_{\text{[fixed objects]}}$$

and

$$\sigma^* = (a_1^*\, a_2^* \ldots a_r^*)(b_1^*\, b_2^* \ldots b_s^*)\ldots(f_1^*)(f_2^*)\ldots(f_t^*).$$

(That it is possible to write them out in this way—with the first cycles in the two products of common length, the second cycles of common length, etc.—is an obvious consequence of their having the same cycle pattern.) Let τ be the element of S_n defined by

$$\tau(a_1) = a_1^*, \tau(a_2) = a_2^*, \ldots, \tau(a_r) = a_r^*, \tau(b_1) = b_1^*, \ldots, \tau(f_t) = f_t^*.$$

(This defines τ satisfactorily since $a_1, a_2, \ldots, f_t$ are the objects $1, 2, \ldots, n$ in some order, as are $a_1^*, a_2^*, \ldots, f_t^*$.) By 57.1, $\tau\sigma\tau^{-1} = \sigma^*$, and therefore σ and σ^* are conjugate in S_n.

This completes proof of the result.

When we with to analyse how (for a particular value of n) the group S_n partitions into conjugacy classes, it is obviously (in view of 57.2) very useful to be able to say how many elements there are with any specified cycle pattern. The answer is in every case obtainable by means of the following general formula.

57.3 The number of elements in S_n with cycle pattern $1^{\lambda_1} \cdot 2^{\lambda_2} \cdot 3^{\lambda_3} \ldots$ is

$$\frac{n!}{1^{\lambda_1} 2^{\lambda_2} 3^{\lambda_3} \ldots \lambda_1! \lambda_2! \lambda_3! \ldots}.$$

Outline of proof. A permutation of cycle pattern $1^{\lambda_1} \cdot 2^{\lambda_2} \cdot 3^{\lambda_3} \ldots$ is produced by filling the n objects into the blanks in the pattern

$$\underbrace{(\cdot)(\cdot)\ldots(\cdot)}_{\lambda_1} \; \underbrace{(\cdot\cdot)(\cdot\cdot)\ldots(\cdot\cdot)}_{\lambda_2} \; \underbrace{(\cdot\cdot\cdot)(\cdot\cdot\cdot)\ldots(\cdot\cdot\cdot)}_{\lambda_3} \ldots \; .$$

One could say that the number of ways of filling these n blanks with the n objects is $n!$, but then one must take account of the fact that for any one permutation σ, there are several ways of filling the blanks to produce σ. This is because of the possibility of varying

(*a*) the starting-point in each cycle that appears in σ, and

(*b*) the order of the λ_j cycles of length j (for each j)

when one writes down σ in d.c. form. The number of permutations of cycle pattern $1^{\lambda_1} \cdot 2^{\lambda_2} \cdot 3^{\lambda_3} \ldots$ is therefore $n!/k$, where k is the number of different ways that there are (on account of the variability of (*a*) and (*b*)) of writing in d.c. form a permutation σ of that cycle pattern.

The variability of (*a*) allows each cycle of length r to be written in r different ways and therefore allows an l-fold variability in the d.c. form expression for σ, where

$$l = \underbrace{1 \times 1 \times \ldots \times 1}_{\lambda_1 \text{ factors}} \times \underbrace{2 \times 2 \times \ldots \times 2}_{\lambda_2 \text{ factors}} \times \underbrace{3 \times 3 \times \ldots \times 3}_{\lambda_3 \text{ factors}} \times \ldots = 1^{\lambda_1} 2^{\lambda_2} 3^{\lambda_3} \ldots \; .$$

The variability of (*b*) means that, for each j, the λ_j cycles of length j can be arranged amongst themselves in $\lambda_j!$ ways and, therefore, that there is altogether an m-fold variability in the d.c. form expression for σ, where $m = \lambda_1! \lambda_2! \lambda_3! \ldots \; .$

Hence the total number, k, of ways of writing σ in d.c. form is given by

$$k = l \times m = 1^{\lambda_1} 2^{\lambda_2} 3^{\lambda_3} \ldots \lambda_1! \lambda_2! \lambda_3! \ldots ;$$

and hence the number of permutations in S_n with cycle pattern $1^{\lambda_1} \cdot 2^{\lambda_2} \cdot 3^{\lambda_3} \ldots$ $(n!/k)$ is as claimed.

As a numerical example of the use of 57.3, let us calculate the number of permutations of cycle pattern $1^2 \cdot 3^2 \cdot 4$ in S_{12}. By 57.3, the number is

$$\frac{12!}{1^2 \times 3^2 \times 4 \times 2! \times 2!},$$

which works out as 3 326 400.

A use of 57.3 of less transient interest is the obtaining of details of how, for a particular value of n, S_n partitions into conjugacy classes. Let us illustrate this first with the case of S_4, which is the subject of the following table. There is one row for each conjugacy class of S_4. Because of 57.2, we can specify each conjugacy class by the common cycle pattern of its elements: thus, for example, in the table the "class $1^2.2$" means the conjugacy class consisting of all the permutations with cycle pattern $1^2.2$. It is the third column of the table (the orders of the conjugacy classes) that 57.3 helps us to fill in. The final column, giving the common period of the elements of each class, is easy to complete in the light of 56.1.

Class	Example element	Order of class	Period of elements
1^4	ι	1	1
$1^2.2$	$(2\,3)$	6	2
2^2	$(1\,3)(2\,4)$	3	2
1.3	$(1\,4\,2)$	8	3
4	$(1\,2\,4\,3)$	6	4

(Total 24)

The corresponding table for S_5 is:

Class	Example element	Order of class	Period of elements
1^5	ι	1	1
$1^3.2$	$(1\,5)$	10	2
1.2^2	$(1\,5)(2\,3)$	15	2
$1^2.3$	$(1\,4\,5)$	20	3
2.3	$(1\,4)(2\,5\,3)$	20	6
1.4	$(2\,4\,5\,3)$	30	4
5	$(1\,3\,4\,5\,2)$	24	5

(Total 120)

In the case of S_4, an interesting easily verified fact is that the set V consisting of the identity and the 3 elements of cycle pattern 2^2 is a subgroup. This subgroup is a four-group, since all its non-identity elements have period 2. Moreover, since V is the union of two of the conjugacy classes of S_4, it is clearly the case that

$$\sigma \in V \text{ and } \tau \in S_4 \ \Rightarrow \tau\sigma\tau^{-1} \in V;$$

i.e. (cf. 49.3) $V \lhd S_4$. This is noteworthy because it turns out that non-trivial proper normal subgroups of symmetric groups are rarities, and the reader may be intrigued to be told (though no attempt will be made to prove it here) that the subgroup V of S_4 is the only instance of a non-trivial normal subgroup of a symmetric group with index greater than 2.

58. Arrangements of the objects $1, 2, \ldots, n$

By an **arrangement** of the objects $1, 2, \ldots, n$, we simply mean a sequence consisting of these objects written down in some particular order. E.g., in the case $n = 7$,

$$3521764 \qquad\qquad (*)$$

is an arrangement of the objects.

The considerations of arrangements in this section may appear to constitute a digression from the theory of symmetric groups and permutations. However, there is an obvious connection, which will be exploited in §59: for each permutation $\sigma \in S_n$ naturally gives rise to a certain arrangement of $1, 2, \ldots, n$, namely the arrangement

$$\sigma(1)\sigma(2)\sigma(3)\ldots\sigma(n),$$

which we shall term the *arrangement associated with* σ and denote by A_σ. Observe that it is the sequence of numbers in the second row of the $2 \times n$ matrix notation for σ.

Returning to the above example arrangement $(*)$, consider how the objects 4 and 7 appear in it. Specifically, notice that 7 appears before 4, whereas in the natural ordering of the positive integers (i.e. $1, 2, 3, 4, 5, 6, 7, \ldots$) 4 comes before 7. We describe this situation by saying that the pair $\{4, 7\}$ is *inverted* in the arrangement $(*)$. It is an easy matter to list all the pairs which are inverted in $(*)$. These pairs are

$$\{1, 2\}, \{1, 3\}, \{1, 5\}, \{2, 3\}, \{2, 5\}, \{4, 5\}, \{4, 6\}, \{4, 7\}, \{6, 7\}$$

making 9 inverted pairs altogether.

Obviously such a count of inverted pairs could be made for any given arrangement of the objects $1, 2, \ldots, n$ (whatever the value of n), and this brings us to the main definition of this section.

For an arbitrary arrangement A of the objects, we define the ε-**value** of A (denoted by $\varepsilon(A)$) to be $(-1)^m$, where m is the total number of inverted pairs in the arrangement A.

So, for example, since (as noted above) there are 9 inverted pairs in the arrangement 3521764 of $1, 2, \ldots, 7$,

$$\varepsilon(3521764) = (-1)^9 = -1.$$

Clearly the value of $\varepsilon(A)$ is in every case 1 or -1, and its value is an indication of whether the number of inverted pairs in A is even or odd. Therefore, if a given arrangement were altered so as to change the number of inverted pairs by precisely 1, the ε-value would be reversed—i.e. changed from $+1$ to -1, or vice versa.

At a first reading it might be thought that the ε-value of an arrangement is a somewhat arcane thing to consider. However, as will be seen before the end of the chapter, it starts us out on a trail which leads to interesting discoveries concerning symmetric groups and their elements.

We conclude the present section with a result which discloses what in retrospect may well be adjudged to be the key fact about ε-values of arrangements.

58.1 Suppose that A is a given arrangement of the objects $1, 2, \ldots, n$, and that arrangement A' is obtained from A by interchanging two objects—say a and b. Then $\varepsilon(A') = -\varepsilon(A)$.

(We can usefully re-express this by saying that the effect of interchanging two objects in an arrangement is to reverse the ε-value.)

Proof. It is fairly easy to see that the assertion is true when the interchanged objects a and b are adjacent in the arrangement A. In this case, when a and b are interchanged, the pair $\{a, b\}$ is the only pair of objects whose order of appearance is disturbed. Hence (depending on whether or not the pair $\{a, b\}$ is inverted in A) the effect of interchanging a and b is either to decrease or to increase the number of inverted pairs by 1. So in any event the number of inverted pairs is changed by precisely 1, and therefore the ε-value is reversed.

We now move on to the less easy case where there are k $(\geqslant 1)$ objects $x_1, x_2, \ldots, x_k$ between a and b in the arrangement A. Thus A is of the form

$$\ldots a\, x_1\, x_2 \ldots x_k\, b \ldots .$$

We observe that the interchange of a and b can be brought about by a sequence of adjacent interchanges (i.e. interchanges of adjacent objects), which we now describe and which the diagram below should make clear. First interchange a and x_1 (i.e., in effect, pass a over x_1); then interchange a and x_2 (i.e. pass a over x_2); and proceed in this way until a has been passed over all the x_js and finally over b. Now interchange b and x_k (i.e. pass b leftwards over x_k); then pass b over x_{k-1}; and continue in this way until b has been passed leftwards over all the x_js (ending with x_1)—at which point b occupies the position a had at first, and vice versa.

$$\ldots a \quad x_1 \quad x_2 \quad \ldots \quad x_{k-1} \quad x_k \quad b \quad \ldots$$

$\Big\downarrow$ $(k+1)$ adjacent interchanges

$$\ldots \quad x_1 \quad x_2 \quad \ldots \quad x_{k-1} \quad x_k \quad b \quad a \quad \ldots$$

$\Big\downarrow$ k adjacent interchanges

$$\ldots \quad b \quad x_1 \quad x_2 \quad \ldots \quad x_{k-1} \quad x_k \quad a \quad \ldots$$

As indicated in the diagram, the first stage (where a is passed rightwards over all the x_js and then b) involves $(k+1)$ adjacent interchanges, and the second stage (where b is passed leftwards over all the x_js) involves k adjacent interchanges. The whole process, therefore, involves altogether $(2k+1)$ adjacent interchanges. Thus the interchange of a and b has been brought about by an odd number of adjacent interchanges, each of which, we know from the first paragraph of the proof, reverses the ε-value. The net result of the odd number of reversals is a reversal of the ε-value.

The stated result is now proved in all cases.

59. The alternating character, and alternating groups

The alternating character on the group S_n is the mapping χ from S_n to the multiplicative group $\{1, -1\}$ defined by

$$\chi(\sigma) = \varepsilon(A_\sigma) \qquad (\sigma \in S_n),$$

where A_σ is the arrangement of $1, 2, \ldots, n$ associated with σ, i.e. the arrangement $\sigma(1)\,\sigma(2)\ldots\sigma(n)$.

A permutation in S_n is described as an **odd** permutation if $\chi(\sigma) = -1$ and as an **even** permutation if $\chi(\sigma) = 1$.

For a numerical illustration of these ideas, suppose that σ is the permutation

$$\begin{pmatrix} 1 & 2 & 3 & 4 & 5 & 6 & 7 \\ 3 & 5 & 2 & 1 & 7 & 6 & 4 \end{pmatrix}$$

in S_7. Then

$$\begin{aligned} \chi(\sigma) &= \varepsilon(A_\sigma) \quad \text{(by the above definition of } \chi\text{)} \\ &= \varepsilon(3521764) \\ &= -1 \quad \text{(as seen in a numerical illustration in §58),} \end{aligned}$$

and hence σ is an odd permutation.

The value of $\chi(\sigma)$ is easily seen in the case $\sigma = \iota$, the identity element of S_n:

59.1 $\chi(\iota) = 1$.

For: the arrangement associated with ι is $1234\ldots n$, in which, obviously, there are no inverted pairs, and hence

$$\chi(\iota) = \varepsilon(1234\ldots n) = (-1)^0 = 1.$$

We shall shortly be in a position to prove the important fact that the alternating character mapping χ is a homomorphism (from the group S_n to the group $\{1, -1\}$). Before that come some preliminary results, the most powerful of which is the next proposition. It is the immediate outcome of applying the major result on arrangements (58.1) to permutations in the obvious way.

59.2 Let $\sigma \in S_n$, and let τ be a transposition in S_n. Then

$$\chi(\tau\sigma) = -\chi(\sigma).$$

Proof. Since σ is a bijection, the objects interchanged by the transposition τ are equal to $\sigma(a)$ and $\sigma(b)$ for certain objects a and b, with, suppose, $a < b$. Moreover, since τ moves only $\sigma(a)$ and $\sigma(b)$, it is clear that $\tau\sigma(c) = \sigma(c)$ for every object c other than a and b. Therefore, the arrangements A_σ and $A_{\tau\sigma}$ are respectively as follows:

$$\sigma(1)\,\sigma(2)\,\sigma(3)\ldots\sigma(a)\ldots\sigma(b)\ldots\sigma(n)$$
$$\sigma(1)\,\sigma(2)\,\sigma(3)\ldots\sigma(b)\ldots\sigma(a)\ldots\sigma(n),$$

one obtainable from the other by interchange of $\sigma(a)$ and $\sigma(b)$. By 58.1, $\varepsilon(A_{\tau\sigma}) = -\varepsilon(A_\sigma)$: i.e. $\chi(\tau\sigma) = -\chi(\sigma)$, as asserted.

From 59.1 and 59.2, it follows that, if τ is any transposition, then

$$\chi(\tau) = \chi(\tau\iota) = -\chi(\iota) = -1.$$

Thus:

59.3 Every transposition is an odd permutation.

A more general result is:

59.4 If $\sigma \in S_n$ and σ is expressible as the product of l transpositions, then $\chi(\sigma) = (-1)^l$.

Proof. Suppose that $\sigma = \tau_1 \tau_2 \ldots \tau_l$, where $\tau_1, \tau_2, \ldots, \tau_l$ are transpositions in S_n. Then

$$\begin{aligned}
\chi(\sigma) &= \chi(\tau_1 \tau_2 \tau_3 \ldots \tau_{l-1} \tau_l) \\
&= (-1) \times \chi(\tau_2 \tau_3 \ldots \tau_{l-1} \tau_l) \quad \text{(by 59.2)} \\
&= (-1)^2 \times \chi(\tau_3 \tau_4 \ldots \tau_{l-1} \tau_l) \quad \text{(by 59.2 again)}, \\
&= \ldots \\
&= (-1)^{l-1} \times \chi(\tau_l) \quad \text{(after } l-1 \text{ such steps)} \\
&= (-1)^l \quad \text{(by 59.3)}.
\end{aligned}$$

This proves the result.

We now come to the earlier announced crucial fact about the alternating character mapping.

59.5 The alternating character $\chi: S_n \to \{1, -1\}$ is a homomorphism.

Proof. Let ρ, σ be arbitrary elements of S_n.

By 54.2, each of ρ and σ can be expressed as a product of transpositions:

say $\rho = \tau_1 \tau_2 \ldots \tau_l$ and $\sigma = \tau_{l+1} \tau_{l+2} \ldots \tau_{l+m}$, where $\tau_1, \tau_2, \ldots, \tau_{l+m}$ are transpositions. Our notation gives us ρ and σ as the products of l transpositions and of m transpositions, respectively; and hence, by 59.4, $\chi(\rho) = (-1)^l$, and $\chi(\sigma) = (-1)^m$. Further, $\rho\sigma = \tau_1 \tau_2 \ldots \tau_{l+m}$, the product of $(l+m)$ transpositions, so that we have

$$\chi(\rho\sigma) = (-1)^{l+m} = (-1)^l \times (-1)^m = \chi(\rho)\chi(\sigma).$$

Since ρ, σ were arbitrary in S_n, this proves that χ is a homomorphism.

The kernel, ker χ, of the homomorphism χ is by definition

$$\{\sigma \in S_n : \chi(\sigma) = 1\}.$$

So ker χ consists precisely of the even permutations in S_n. It follows at once (by 49.4) that the even permutations in S_n form a normal subgroup of S_n. This normal subgroup comprising all the even permutations in S_n is called the **alternating group** of degree n, and we shall denote it by A_n.

To ascertain the order of A_n, we prove the following general proposition about the possible proportions of even permutations in an arbitrary subgroup of S_n.

59.6 Suppose that G is a subgroup of S_n ($n \in \mathbb{N}$). Then either G consists entirely of even permutations (i.e. $G \subseteq A_n$), or else the even permutations in G form a subgroup of index 2 in G.

Proof. Let K be the set of even permutations in G. It may be the case that K is the whole of G. Consider henceforth the remaining case where G contains at least one odd permutation, and let ρ denote an odd permutation in G. It suffices to prove that in this case K is a subgroup of G with $|G:K| = 2$.

Let $\chi^* = \chi|_G$, the restriction of χ to G as domain. (See the last paragraph of §23.) Clearly χ^* is a homomorphism, and its kernel is K. So K is a (normal) subgroup of G. Further, by the first isomorphism theorem (52.1),

$$G/K \cong \text{im } \chi^*.$$

It follows that $|G/K| = |\text{im } \chi^*|$, i.e. $|G:K| = |\text{im } \chi^*|$. But, since ι and ρ belong to G and their images under χ^* are 1 and -1, respectively, it is evident that im χ^* is the whole of $\{1, -1\}$. Thus $|\text{im } \chi^*| = 2$, and therefore $|G:K| = 2$. The stated result follows.

A corollary is:

59.7 For $n \geqslant 2$, $|S_n : A_n| = 2$, and so $|A_n| = \frac{1}{2}(n!)$.

Proof. Suppose $n \geqslant 2$. Then (cf. 59.3) S_n does contain at least one odd permutation—e.g. the transposition $(1\ 2)$. So it follows from 59.6 that $|S_n : A_n| = 2$, as asserted.

Remarks

(a) It is necessary to exclude the case $n = 1$ in the statement of 59.7. Of course, the symmetric group S_1 of degree 1 is a trivial group (of order 1), and so there is seldom if ever any point in considering S_1 in permutation group theory. However, for the record, the one element, ι, of S_1 is an even permutation, and thus the alternating group A_1 is the whole of S_1.

(b) The alternating groups A_2 and A_3 are also of limited interest. On calculating their orders, one sees that they are trivial and cyclic of order 3, respectively (cf. 44.2).

(c) The group of rotational symmetries of a regular icosahedron can be shown to be isomorphic to A_5 which, for this reason, is often called the *icosahedral group*.

(d) The perception of the distinction between odd and even permutations and the discovering of the existence and order of the subgroup A_n of S_n certainly constitute significant advances so far as the theory of permutation groups and its applications are concerned. But in addition there is, as will be indicated in §60, a reason (quite apart from permutation groups specifically) for regarding the groups A_n for $n \geq 5$ as having a special status in the general theory of the structure of finite groups.

The final objective of this section is to show how one can tell from its cycle pattern whether a given permutation is even or odd. As will be demonstrated, this is possible as soon as one is aware of the following fact about cycles.

59.8 A cycle of even length is an odd permutation, and a cycle of odd length is an even permutation.

Proof. Consider an arbitrary cycle $\alpha = (a_1 a_2 \ldots a_r)$ of length r (≥ 2) in S_n. The key to the proof is the observation (easily verified once it comes to one's attention) that α is equal to

$$(a_1 a_r)(a_1 a_{r-1}) \ldots (a_1 a_4)(a_1 a_3)(a_1 a_2).$$

a product of $(r-1)$ transpositions. Hence, by 59.4,

$$\chi(\alpha) = (-1)^{r-1},$$

and it can be seen that the stated result follows from this.

It is because χ is a homomorphism that 59.8 enables us to determine the evenness or oddness of any permutation given in d.c. form. A specific example will make this clear: let us illustrate with the permutation

$$\sigma = (1\ 3)(2\ 14)(4\ 13\ 8)(5\ 11\ 10\ 6\ 9\ 12) \qquad (\in S_{14}).$$

(σ fixes 7, but in this context we can ignore fixed objects.) By 59.8, the χ-values of the cycles $(1\ 3)$, $(2\ 14)$, $(4\ 13\ 8)$ and $(5\ 11\ 10\ 6\ 9\ 12)$ whose product is σ are respectively

$$-1, -1, 1, \text{ and } -1.$$

Hence, since χ is a homomorphism,

$$\chi(\sigma) = (-1) \times (-1) \times 1 \times (-1) = -1;$$

i.e. σ is an odd permutation.

A little thought will soon make it clear that the general result obtainable through the approach just illustrated is:

59.9 A permutation in S_n is odd iff there is altogether an odd number of cycles of even length in its d.c. form.

Looking back at the tables of conjugacy classes of S_4 and S_5 towards the end of §57, one can now easily see which members of S_4 belong to A_4 and which members of S_5 belong to A_5. From the S_4 table one sees that the group A_4 (which has order 12) consists of the identity element, the 3 permutations of cycle pattern 2^2, and the 8 elements of cycle pattern $1 \cdot 3$. The corresponding story for A_5 (which we shall refer back to in §60) is:

59.10 The group A_5 (which has order 60) consists of the following elements of S_5: the identity, the 15 elements of cycle pattern $1 \cdot 2^2$, the 20 elements of cycle pattern $1^2 \cdot 3$, and the 24 elements of cycle pattern 5.

60. The simplicity of A_5

A **simple group** means a non-trivial group which has no non-trivial proper normal subgroup.

To see the point of this definition, consider a finite group G which *does* have a non-trivial proper normal subgroup K. Because $K \lhd G$, the group G/K exists; and both of the groups K and G/K are smaller than G. It must be emphasised that it is *not* true that the structures of the two groups K and G/K determine the structure of G. Nevertheless, knowledge of the structures of the groups K and G/K will in general help us towards some purposeful statements about the group G in terms of two smaller groups. Clearly, however, this approach is not open to us in the case of a simple group, and so we may liken simple groups to atoms or fundamental building-bricks in terms of which the structures of other groups may be, to some extent, described. There is, therefore, great interest in simple groups in group theory.

Since every subgroup of an abelian group is normal in that group, a simple abelian group is a non-trivial group with no non-trivial proper subgroups whatsoever. It is not difficult to show (cf. exercise 23 on page 88) that the groups with this property are precisely the cyclic groups of prime order. Thus:

60.1 The abelian simple groups are the cyclic groups of prime order.

The search for non-abelian simple groups is an infinitely harder task. It is a task to which a formidable effort has been devoted by group-theorists in recent decades—very productively, let it be added. Here we shall prove just one modest, but very interesting, result on this theme—that A_5 (which has order 60) is one example of a non-abelian simple group. (It is certainly non-abelian: for example, the cycles (1 2 3) and (1 2 4), which by 59.8 belong to A_5, do not commute with each other.)

As a preliminary we shall investigate in general how an alternating group partitions into conjugacy classes. For this purpose it is necessary to distinguish carefully (for arbitrary $\sigma \in A_n$) between

$$C_{A_n}(\sigma)\text{—the centralizer of } \sigma \text{ in } A_n, \text{ i.e. } \{\rho \in A_n : \rho\sigma = \sigma\rho\},$$

and

$$C_{S_n}(\sigma)\text{—the centralizer of } \sigma \text{ in } S_n, \text{ i.e. } \{\rho \in S_n : \rho\sigma = \sigma\rho\},$$

and between

$$\mathrm{cl}_{A_n}(\sigma)\text{—the conjugacy class of } \sigma \text{ in } A_n, \text{ i.e. } \{\rho\sigma\rho^{-1} : \rho \in A_n\},$$

and

$$\mathrm{cl}_{S_n}(\sigma)\text{—the conjugacy class of } \sigma \text{ in } S_n, \text{ i.e. } \{\rho\sigma\rho^{-1} : \rho \in S_n\}.$$

We know (cf. 57.2) that $\mathrm{cl}_{S_n}(\sigma)$ consists precisely of the permutations in S_n with the same cycle pattern as σ, and it is readily seen that:

60.2 $\mathrm{cl}_{A_n}(\sigma) \subseteq \mathrm{cl}_{S_n}(\sigma)$ $(\sigma \in A_n)$,

since $\{\rho\sigma\rho^{-1} : \rho \in A_n\} \subseteq \{\rho\sigma\rho^{-1} : \rho \in S_n\}$, obviously, as A_n is a subset of S_n. However, it does happen in certain instances that $\mathrm{cl}_{A_n}(\sigma)$ is a proper subset of $\mathrm{cl}_{S_n}(\sigma)$, and the next result gives details of this. Attention here is restricted to $n \geqslant 2$, since the proof to be given does not cover the case $n = 1$ and since that case is of zero interest.

60.3 Let $\sigma \in A_n (n \geqslant 2)$. If there is an odd permutation in S_n that commutes with σ, then $\mathrm{cl}_{A_n}(\sigma) = \mathrm{cl}_{S_n}(\sigma)$. Otherwise, $\mathrm{cl}_{A_n}(\sigma)$ consists of precisely half the elements of $\mathrm{cl}_{S_n}(\sigma)$.

Proof. We begin by remarking that we must have one or other of the following two cases.

Case 1: there is an odd permutation in S_n that commutes with σ, i.e. there is an odd permutation in $C_{S_n}(\sigma)$. In this event (by 59.6) precisely half the elements of $C_{S_n}(\sigma)$ are even permutations.

Case 2: there is no odd permutation in S_n that commutes with σ, i.e. all the elements of $C_{S_n}(\sigma)$ are even permutations.

Since, clearly, $C_{A_n}(\sigma)$ consists of the even permutations in S_n that commute with σ, it is apparent that

$$|C_{A_n}(\sigma)| = \begin{cases} \frac{1}{2}|C_{S_n}(\sigma)| & \text{in case 1,} \\ |C_{S_n}(\sigma)| & \text{in case 2.} \end{cases}$$

Now, in any event,

$$|cl_{A_n}(\sigma)| = \frac{|A_n|}{|C_{A_n}(\sigma)|} \qquad \text{(by 48.2)}$$

$$= \frac{\frac{1}{2}|S_n|}{|C_{A_n}(\sigma)|} \qquad \text{(cf. 59.7).}$$

So, in case 1,

$$|cl_{A_n}(\sigma)| = \frac{\frac{1}{2}|S_n|}{\frac{1}{2}|C_{S_n}(\sigma)|} = \frac{|S_n|}{|C_{S_n}(\sigma)|} = |S_n : C_{S_n}(\sigma)| = |cl_{S_n}(\sigma)|,$$

while, in case 2,

$$|cl_{A_n}(\sigma)| = \frac{\frac{1}{2}|S_n|}{|C_{S_n}(\sigma)|} = \frac{1}{2}|S_n : C_{S_n}(\sigma)| = \frac{1}{2}|cl_{S_n}(\sigma)|;$$

and, in view of 60.2, these conclusions justify what is stated in the present proposition.

Another way of expressing what 60.3 tells us is this: if X is a conjugacy class in S_n whose members are even permutations, then either (1) X is also a conjugacy class in A_n—which happens if the centralizer of an element of X contains an odd permutation, or (2) X splits into two conjugacy classes of A_n, each of order $\frac{1}{2}|X|$—which happens if the centralizer of each element of X consists entirely of even permutations.

Let us now apply this knowledge to A_5, whose membership was listed in terms of S_5-conjugacy classes in 59.10. Obviously, an S_5-conjugacy class of odd order cannot split into two equal-sized subsets; and so it is immediately apparent that the 15 elements of cycle pattern 1.2^2 must form a single conjugacy class in A_5, as of course does the identity element. On the other hand, it is obvious (since the order of a conjugacy class in a group G must divide $|G|$) that the 24 elements of cycle pattern 5 cannot form a single conjugacy class in A_5 (which has order 60); and so these elements must form two A_5-conjugacy classes, each of order 12. That leaves the set of 20 elements of cycle pattern $1^2.3$. One such element is the cycle $\alpha = (1\,2\,3)$; and it is not difficult to spot that $(4\,5)$ is an odd permutation commuting with α. So, in fact, by 60.3 the 20 elements of cycle pattern $1^2.3$ form a single conjugacy class in A_5.

To sum up, the following conclusion has been reached.

60.4 The group A_5 partitions into 5 conjugacy classes, namely:

(1) $\{\imath\}$,
(2) the set of 15 elements of cycle pattern $1 \cdot 2^2$,
(3) the set of 20 elements of cycle pattern $1^2 \cdot 3$,
(4) a set comprising 12 of the elements of cycle pattern 5, and
(5) the set consisting of the other 12 elements of cycle pattern 5.

The result which is the climax of §60 can now be proved, namely:

60.5 The group A_5 is simple.

Proof. Suppose that A_5 is not simple. Then A_5 contains a normal subgroup K whose order lies strictly between 1 and 60 and is (by Lagrange's theorem) a divisor of 60.

Because $K \lhd A_5$,

$$\sigma \in K \Rightarrow \tau \sigma \tau^{-1} \in K \text{ for all } \tau \in A_5 \qquad \text{(cf. 49.3),}$$

i.e. $\sigma \in K \Rightarrow$ every element of $\mathrm{cl}_{A_5}(\sigma)$ lies in K.

Hence it is apparent that K must be the union of some entire conjugacy classes of A_5; and of course one of these is $\{\imath\}$, since the subgroup K must contain the identity element. Different conjugacy classes being disjoint, it follows that $|K|$ is equal to the total of the orders of certain conjugacy classes of A_5 including $\{\imath\}$. However, from 60.4 one can soon check that no such total is equal to a divisor of 60 strictly between 1 and 60, and thus we have a contradiction of what was said earlier about $|K|$. From this contradiction the stated result follows.

Two supplementary facts are worth mentioning which help to put 60.5 in perspective. The first is that A_5 is the *smallest* of the non-abelian simple groups. (The proof that there is no non-abelian simple group with order $\leqslant 59$ could be set as a not too difficult exercise at a slightly more advanced stage in the study of group theory than is reached here.) The second supplementary fact is a generalization of 60.5—that A_n is simple for *every* $n \geqslant 5$. Because of the significance of simple groups (as outlined at the beginning of §60), it follows that alternating groups are of interest in group theory well beyond that part of the subject which deals purely with permutations. (There exists an elementary, albeit somewhat long, proof of the simplicity of A_n for all $n \geqslant 5$. A version of this elementary proof may be found in [3].)

At this point the section of the book devoted to groups draws to a close, since to go further would be at variance with the book's aim to provide an introduction, of reasonable length, to abstract algebra as a whole. The last four chapters have introduced the student to the basic concepts of group theory such as period of an element, cyclic groups, subgroups, cosets, homomorphisms and their kernels, quotient groups, and matters of fundamental importance in permutation group theory. An understanding of these things equips the student to advance to higher levels in the study of groups if he wishes. It will also help him in the study of other species of algebraic systems, since concepts closely analogous to those of elementary group theory often loom large in the study of other systems—a point which will emerge clearly in the discussion of rings in the next chapter.

The further theory of groups is a particularly attractive part of mathematics. In it one comes across theorems of great power and elegance on group structure (e.g. theorems about the existence of and relationships between subgroups of a group—both general theorems and theorems about special classes of groups that one can identify as in some sense nice). However, many significant problems about groups are very formidable problems, and it should not be imagined that complete solutions to all these problems are known to mankind.

Students keen to read more about group theory will find plenty to interest them in textbooks on the subject such as those by I. D. Macdonald, W. R. Scott, and J. Rose ([4], [5], and [6], respectively).

EXERCISES ON CHAPTER NINE

1. In S_6, let

$$\sigma = \begin{pmatrix} 1 & 2 & 3 & 4 & 5 & 6 \\ 5 & 4 & 1 & 3 & 6 & 2 \end{pmatrix}, \qquad \tau = \begin{pmatrix} 1 & 2 & 3 & 4 & 5 & 6 \\ 6 & 5 & 2 & 4 & 3 & 1 \end{pmatrix}.$$

Write down, in the same form, each of the permutations $\sigma\tau, \tau\sigma, \sigma^{-1}, \tau^{-1}, \tau^2$.

2. Express in d.c. form the permutations σ and τ given in exercise 1.

3. Write down the cube and the inverse of the cycle $(1\,5\,3\,2\,4)$.

4. In S_{10}, let

$$\sigma = (1\,4)(2\,9\,8\,6)(3\,10\,5\,7), \qquad \tau = (1\,7\,5)(2\,6)(3\,9)(8\,10)(4).$$

(a) Write down the cycle patterns of σ and τ.

(b) Without translating σ or τ into any other form, find in d.c. form each of the permutations $\sigma\tau, \tau\sigma, \sigma^{-1}, \tau^{-1}, \sigma^2, \sigma^3, (\sigma\tau)^3$.

(c) What are the periods of σ, τ, and $\sigma\tau$?

5. In S_8, let

$$\sigma = \begin{pmatrix} 1 & 2 & 3 & 4 & 5 & 6 & 7 & 8 \\ 3 & 8 & 4 & 6 & 5 & 1 & 2 & 7 \end{pmatrix}, \qquad \tau = \begin{pmatrix} 1 & 2 & 3 & 4 & 5 & 6 & 7 & 8 \\ 7 & 5 & 3 & 1 & 8 & 6 & 4 & 2 \end{pmatrix}.$$

(a) Express each of the permutations $\sigma, \tau, \sigma\tau, \tau\sigma\tau^{-1}$ in d.c. form. (Be sure to use 57.1 in the last case.)

(b) State the periods of σ and τ.

(c) Find the number of conjugates of τ in S_8.

(d) Find a permutation $\gamma \in S_8$ such that $\gamma^2 = \tau$ and γ moves all the objects $1, 2, 3, \ldots, 8$.

6. In S_7, let

$$\sigma = (1\,4)(2\,7\,5\,3)(6), \qquad \tau = (2\,4)(3\,6\,7\,5)(1),$$

so that σ and τ have the same cycle pattern and are therefore conjugate in S_7. Write down a permutation $\gamma \in S_7$ such that $\gamma\sigma\gamma^{-1} = \tau$.

7. What cycle patterns are possible for a permutation in S_7 that fixes none of the objects? Hence find the total number of such permutations in S_7.

8. Find (a) the number of elements of period 3 in S_7, (b) the number of elements of period 6 in S_8.

9. Show that in S_6 the elements of period 3 partition into two equal-sized conjugacy classes. Show also that if σ is any element of period 3 in S_6, $C_{S_6}(\sigma)$ contains an element of period 6.

10. Let α be a cycle in S_n (n arbitrary). Suppose that there is an object a such that $\alpha(a) \neq a$ but $\alpha^r(a) = a$ for a certain positive integer r. Show that $\alpha^r = \iota$.

11. A certain permutation σ in S_9 has cycle pattern $1^2 . 2^2 . 3$, and the permutation γ satisfies $\gamma^2 = \sigma$. How many different cycle patterns might γ have?

12. In S_9, let $\sigma = (1\,3)(2\,9)(4\,7)(5\,8\,6)$. Write down a permutation of cycle pattern 3^3 that commutes with σ. (Note that $\tau\sigma = \sigma\tau$ is equivalent to $\tau\sigma\tau^{-1} = \sigma$.)

13. Let α be a cycle of length n in S_n (n arbitrary). Show, by considering $|C_{S_n}(\alpha)|$, that $C_{S_n}(\alpha) = \langle \alpha \rangle$.

14. By counting inverted pairs, find the ε-value of the arrangement 25341.

15. In S_{10}, let

$$\sigma = \begin{pmatrix} 1 & 2 & 3 & 4 & 5 & 6 & 7 & 8 & 9 & 10 \\ 5 & 6 & 4 & 3 & 10 & 2 & 1 & 8 & 7 & 9 \end{pmatrix}, \quad \tau = \begin{pmatrix} 1 & 2 & 3 & 4 & 5 & 6 & 7 & 8 & 9 & 10 \\ 3 & 4 & 10 & 2 & 7 & 1 & 6 & 9 & 8 & 5 \end{pmatrix}.$$

Express each of σ and τ in d.c. form, and hence determine whether each of σ, τ is an even or odd permutation.

16. If σ, τ are respectively odd and even permutations in S_n, which of the permutations

$$\sigma\tau, \sigma^2, \tau^2, \sigma^3, \sigma^{-1}, \tau^{-1}, \tau\sigma\tau^{-1}$$

belong to A_n?

17. (a) What is the smallest value of n such that there is an element of period 18 in S_n? (b) What is the smallest value of n such that there is an element of period 18 in A_n?

18. Let $\sigma = (1\,2\,3)(4\,5\,6)$. Find (a) an odd permutation, (b) a non-identity even permutation in $C_{S_6}(\sigma)$.

19. Find the details of how A_6 partitions into conjugacy classes.

20. Verify that if a, b are objects and $1, a, b$ are all different, then

$$(1\ a)(1\ b)(1\ a) = (a\ b).$$

Deduce that (a) for $n \geqslant 2$, every element of S_n can be expressed in terms of the transpositions of the form $(1\ c)$ ($c \neq 1$), and (b) for $n \geqslant 3$, every element of A_n can be expressed as a product of cycles of length 3.

21. This is a nice application of 59.6 to prove that a group of order $2n$, where n is odd, contains a subgroup of order n. Let G be such a group, and consider the isomorphic group $\theta(G)$, where θ is the isomorphism introduced in the proof of Cayley's theorem (47.9). By exercise 13 on chapter 6, G contains an element c of period 2. Show that in $\theta(G)$ the permutation λ_c (which must also have period 2 and is to be regarded as a permutation of the n objects in G) has cycle pattern 2^n. Deduce that $\theta(G)$ contains a subgroup of order n, and therefore so does G.

22. Deduce from exercise 21 that if G is a simple group of even order greater than 2, then $4\,||G|$. (Note: there is a famous theorem, the Feit–Thomson theorem, proof of which was first published in 1963, which tells us that every non-abelian simple group has even order.)

23. A subgroup G of S_n is described as *transitive* iff for every ordered pair (a, b) of objects there exists $\sigma \in G$ such that $\sigma(a) = b$.
(a) Show that, for every $n \geqslant 3$, A_n is transitive.
(b) Show that if G is a transitive abelian subgroup of S_n, then each non-identity element of G moves all the objects.

24. Let G be a subgroup of S_n.
(a) Show that the relation $\sim$ defined on the set of objects by

$$a \sim b \Leftrightarrow \exists \sigma \in G \text{ s.t. } \sigma(a) = b \qquad (a, b \in \{1, \ldots, n\})$$

is an equivalence relation. The $\sim$-class of object a is termed the *orbit* of a. Notice that the number of objects in the orbit of a is (by exercise 17 on chapter 7) equal to $|G:G_a|$, where G_a is the stabilizer of a in G, i.e. $\{\sigma \in G : \sigma(a) = a\}$.

(b) Let a be an object, and let $\sigma \in G$. Show that $G_{\sigma(a)}$, the stabilizer of $\sigma(a)$ in G, is $\sigma G_a \sigma^{-1}$ (G_a being the stabilizer of a).

(c) For each $\gamma \in G$, let $F(\gamma)$ denote the number of objects fixed by γ. By counting the number of pairs of the form (γ, a) with a an object and γ a permutation in G fixing a, show that

$$\sum_{\gamma \in G} F(\gamma) = |G| \times k,$$

where k is the number of different orbits under the action of G, i.e. the number of different $\sim$-classes that there are, where $\sim$ is the equivalence relation introduced in part (a).

25. (i) Show that, although $6 \,|\, |A_4|$, A_4 contains no subgroup of order 6. Is A_4 a simple group?

(ii) Part (i) reveals that what is sometimes termed "the converse of Lagrange's theorem"—i.e. the statement that if d ($\in \mathbb{N}$) divides the order of the finite group G, then G must contain a subgroup of order d—is false. A more sophisticated suggestion along the same lines that might be put forward is: "If d ($\in \mathbb{N}$) divides the order of the finite group G and if d and $|G|/d$ are relatively prime, then G contains a subgroup of order d". Prove that this suggestion is also false by making use of the result of exercise 28 on chapter 8 (that every group of order 15 is cyclic).

CHAPTER TEN

THE FUNDAMENTALS OF RING THEORY

61. Introduction

In this chapter we turn our attention to rings, i.e. algebraic systems with two binary operations (addition and multiplication) satisfying certain postulates. For part of the time our concern will be with special kinds of rings — e.g. fields, to which the reader was introduced towards the end of Chapter 3.

Students should find much of the material in this chapter rapidly accessible because it deals with features of ring theory which are obviously analogous to, or connected with, features of group theory developed in Chapters 6–8. For example, after disposing of some very elementary matters in §§ 62 and 63, we shall go on, in §§ 64–68, to define and discuss subrings, homomorphisms from one ring to another, and kernels of such homomorphisms. A subring which is the kernel of a homomorphism between rings is what we call an *ideal* (cf. a normal subgroup in group theory), and it turns out that, just as in group theory the set of cosets of a normal subgroup can be turned into a group, so in ring theory the set of cosets of an ideal can be turned into a ring. Details of that particular matter come in § 68.

In § 69 we shall use the idea of the period of a group element to focus on something fundamental to field theory. It will be seen, in particular, that the fields $\mathbb{Q}$ and $\mathbb{Z}_p$ (p a prime number) play a special role in the theory of fields because every field contains a subfield isomorphic either to $\mathbb{Q}$ or to $\mathbb{Z}_p$ for some prime number p. Finally, in § 70, we shall gain an extra dimension of insight into the relationship between $\mathbb{Q}$ and $\mathbb{Z}$ by seeing that the construction of $\mathbb{Q}$ from $\mathbb{Z}$ is but one case of something very general.

62. The definition of a ring and its elementary consequences

Definition. A non-empty set R is called a **ring** when there are defined on R two binary operations, addition and multiplication, and the following postulates are satisfied:

A: R is an abelian group with respect to addition;

M0: R is closed under multiplication;

M1: multiplication is associative on R;

$$D: \forall x, y, z \in R, \left\{ \begin{array}{l} x(y + z) = xy + xz \\ \text{and } (y + z)x = yx + zx \end{array} \right\} \quad (\textit{distributive laws}).$$

The postulates A, M0, M1, D are called the **ring axioms**.

Examples of rings include the following systems: $\mathbb{Z}$; $\mathbb{R}$; $\mathbb{C}$; the system of all $n \times n$ matrices with entries in $\mathbb{R}$ (n being a given positive integer); the system $\mathbb{Z}_m$ of congruence classes modulo m (m being a given positive integer, and addition and multiplication being defined as in § 18).

Remarks. (*a*) Pedantically speaking, a ring is not just a set R but rather an ordered triple $(R, +, .)$ consisting of a set R and two binary operations defined on that set. Nevertheless (cf. the remark following the definition of a semigroup in § 29) it is commonplace to use phraseology such as "the ring R".

(*b*) Clearly, with respect to multiplication, the set of elements in a ring is a semigroup.

(*c*) Ring axiom A tells us that in every ring there is a zero and each element x has a negative, $-x$; and it tells us how zero and negatives behave in addition in a ring. We now show that it is a consequence of the ring axioms that, in multiplication in a ring, zero and negatives behave in a definite (and natural-looking) way.

62.1 In a ring R, $\quad 0 . x = x . 0 = 0$ for every $x \in R$.

Proof. Let x be an arbitrary element of the ring R. We shall prove that $0 . x = 0$. (The proof that $x . 0 = 0$ is similar.)

Since 0 is the identity element in the group $(R, +)$, $0 + 0 = 0$; and so

$$x(0 + 0) = x . 0$$

i.e. $\qquad\qquad\qquad x . 0 + x . 0 = x . 0 \quad \text{(by D)}$

$$= x . 0 + 0.$$

By cancellation in the additive group $(R, +)$ it follows that $x . 0 = 0$.

62.2 Let x, y be elements of a ring R. Then

(i) $(-x)y = x(-y) = -(xy)$, (ii) $(-x)(-y) = xy$.

Proof. (i) $\{x + (-x)\}y = 0 . y$

$$= 0 \qquad \text{(by 62.1)}$$

i.e. by D, $\quad xy + (-x)y = 0$.

By the additive version of 35.3, it follows that $(-x)y = -(xy)$.

It can similarly be proved that $x(-y) = -(xy)$.

(ii) $(-x)(-y) = -\{x(-y)\}$ (by part (i))

$\qquad\qquad\quad = -\{-(xy)\}$ (by part (i))

$\qquad\qquad\quad = xy$ (by the additive version of 35.5).

63. Special types of ring and ring elements

(1) *Commutative rings*

A **commutative ring** means a ring in which the multiplication is commutative. The most familiar example is $\mathbb{Z}$.

(2) *Rings with unity*

In a ring R, a **unity** (or identity) is a nonzero element 1 with the property that $1x = x1 = x$ for every $x \in R$. Thus a unity in R is an identity, different from zero, in the semigroup $(R, .)$.

By 31.1, it follows that no ring can possess more than one unity.

A ring with a unity is often referred to colloquially as "a ring with a one".

The ring of 2×2 real matrices is an example of a non-commutative ring with a one, the "one" (or unity) being the matrix $I = \begin{bmatrix} 1 & 0 \\ 0 & 1 \end{bmatrix}$.

The set of even integers (with ordinary addition and multiplication) provides an example of a ring without a unity.

(3) *Division rings*

A **division ring** is a (not necessarily commutative) ring with a unity, 1, in which every nonzero element has a multiplicative inverse (i.e. corresponding to each nonzero element x of the ring there is an element x^{-1}, also in the ring, such that $x^{-1}x = xx^{-1} = 1$).

An example of a division ring which is non-commutative is the ring of all complex matrices of the form $\begin{bmatrix} a & b \\ -\bar{b} & \bar{a} \end{bmatrix}$ $(a, b \in \mathbb{C})$. It is easy but tedious to check that the set of all such matrices actually is a non-commutative ring with a one; the point of special interest is that each nonzero element $\begin{bmatrix} a & b \\ -\bar{b} & \bar{a} \end{bmatrix}$ $(a, b \in \mathbb{C}$ and not both zero) of the ring has a multiplicative inverse in the ring, viz. $\begin{bmatrix} c & d \\ -\bar{d} & \bar{c} \end{bmatrix}$, where

$$c = \bar{a}/(|a|^2 + |b|^2), \qquad d = -b/(|a|^2 + |b|^2).$$

This division ring, which can be exhibited in a variety of isomorphic versions, is called the *ring of quaternions*.

(4) *Fields*
Arriving on familiar territory by an unfamiliar route, we define a **field** to be a commutative division ring, i.e. a field is a non-empty set F on which are defined two binary operations, viz. addition and multiplication, the following postulates all being satisfied:

A, M0, M1, D as in the definition of a ring (with F replacing R);

M2: multiplication is commutative on F;

M3: there is a unity in F;

M4: each nonzero element of F has a multiplicative inverse in F.

(It is easily seen that this definition of a field coincides with that given in § 19.)

The very well-known algebraic systems $\mathbb{Q}$, $\mathbb{R}$, $\mathbb{C}$ are all fields. And (cf. 19.2) the system $\mathbb{Z}_p$ is a field when p is prime.

Notational remarks. (a) When we are working with elements a, b of a field F and $b \neq 0$, the notation $\dfrac{a}{b}$ (or a/b) is acceptable for ab^{-1}. There is no ambiguity in "a/b" since ab^{-1} and $b^{-1}a$ mean the same thing, the multiplication defined on F being commutative.

(b) In this chapter, where the aim is clear exposition of fundamentals, we shall persist with the notation hitherto used for the elements of the field $\mathbb{Z}_p$ (p prime) — namely $C_0, C_1, C_2, \ldots, C_{p-1}$. However, in more advanced work it is customary to denote the elements of $\mathbb{Z}_p$ simply by $0, 1, 2, \ldots, p - 1$ (i.e. to abbreviate C_x just to x) provided there is no risk of confusion between integers and the member objects of $\mathbb{Z}_p$. This is particularly advantageous in the cases of $\mathbb{Z}_2$ (whose elements can be denoted simply by 0 and 1) and of $\mathbb{Z}_3$ (whose elements can be denoted simply by 0, 1 and 2 or by 0, 1 and -1, "-1" and "2" being the same thing in $\mathbb{Z}_3$ since the congruence classes C_2 and $C_{(-1)}$ (mod 3) coincide). The simpler notation for elements of $\mathbb{Z}_p$ will be employed in Chapter 11 of this book, and students should feel free to use the simpler notation in their own work as soon as they are absolutely secure in their understanding of what $\mathbb{Z}_p$ and its elements are.

(5) *Integral domains*
Preliminary definition. In a commutative ring R, a nonzero element x is called a **divisor of zero** iff there exists $y \in R - \{0\}$ such that $xy = 0$.

Thus the presence of a divisor of zero in a commutative ring means that the product of two elements of the ring may be zero, even though neither factor is zero: e.g. in the commutative ring $\mathbb{Z}_6$, we have $C_2 C_3 = C_0$ (C_0 being the zero of the ring), while neither C_2 nor C_3 is zero: this shows that C_2 and C_3 are divisors of zero in the ring $\mathbb{Z}_6$.

Main definition. An **integral domain** is a commutative ring with a one in which there are no divisors of zero; i.e. an integral domain is a commutative ring R with a one in which there holds the additional axiom

         I: For $x, y \in R$, $xy = 0 \Rightarrow x = 0$ or $y = 0$ (or both).

The ring $\mathbb{Z}$ is a familiar example of an integral domain. Moreover, $\mathbb{Z}$ is an example of an integral domain which is not a field. This remark should be set alongside the fact that:

63.1   Every field is an integral domain.

Proof. Let F be a field. Since F is certainly a commutative ring with a one, the result will follow if we can show that the axiom I holds in the case $R = F$.
Suppose that $xy = 0$ $(x, y \in F)$. If $x \neq 0$, x^{-1} exists in F and we have

$$y = 1y = x^{-1}xy = x^{-1}0 = 0 \qquad \text{(using 62.1 at the last step)}.$$

The only remaining case is $x = 0$.
     Therefore, for $x, y \in F$, $xy = 0 \Rightarrow x = 0$ or $y = 0$.
     Thus I holds for F; and the result follows.

It is helpful to note that, by the law of contraposition, the axiom I is equivalent to

$$x \neq 0 \quad \text{and} \quad y \neq 0 \Rightarrow xy \neq 0 \quad (x, y \in R)$$

i.e. $R - \{0\}$ is closed under multiplication.

This version of I makes it easy to see that:

63.2   If R is an integral domain, then $(R - \{0\}, .)$ is a semigroup.
     (For, if R is an integral domain, then (1) $R - \{0\} \neq \varnothing$ since $1 \in R - \{0\}$, (2) $R - \{0\}$ is closed under multiplication (cf. above version of I), and (3) multiplication is associative on $R - \{0\}$ since multiplication is associative on R).
     An analogous observation about fields, more conveniently made here than in subsection (4), is:

63.3   If F is a field, then $(F - \{0\}, .)$ is a group. [We refer to this group as the *multiplicative group* of F.]

Proof. Let F be a field. We show that the group axioms hold for $F - \{0\}$.
     G0: By 63.1 and 63.2, $F - \{0\}$ is closed under multiplication.
     G1: Holds, since multiplication is associative on F.
     G2: Holds because (since $1 \neq 0$) $1 \in F - \{0\}$.
     G3: Corresponding to each $x \in F - \{0\}$, there is (by the definition of a field) an element $x^{-1} \in F$ such that $x^{-1}x = 1$; and in every case $x^{-1} \neq 0$ (since $x^{-1} = 0$ would imply $1 = x^{-1}x = 0x = 0$). Thus each element of $F - \{0\}$ has an inverse in $F - \{0\}$, i.e. G3 holds.

The result follows.

Our final remark on integral domains is that in such systems there holds a cancellation law to the effect that one can cancel a nonzero element from both sides of an equation:

63.4 Suppose that elements a, b, c of an integral domain R satisfy $ab = ac$ and $a \neq 0$. Then $b = c$.

Proof. $a(b - c) = a\{b + (-c)\}$ (by meaning of subtraction in an additive
$\qquad\qquad\qquad\qquad\qquad$ group — see remark (b) in § 33)

$\qquad\quad = ab + a(-c)$ (by ring axiom D)

$\qquad\quad = ab - ac$ (by 62.2)

$\qquad\quad = 0.$

Therefore, since R has no divisors of zero, one at least of $a, b - c$ must be zero. However, $a \neq 0$. Hence $b - c = 0$, and hence $b = c$.

(6)   *Units in a ring with a one*
Definition. Let R be a ring with a one. A **unit** in R means an element of R with a multiplicative inverse in R (i.e. the element x of R is a unit iff x has an inverse in the semigroup $(R, .)$).

For example, in $\mathbb{Z}$ the units are 1 and -1; in a field every nonzero element is a unit.

A generalization of 63.3 is:

63.5 Let R be a ring with a one, and let U be the set of units in R. Then U is a group with respect to multiplication.

The proof of 63.5 is left as an exercise.

64. Subrings and subfields

Definition 1. Let R be a ring. [This means that the system $(R, +, .)$ is a ring.] A subset S of R is called a **subring** of R iff the system $(S, +, .)$ is a ring, i.e. iff S is a ring with respect to the addition and multiplication which are defined on R and therefore on S (see remark (c) in § 27).

Definition 2. Let F be a field. [This means that the system $(F, +, .)$ is a field.] A subset K of F is called a **subfield** of F iff the system $(K, +, .)$ is a field.

Familiar illustrations: $\mathbb{Z}$ is a subring of $\mathbb{R}$; $\mathbb{R}$ is a subfield of $\mathbb{C}$.
Notice that, clearly:

64.1 In any ring R, R itself and $\{0\}$ are subrings of R; and, in any field F, F is a subfield of F.

Tests for a subring or subfield

64.2 Let R be a ring and let S be a subset of R. Then S is a subring of R iff the following conditions all hold:

(1) $S \neq \varnothing$;
(2) S is closed under subtraction;
(3) S is closed under multiplication.

Proof. (i) If S is a subring, then the conditions (1), (2), (3) all hold, by virtue, respectively, of the ring axioms A, A, M0 as they apply to S.

(ii) Conversely, suppose that the conditions (1), (2), (3) all hold.

By (1), (2) and the additive version of 36.4, it follows that S is a group with respect to addition; and the group $(S, +)$ is abelian since the addition defined on R is commutative. Hence A holds for S. The condition (3) tells us that M0 holds for S. Finally, M1 and D hold for S since they hold for the whole set R. Therefore $(S, +, .)$ is a ring, i.e. S is a subring of R.

Proof of the stated result is now complete.

Formulation of a test for a subfield of a given field is more subtle. We leave as an exercise proof of the following result.

64.3 Let F be a field and K a subset of F. Then K is a subfield of F iff the following conditions all hold: (1) $K - \{0\} \neq \varnothing$;

(2) K is closed under subtraction;
(3) K is closed under multiplication;
(4) $x \in K - \{0\} \Rightarrow x^{-1} \in K$.

Unity elements in relation to subrings/subfields

Consider the set S of matrices of the form $\begin{bmatrix} x & 0 \\ 0 & 0 \end{bmatrix}$, with $x \in \mathbb{R}$. By means of 64.2, it is easily seen that S is a subring of the ring R of all 2×2 real matrices. Clearly, the subring S has a unity, viz. $\begin{bmatrix} 1 & 0 \\ 0 & 0 \end{bmatrix}$; and this unity is not the same as the unity of the whole ring R. The moral is that, in a ring with a unity, the unity of a subring (if it possesses one) need not be the same as the unity of the whole ring.

It is easy to prove that, in a field F, the unity of every subfield must coincide with that of F and, more generally, that if D is an integral domain contained as a subring in a larger integral domain E, then the unities of D and E must coincide.

65. Ring homomorphisms

Definition. Let R and R_1 be rings. A **(ring) homomorphism** from R to R_1

means a mapping $\theta: R \rightarrow R_1$ such that

$$\forall x, y \in R, \text{ (1) } \theta(x + y) = \theta(x) + \theta(y) \quad \text{and} \quad \text{(2) } \theta(xy) = \theta(x)\theta(y).$$

Thus, as in group theory, a homomorphism is a mapping which, so to speak, respects the binary operations in the systems concerned.

Let $\theta: R \rightarrow R_1$ be a ring homomorphism. If we temporarily forget the multiplication defined on R and R_1, we see that θ is a group homomorphism from the additive group $(R, +)$ to the additive group $(R_1, +)$. By 47.1 and 47.2, it follows that:

65.1 If $\theta: R \rightarrow R_1$ is a ring homomorphism (R, R_1 being rings), then
(i) $\theta(0) = 0_1$ (where 0, 0_1 are the zeros of R, R_1, respectively),
(ii) $\forall x \in R$, $\theta(-x) = -\{\theta(x)\}$.

In the same vein, we note that the kernel of a ring homomorphism $\theta: R \rightarrow R_1$ already has a meaning, viz.

$$\ker \theta = \{x \in R : \theta(x) = 0_1\} \quad \text{(where } 0_1 \text{ is the zero of } R_1\text{)};$$

and 47.8 tells us that:

65.2 If $\theta: R \rightarrow R_1$ is a ring homomorphism, then θ is injective iff $\ker \theta = \{0\}$.

To obtain other facts about ring homomorphisms and their kernels, we must bring back into our thinking the multiplication defined on the rings concerned.

65.3 Let $\theta: R \rightarrow R_1$ be a ring homomorphism (R and R_1 being rings). Then:
(i) for every subring S of R, $\theta(S)$ is a subring of R_1; and, in particular [the case $S = R$], im θ is a subring of R_1;
(ii) $\ker \theta$ is a subring of R.

Proof. In each part we use 64.2.
(i) Let S be a subring of R.
(1) Since $S \neq \varnothing$, $\theta(S) \neq \varnothing$.
(2) and (3). Let x, y be arbitrary elements of $\theta(S)$. Then $x = \theta(s)$ and $y = \theta(t)$ for some $s, t \in S$. Hence

$$
\begin{aligned}
x - y &= \theta(s) - \theta(t) \\
&= \theta(s) + \theta(-t) \quad \text{(by 65.1(ii))} \\
&= \theta(s - t) \quad \text{(since } \theta \text{ is a homomorphism)}.
\end{aligned}
$$

Since S, being a subring, is closed under $-$, $s - t \in S$. Hence $x - y \in \theta(S)$.

Similarly, $xy = \theta(s)\theta(t) = \theta(st)$, and so, since S is closed under multiplication (so that $st \in S$), $xy \in \theta(S)$.

This proves that $\theta(S)$ is closed under subtraction and multiplication.

Part (i) now follows by 64.2.

(ii) (1) By 65.1(i), $\theta(0) = 0_1$ ($=$ the zero of R_1), i.e. $0 \in \ker \theta$. Therefore $\ker \theta \neq \varnothing$.

(2) and (3) Let $x, y \in \ker \theta$. Then $\theta(x) = \theta(y) = 0_1$. Hence

$$\begin{aligned}
\theta(x - y) &= \theta(x) + \theta(-y) \quad \text{(since θ is a homomorphism)} \\
&= \theta(x) - \theta(y) \quad \text{(by 65.1(ii))} \\
&= 0_1 - 0_1 \\
&= 0_1
\end{aligned}$$

and $\theta(xy) = \theta(x)\theta(y) = 0_1 0_1 = 0_1$. So both $x - y$ and xy belong to $\ker \theta$.

This proves that $\ker \theta$ is closed under subtraction and multiplication.

It follows by 64.2 that $\ker \theta$ is a subring of R.

Remarks. (*a*) Ring homomorphisms of special character are given names like those given correspondingly to special group homomorphisms: i.e.

    (1) an injective ring homomorphism is called a **monomorphism**;

    (2) a surjective ring homomorphism is called an **epimorphism**;

    (3) a bijective ring homomorphism is called an **isomorphism**.

(*b*) On reflection, it can be apreciated that the existence of a ring isomorphism from R to R_1 (where R and R_1 are rings) tells us that R and R_1 are, in a significant sense, the same — that they have the same algebraic structure. When a ring isomorphism from the ring R to the R_1 exists, we say that R and R_1 are **isomorphic** to each other; and we indicate this by writing $R \cong R_1$.

66. Ideals

We now define a special kind of subring that we call an *ideal*. As indicated in § 61, ideals are to ring theory what normal subgroups are to group theory.

Definition. Let S be a subring of the ring R. Then S is called an **ideal** of R iff the condition

$$s \in S \text{ and } x \in R \Rightarrow \text{both } xs \text{ and } sx \text{ belong to } S$$

holds.

Thus a subring S of the ring R is an ideal of R iff it has a subtle kind of closure property — namely that, when we take an element of the subring S

and multiply it (on left or on right) by any element of R (either in S or in $R - S$), the resulting product is always in S.

Simple illustration. In $\mathbb{Z}$, the subring comprising all the even integers is an ideal: for if we multiply an even integer by any integer whatever, we get for the product an even integer.

It is clear from the definition of an ideal that:

66.1 In every ring R, $\{0\}$ and R are ideals of R.

The next result is a simple but useful observation that follows almost immediately from the definition of an ideal.

66.2 Let R be a ring with a one, 1; and let S be an ideal of R. If S contains a unit of R (in particular if $1 \in S$), then S is the whole of R.

Proof. Suppose that S contains the unit u of R (with inverse u^{-1} in R).

Let x be an arbitrary element of R. Since S is an ideal and since $u \in S$ while $xu^{-1} \in R$, it follows from the definition of an ideal that $(xu^{-1})u \in S$, i.e. $x \in S$.

Since x was arbitrary, this proves that S is the whole of R; and the result follows.

A corollary is that a field has no ideals except the obvious ones that 66.1 points out to us, i.e.

66.3 In a field F, the only ideals are F and $\{0\}$.

Proof. If S is an ideal of the field F and $S \neq \{0\}$, then S contains a nonzero element of F, i.e. S contains a unit of F, and hence, by 66.2, $S = F$.

The following result presents the standard simple test for an ideal not already known to be a subring.

66.4 Let R be a ring and let S be a subset of R. Then S is an ideal of R iff the following conditions all hold:
   (1) $S \neq \varnothing$;
   (2) S is closed under subtraction;
   (3) $s \in S$ and $x \in R \Rightarrow xs$ and sx belong to S.

Proof. (i) If S is an ideal of R, then, because S is a subring, the conditions (1) and (2) hold, and because, further, S is an ideal, condition (3) also holds.

(ii) Conversely, suppose that conditions (1), (2), (3) all hold.

Condition (3) tells us that, in particular, S is closed under multiplication; and, by this and conditions (1) and (2), it follows that S is a subring. By condition (3) it now follows that S is an ideal.

Proof of the stated results is now complete.

Worked example. Let R be the ring of all matrices of the form $\begin{bmatrix} x & y \\ 0 & z \end{bmatrix}$ with $x, y, z \in \mathbb{R}$. (Incidentally, one would prove that R is a ring by using 64.2 to prove that R is a subring of the ring of all 2×2 real matrices.) Let S be the subset of R comprising the matrices of the form $\begin{bmatrix} x & y \\ 0 & 0 \end{bmatrix}$ $(x, y \in \mathbb{R})$. Show that S is an ideal of R.

Solution. (1) $S \neq \varnothing$, clearly.

(2) Let $A = \begin{bmatrix} x_1 & y_1 \\ 0 & 0 \end{bmatrix}$, $B = \begin{bmatrix} x_2 & y_2 \\ 0 & 0 \end{bmatrix}$ be two arbitrary members of S $(x_1, x_2, y_1, y_2 \in \mathbb{R})$. Then

$$A - B = \begin{bmatrix} x_1 - x_2 & y_1 - y_2 \\ 0 & 0 \end{bmatrix}, \text{ which belongs to } S.$$

Hence S is closed under subtraction.

(3) Let $A = \begin{bmatrix} x & y \\ 0 & 0 \end{bmatrix}$ be an arbitrary member of S and $X = \begin{bmatrix} a & b \\ 0 & c \end{bmatrix}$ be an arbitrary member of R $(x, y, a, b, c \in \mathbb{R})$. Then

$$AX = \begin{bmatrix} xa & xb + yc \\ 0 & 0 \end{bmatrix} \text{ and } XA = \begin{bmatrix} xa & ya \\ 0 & 0 \end{bmatrix},$$

and both of these matrices belong to S. Thus $A \in S$ and $X \in R \Rightarrow AX$ and XA belong to S.

It follows (by 66.4) that S is an ideal of R.

The next theorem is the first to give substance to the claim that ideals are the analogues in ring theory of normal subgroups in group theory.

66.5 Let $\theta: R \to R_1$ be a ring homomorphism (R, R_1 being rings). Then $\ker \theta$ is an ideal of R.

Proof. By 65.3(ii), $\ker \theta$ is a subring of R.

Let s be an arbitrary element of $\ker \theta$ (so that $\theta(s) = 0_1$, the zero of R_1), and let x be an arbitrary element of R. Then

$$\theta(xs) = \theta(x)\theta(s) = \theta(x).\, 0_1 = 0_1, \text{ and } \theta(sx) = \theta(s)\theta(x) = 0_1.\, \theta(x) = 0_1,$$

so that both xs and sx belong to $\ker \theta$.

Thus $s \in \ker \theta$ and $x \in R \Rightarrow$ both xs and sx belong to $\ker \theta$.

It follows that ker θ is an ideal of R.

Later (in § 68), when we construct the ring-theory analogues of quotient groups, it will be apparent that, conversely, every ideal of a ring R is the kernel of at least one homomorphism with domain R.

The final item in this section is a means by which a new ideal of a ring can be constructed from given ideals.

Suppose that R is a ring. In § 41 we defined the product of two subsets of a multiplicative group. The corresponding notion in an additive group is the *sum* of two subsets of the group. So if A and B are subsets of R (and are therefore contained in the additive group $(R, +)$), "the sum $A + B$" already has a meaning, namely

$$A + B = \{a + b : a \in A, b \in B\}$$
$[=$ the set of elements of R expressible as the sum
of an element of A and an element of $B.]$

This is relevant to the discussion of ideals because, as the following proposition reveals along with some supplementary details, the sum of two ideals of R is also an ideal of R.

66.6 Let I and J be ideals of a ring R. Then $I + J$ is an ideal of R that contains both I and J as subsets, and it is the smallest ideal of R that contains both I and J. [Here "smallest" is used in what might be called the set-theoretic sense: in describing $I + J$ as the *smallest* ideal of R that contains both I and J, what we mean is that if K is any such ideal of R, then $I + J$ is *contained in K*.]

Proof. (1) $I + J$ is certainly non-empty: e.g. since $0 \in I$ and $0 \in J$, $I + J$ contains $0 + 0$ ($= 0$).

(2) $I + J$ is closed under subtraction, since

$x, y \in I + J \Rightarrow x = i_1 + j_1$ and $y = i_2 + j_2$ for some $i_1, i_2 \in I$, $j_1, j_2 \in J$
$\qquad \Rightarrow x - y = (i_1 - i_2) + (j_1 - j_2)$ [with $i_1, \ldots, j_2$ as above]
$\qquad \Rightarrow x - y \in I + J$ (since $i_1 - i_2 \in I$ and $j_1 - j_2 \in J$, the ideals I
$\qquad\qquad\qquad\qquad$ and J being closed under subtraction).

(3) $x \in R$ and $s \in I + J$
$\qquad \Rightarrow x \in R$ and $s = i + j$ for some $i \in I$, $j \in J$
$\qquad \Rightarrow xs = x(i + j) = xi + xj$ [with $i \in I$, $j \in J$]
$\qquad \Rightarrow xs \in I + J$ (since, because I and J are ideals, $xi \in I$ and $xj \in J$);

and similarly $s \in R$ and $x \in I + J \Rightarrow sx \in I + J$.

It now follows, by 66.4, that $I + J$ is an ideal of R.

Each element i of I belongs to $I + J$ (as is seen from the equation $i = i + 0$ and the fact that $0 \in J$). So $I \subseteq I + J$. Similarly $J \subseteq I + J$. So $I + J$ is an ideal of R that contains both I and J. To see that it is the smallest such ideal, suppose that K is any ideal of R that contains I and J. Then, because K is closed under addition, K contains every element of the form $i + j$ with $i \in I$, $j \in J$; and so $K \supseteq I + J$.

Proof of the stated result is now complete.

The idea of the sum of two ideals will be used to prove an important theorem (68.3) later in this chapter and will again be used in Chapter 11.

67. Principal ideals in a commutative ring with a one

Throughout this section R will denote a *commutative* ring with a one.

For each $a \in R$, we denote by (a) the subset $\{ra : r \in R\}$ of R, i.e. the set of all elements obtainable by multiplying the particular element a by an element of R.

67.1 For each $a \in R$, (a) is an ideal of R. Moreover, $a \in (a)$, and (a) is the smallest ideal of R to which a belongs — smallest in the sense that it is *contained in* every ideal of R to which a belongs [cf. remark on "smallest" in the statement of 66.6].

Proof. Let a be an arbitrary element of R.

We use 66.4 to prove that (a) is an ideal of R.

(1) $(a) \neq \varnothing$ since, for example, $1a \in (a)$, i.e. $a \in (a)$.

(2) Let x, y be arbitrary elements of (a). Then $x = ra$ and $y = sa$ for some $r, s \in R$. Hence $x - y = ra - sa = (r - s)a$, and so $x - y \in (a)$.

This shows that (a) is closed under subtraction.

(3) If s is an element of (a) and $x \in R$, then $s = ra$ for some $r \in R$ and so $xs = x(ra) = (xr)a$, which belongs to (a).

Thus $s \in (a)$ and $x \in R \Rightarrow xs \ (= sx) \in (a)$.

It follows by 66.4 that (a) is an ideal of R; and it has already been remarked (under (1) above) that $a \in (a)$.

Let now J be any ideal of R such that $a \in J$. By the definition of an ideal, $ra \in J$ for every $r \in R$. So every element of (a) belongs to J, i.e. $(a) \subseteq J$.

This shows that (a) is the smallest ideal of R to which a belongs and completes the proof of the theorem.

Nomenclature. For each $a \in R$, the ideal (a) of R is called the **principal ideal of R generated by** a.

Simple illustration. In $\mathbb{Z}$, for each integer m, (m) is the ideal of $\mathbb{Z}$ comprising all the integral multiples of m. As the next theorem shows, such ideals are the only ideals of $\mathbb{Z}$.

67.2 Every ideal of $\mathbb{Z}$ is a principal ideal, i.e. equals (m) for some integer m.

Foreword to proof. A point of some subtlety arises in the proof of 67.2 given below — namely that, for $k, m \in \mathbb{Z}$, the kth multiple of m in the additive group $(\mathbb{Z}, +)$, in the sense of kth "power" of m under the binary operation $+$, is the same as the product km in the ring $\mathbb{Z}$. The student should reflect on this delicate point before proceeding.

Proof of 67.2. Let J be an arbitary ideal of $\mathbb{Z}$. Then J is a subring of $\mathbb{Z}$ and, in particular, $(J, +)$ is a subgroup of the additive group $(\mathbb{Z}, +)$. But the additive group $(\mathbb{Z}, +)$ is cyclic (1 being a generator) and hence, by 39.4, $(J, +)$ is a cyclic group — generated, say, by m ($m \in \mathbb{Z}$). Therefore (cf. foreword)

$$J = \{km : k \in \mathbb{Z}\} = (m)$$

This proves the stated result.

Definition. An integral domain in which every ideal is a principal ideal is called a **principal ideal domain**.

Thus, since $\mathbb{Z}$ is an integral domain, 67.2 tells us that $\mathbb{Z}$ is a principal ideal domain.

It turns out that every principal ideal domain has nice factorization properties akin to those of $\mathbb{Z}$. One particular case of this will be demonstrated in Chapter 11 — the case of the ring of polynomials with coefficients in an arbitrary field (a ring which, as will be duly proved in Chapter 11, is, like $\mathbb{Z}$, a very naturally occurring example of a principal ideal domain).

68. Factor rings

Throughout this section R will denote a ring (not necessarily commutative, not necessarily possessing a unity), and J will denote an ideal of R. We are about to carry out the ring theory analogue of the construction of a quotient group.

To begin with, let us temporarily forget that multiplication is defined on R. Then we are left with an additive group $(R, +)$ in which $(J, +)$ is a subgroup — indeed a normal subgroup since $(R, +)$ is an abelian group. By 50.3 the additive quotient group $(R/J, +)$ exists; and it is an abelian group (since, obviously, every quotient group of an abelian group is also abelian). The results of § 50, when translated into additive notation, give us certain information about the group $(R/J, +)$, including:

(a) each element is a coset of J in R, i.e. an object of the form

$$x + J (= \{x + j : j \in J\}) \text{ for some } x \in R;$$

(b) the rule of addition of such cosets is

$$(x + J) + (y + J) = (x + y) + J \quad (x, y \in R);$$

(c) in the group $(R/J, +)$ the zero is J and the negative of $x + J$ is $(-x) + J$ (for each $x \in R$).

Taking R/J to mean the set of elements of the group $(R/J, +)$, let us now remember the multiplication on R and propose the rule

$$(x + J)(y + J) = xy + J \quad (x, y \in R) \tag{1}$$

as a definition of multiplication on R/J. Since each coset has, in general, several labels of the form $x + J$ $(x \in R)$, the question of consistency arises here (cf. § 18). The question is favourably settled by the following lemma.

68.1 Suppose that $x + J = x' + J$ $(= \xi)$ and $y + J = y' + J$ $(= \eta)$, where $x, x', y, y' \in R$. Then $xy + J$ and $x'y' + J$ (the two meanings that (1) indicates for $\xi\eta$) are equal.

Proof. By the additive version of 42.6, since $x + J = x' + J$, it follows that $x - x' \in J$. Similarly, $y - y' \in J$. Hence, since J is an ideal, $(x - x')y \in J$ and $x'(y - y') \in J$. So, since, in particular, J is closed under addition,

$$(x - x')y + x'(y - y') \in J, \text{ i.e. } xy - x'y' \in J, \text{ i.e. } xy + J = x'y' + J.$$

We can now proceed, knowing that the equation (1) provides a consistent definition of multiplication on the set R/J, so that we have an algebraic system $(R/J, +, .)$.

68.2 (i) The system $(R/J, +, .)$ is a ring (which we henceforth call "the ring R/J").

(ii) The mapping $v: R \to R/J$ given by $x \mapsto x + J$ $(x \in R)$ is a ring epimorphism with kernel J.

(iii) If R is commutative, so is R/J.

(iv) If R has a unity, 1, then, provided $J \subset R$, R/J has a unity, viz. $1 + J$.

Proof. (i) We check that the ring axioms hold for $(R/J, +, .)$.

A: Before introducing multiplication, we saw that R/J is an abelian group with respect to addition.

M0: It is clear from the definition of multiplication (equation (1)) that R/J is closed under multiplication.

M1: For all $x, y, z \in R$, we have, using the definition of multiplication on R/J and the fact that the multiplication on R is associative:

$$(x + J)\{(y + J)(z + J)\} = (x + J)(yz + J) = xyz + J$$

and

$$\{(x + J)(y + J)\}(z + J) = (xy + J)(z + J) = xyz + J.$$

Hence the multiplication defined on R/J is associative.

D: The distribution laws may be checked using the same approach as has just been used to check M1.

This done, it follows that $(R/J, +, .)$ is a ring.

(ii) It is a formality to verify that v is a ring homomorphism. It is surjective, since the arbitrary element $x + J$ of R/J $(x \in R)$ is the image under v of x. So v is a ring epimorphism. Further, for $x \in R$,

$$x \in \ker v \Leftrightarrow v(x) = \text{zero of } R/J$$
$$\Leftrightarrow x + J = J$$
$$\Leftrightarrow x \in J \quad \text{(by additive version of 42.6)};$$

and hence $\ker v = J$.

(iii) If R is commutative, then, for all $x, y \in R$,

$$(x + J)(y + J) = xy + J = yx + J = (y + J)(x + J)$$

and this shows that R/J is commutative.

(iv) Suppose that R has a unity 1, and that $J \subset R$.

By 66.2, $1 \notin J$ and so $1 + J \neq J$, i.e. $1 + J$ is a nonzero element of R/J. Moreover, for all $x \in R$,

$$(1 + J)(x + J) = 1x + J = x + J = x1 + J = (x + J)(1 + J)$$

and hence $1 + J$ is a unity in R/J. The final part of the theorem follows.

Nomenclature. The ring R/J is called the **factor ring** of R modulo J. The epimorphism v detailed in 68.2(ii) is called the **natural epimorphism** from R onto R/J.

Illustration of a factor ring. Consider the factor ring $\mathbb{Z}/(m)$, where m is a positive integer. Notice that the typical element $x + (m)$ of $\mathbb{Z}/(m)$ $(x \in \mathbb{Z})$ is what we have previously called C_x, the congruence class of x modulo m: this is so because, for $t \in \mathbb{Z}$,

$$t \in x + (m) \Leftrightarrow t - x \in (m) \quad \text{(by additive version of 42.6)}$$
$$\Leftrightarrow m | (t - x) \quad \text{(by definition of } (m))$$
$$\Leftrightarrow t \equiv x \, (\text{mod } m)$$
$$\Leftrightarrow t \in C_x.$$

Thus $\mathbb{Z}/(m)$ is a ring whose elements are the congruence classes modulo m; and in this ring the rules of addition and multiplication are

$$(x + (m)) + (y + (m)) = (x + y) + (m), \ (x + (m))(y + (m)) = xy + (m),$$

i.e.

$$C_x + C_y = C_{x+y}, \ C_x C_y = C_{xy}$$

for all $x, y \in \mathbb{Z}$. Thus the factor ring $\mathbb{Z}/(m)$ is nothing other than our old friend $\mathbb{Z}_m$!

Interestingly, therefore, the construction of a general factor ring R/J is a generalization of the construction of the algebraic system $\mathbb{Z}_m$.

Next comes an important theorem about factor rings (68.3), which will be exploited in Chapter 11 to construct new fields. To understand 68.3, it is necessary first to absorb one further definition: an ideal J of the ring R is a **maximal** ideal of R iff (1) $J \subset R$ and (2) there is no ideal K of R such that $J \subset K \subset R$ [and therefore any ideal of R known to contain J strictly must be the whole of R].

68.3 Suppose that R is commutative and has a one and that J is a maximal ideal of R. Then R/J is a field.

Proof. Since $J \subset R$ [cf. condition (1) in the definition of "maximal ideal"], it follows from 68.2(iii, iv) that R/J is a commutative ring with a one. It remains to prove that each nonzero element of R/J has a multiplicative inverse in R/J.

Let then $x + J$ (where $x \in R$) be an arbitrary nonzero element of R/J; "$x + J$ nonzero" means that $x + J \neq J$, i.e. that $x \notin J$.

Let $K = J + (x)$, which (cf. 66.6) is an ideal of R that contains J and also contains x [$= 0 + 1x$], which does not belong to J. Therefore K is an ideal of R that strictly contains J; and so, since J is a maximal ideal of R, K must be the whole of R. Hence $1 \in K$, and therefore $1 = j + t$ for some $j \in J$ and $t \in (x)$, i.e.

$$1 = j + rx \text{ for some } j \in J \text{ and some } r \in R.$$

From this it follows that $1 - rx \in J$; and hence

$$rx + J = 1 + J,$$
$$\text{i.e. } (r + J)(x + J) = \text{the unity of } R/J$$

—which shows that, in the commutative ring R/J, $x + J$ has $r + J$ as a multiplicative inverse.

It now follows that R/J is a field.

We conclude the section by stating the first isomorphism theorem for rings (cf. 52.1).

68.4 Let $\theta: R \to R_1$ be a ring homomorphism (R, R_1 being rings). Then

$$\operatorname{im} \theta \cong R/\ker \theta.$$

The proof is closely analogous to that of 52.1. It consists of establishing that the rule

$$\phi(x + \ker \theta) = \theta(x) \quad (x \in R)$$

consistently defines a mapping ϕ from $R/\ker \theta$ to R_1 and proving then that ϕ is a ring monomorphism with image set equal to $\operatorname{im} \theta$.

69. Characteristic of an integral domain or field

Throughout this section D will denote an integral domain. By 63.1, all that we are about to say will apply in particular when D is a field.

Notice first that, for $x \in D$ and $n \in \mathbb{Z}$, nx already has a meaning, viz. the nth multiple (additive analogue of nth power) of x in the group $(D, +)$. In particular, if $n \in \mathbb{N}$ and $x \in D$, nx means $x + x + \dots$ to n terms. Observe also that, for each $x \in D$, the period of x in the additive group $(D, +)$ is

(i) infinite iff $nx \neq 0$ for all $n \in \mathbb{N}$,

(ii) the positive integer k iff $kx = 0$ while $nx \neq 0$ for $1 \leqslant n \leqslant k - 1$.

We focus attention on the period of the unity, 1, of D.

Definition. (1) We say that the integral domain D has **characteristic zero** iff 1 has infinite period in the group $(D, +)$.

(2) We say that D has **characteristic** p ($\in \mathbb{N} - \{1\}$) iff 1 has finite period p in the group $(D, +)$.

Illustrations. (i) The integral domain $\mathbb{Z}$ has characteristic zero (since, $\forall n \in \mathbb{N}$, $n1 \neq 0$).

(ii) The field $\mathbb{Z}_5$ has characteristic 5. Its unity is C_1 and, for $n \in \mathbb{N}$,

$$nC_1 = C_1 + C_1 + \dots \text{ (to } n \text{ terms)} = C_n$$

so that nC_1 is zero when $n = 5$ and nonzero when $1 \leqslant n \leqslant 4$. It is easily seen that, more generally, for every prime number p the field $\mathbb{Z}_p$ has characteristic p.

The following two theorems give basic information about the nonzero characteristic p case and the characteristic zero case, respectively.

69.1 Suppose that the integral domain D has nonzero characteristic p. Then

(i) p must be prime,

(ii) p is the period of every nonzero element in the group $(D, +)$, and

(iii) for every $x \in D$, $nx = 0$ if n is an integer divisible by p.

Proof. (i) As a preliminary, note that, for $m, n \in \mathbb{N}$,

$$(m1)(n1) = (1 + 1 + 1 + \ldots \text{ to } m \text{ terms}) . (n1)$$
$$= n1 + n1 + n1 + \ldots \text{ to } m \text{ terms} \quad \text{(by distributive law)}$$
$$= m(n1)$$
$$= (mn)1 \quad \text{(by additive version of 35.10(ii))}.$$

Suppose that p is composite, so that $p = kl$, where $1 < k, l < p$. Then, by the meaning of "characteristic", $p1 = 0$, i.e. $(kl)1 = 0$, i.e. (by the above preliminary) $(k1)(l1) = 0$; and so, since D is an integral domain, $k1 = 0$ or $l1 = 0$. In either case, we have a contradiction, since the pth multiple of 1 is the lowest positive multiple of 1 that equals zero.

It follows that p must be prime.

(ii) Let x be an arbitrary element of $D - \{0\}$. Since

$$px = x + x + \ldots \text{ to } p \text{ terms} = (1 + 1 + \ldots \text{ to } p \text{ terms})x = (p1)x = 0x = 0,$$

it follows (cf. 38.4(ii)) that the period of x is a factor of p. Hence, since p is prime, the period of x is 1 or p. But it is not 1, since $x \neq 0$. Therefore the period of x is p.

This proves part (ii).

Part (iii) is trivial in the case $x = 0$ and follows from part (ii) in the case $x \neq 0$ by the additive notation version of 38.4(ii).

69.2 Suppose that the integral domain D has characteristic zero. Then, in the group $(D, +)$, every nonzero element has infinite period.

Proof. Let x be an arbitrary element of $D - \{0\}$. For every $n \in \mathbb{N}$, $nx = x + x + \ldots \text{ to } n \text{ terms} = (1 + 1 + \ldots \text{ to } n \text{ terms})x = (n1)x$. But $x \neq 0$; and, since D has characteristic zero, $n1 \neq 0$ (for every $n \in \mathbb{N}$). So, since D is an integral domain, it follows that, $\forall n \in \mathbb{N}$, $nx \neq 0$; i.e. x has infinite period in the group $(D, +)$. This proves the result.

Notes. (*a*) Because of 38.3 it is clear that:

69.3 A finite integral domain cannot have characteristic zero.

In fact (as one of the exercises at the end of this chapter reveals) there cannot be a finite integral domain which is not a field. So nothing is lost in restating 69.3 to read, "A finite *field* cannot have characteristic zero."

(*b*) Work in any system of characteristic 2 is in one respect unusually simple: for in such a system subtraction is the same as addition. (If D is an integral domain of characteristic 2 and $x, y \in D$, then (cf. 69.1(iii)) $2y = 0$, i.e. $y + y = 0$, so that $-y = y$ and therefore $x - y = x + y$.) On the other hand, beware of the fact that one cannot halve any nonzero element t in a system of characteristic 2: there cannot be an element s in the system such that $2s = t$, because, no matter what s is, $2s$ is bound to be zero.

Prime subfield of a field

Let F be a field. In the light of the above discussion (and 69.1(i) in particular), we recognize that either (1) F has characteristic zero or (2) F has characteristic p for some prime p. We discuss each of these cases in turn, showing, without going into every detail, that in each case F has an easily picked out subfield isomorphic to a well-known field.

Case 1: F has characteristic zero. In this case, we can define (consistently!) a mapping $\theta : \mathbb{Q} \rightarrow F$ by

$$m/n \mapsto (m1)(n1)^{-1} \ (m, n \in \mathbb{Z}, n \neq 0)$$

and we can prove that this mapping θ is a ring monomorphism. By the ring theory analogue of 47.6, it follows that im θ is a subfield of F isomorphic to $\mathbb{Q}$. Moreover this subfield is contained in every subfield of F: for, if K is any subfield of F, then, because $1 \in K$, K contains all the elements $k1$ $(k \in \mathbb{Z})$ and hence contains all the elements $(m1)(n1)^{-1}$ $(m, n \in \mathbb{Z}, n \neq 0)$, i.e. contains im θ.

Case 2: F has characteristic p, where p is prime. In this case, we can define (consistently!) a mapping $\phi : \mathbb{Z}_p \rightarrow F$ by

$$C_n \mapsto n1 \quad (n \in \mathbb{Z})$$

where C_n denotes the congruence class of n modulo p; and we can prove that this mapping ϕ is a ring monomorphism. It follows that im ϕ is a subfield of F isomorphic to $\mathbb{Z}_p$. Note that the subfield im ϕ consists of the p elements $0, 1, 2.1, \ldots, (p-1)1$. Moreover the subfield im ϕ is contained in every subfield of F: for, if K is any subfield of F, then, because $1 \in K$, it is clear that K must contain all the elements of im ϕ.

Summing up what we have found in the two cases, we have the following theorem.

69.4 In every field F, there is a minimal subfield P (i.e. a subfield P contained in every subfield of F), P being isomorphic to $\mathbb{Q}$ if F has characteristic zero and to $\mathbb{Z}_p$ if F has prime characteristic p.

(*Note:* A field F cannot have more than one such minimal subfield: for, if K, L are both minimal subfields of the field F, then $K \subseteq L$ and $L \subseteq K$, so that $K = L$.)

The unique minimal subfield in a field F is called the **prime subfield** of F. Notice finally that:

69.5 For any field F, every subfield of F has the same characteristic as F.

(Quite apart from considerations of prime subfields, 69.5 is obvious because the characteristic of a field is a property of its unity and because (cf. concluding remark in § 64) every subfield of a field F has the same unity as F.)

70. The field of fractions of an integral domain

The integral domain $\mathbb{Z}$, which is not a field, is a subring of $\mathbb{Q}$, which *is* a field. Moreover, the elements of $\mathbb{Q}$ are "fractions" (or quotients) that can be formed using elements of $\mathbb{Z}$: each element of $\mathbb{Q}$ is expressible as $a/b\,[=ab^{-1}]$ for some $a, b \in \mathbb{Z}$ with $b \neq 0$. The purpose of this section is to exhibit a general pattern of which the relationship of $\mathbb{Q}$ to $\mathbb{Z}$ is one particular case. It will be seen that starting with any integral domain D, one can construct a field F that contains D, each element of the constructed field F being expressible as a "fraction" or quotient a/b [meaning ab^{-1}] for some $a, b \in D$ with $b \neq 0$.

Execution of the general construction is in fact a very mechanical business, once sufficient analysis has been done of how exactly, in terms of the ideas which abstract algebra deals with, $\mathbb{Q}$ is related to $\mathbb{Z}$. To find the right starting-point, it is necessary to realize that a rational number does not simply correspond to a single (numerator, denominator) ordered pair of integers, but rather corresponds to a whole set of (numerator, denominator) ordered pairs of integers: e.g. what we call "a half" is represented as $\frac{1}{2}$ or as $\frac{2}{4}$ or as $\frac{3}{6}$ or as $\frac{-7}{-14}$, etc., etc., and so corresponds not merely to the ordered pair $(1, 2)$ but to a set of ordered pairs that includes $(1, 2), (2, 4), (3, 6), (-7, -14)$ and indeed all objects of the form $(n, 2n)$ with $n \in \mathbb{Z} - \{0\}$. A key insight is that this set of ordered pairs (a subset of $\mathbb{Z} \times (\mathbb{Z} - \{0\})$) is an equivalence class under a certain relation $\sim$ on $\mathbb{Z} \times (\mathbb{Z} - \{0\})$—namely the relation $\sim$ defined by

$(a, b) \sim (c, d)$ iff (a, b) and (c, d) arise as (numerator, denominator) pairs for the same rational number,

i.e. $(a, b) \sim (c, d)$ iff a/b and c/d are the same rational number,

i.e. $(a, b) \sim (c, d)$ iff $ad = bc$.

Here a, b, c, d denote integers with $b, d \neq 0$.

To achieve generalization, therefore, one is drawn to define a relation $\sim$

on $D \times (D - \{0\})$, where D is an arbitrary given integral domain, by

$$(a, b) \sim (c, d) \Leftrightarrow ad = bc \quad (a, b, c, d \in D; b, d \neq 0)$$

and to expect that this will be an equivalence relation and that, if indeed it is, the $\sim$-classes will form a set of objects that can be turned into a field (by defining addition and multiplication so as to copy, in effect, the way rational numbers get added and multiplied). To someone given no clue as to what was going on, the details might seem impenetrably obscure and full of unlikely initiatives requiring the inventiveness of a genius. However, armed with the explanation that everything that is done is a matter simply of analysing and copying the relationship of $\mathbb{Q}$ to $\mathbb{Z}$, the student should see that no inventiveness at all is required to carry through the construction procedure and that at every point the way ahead, far from being obscure, is clear as could be.

There is one major side-issue which calls for comment at this preliminary stage. Our aim is to produce a field F containing an arbitrary given integral domain D, but if the word "containing" is to be taken with unyielding literalness, we face a tall order indeed. Instead we should content ourselves with constructing a field containing a subring D^* *isomorphic to D*. Since D^* can sensibly be thought of as a copy of D, it is then possible to identify D^* with D to give us $F \supseteq D$. To consider the same point in a slightly different way, observe that when we know of an isomorphism $\theta : D \rightarrow D^*$ ($\subseteq F$), we can, for each $x \in D$, re-label the object $\theta(x)$ in D^* simply as x, so bringing about a situation where F contains a subring D^* consisting of elements which are labelled by elements of D and behave in all algebraic operations just like the elements of D—a situation in which it becomes unhelpfully pedantic to persevere with any distinction between D and D^* (or between the statements "F contains a copy D^* of D" and "F contains D"). There is a technical term used in this type of context: we say that D is *imbedded* in F when F contains a subsystem D^* isomorphic to D (and when, consequently, it is open to us to identify D^* with D to give us $D \subseteq F$).

To have one system imbedded, in the sense just described, in a constructed larger system is not at all uncommon. For example, to introduce complex numbers in a logically watertight way, an obvious procedure to adopt is to take $\mathbb{C}$ to be the set $\mathbb{R}^2$ of all ordered pairs of real numbers [the ordered pair (x, y) being what will ultimately be labelled $x + iy$] and to define addition and multiplication on this set of ordered pairs in accordance with what one intends the system $\mathbb{C}$ to be. (For example, multiplication is defined by $(a, b)(c, d) = (ac - bd, ad + bc)$.) In the resulting field $(\mathbb{C}, +, .)$ the objects of the form $(x, 0)$ form a subfield $\mathbb{R}^*$ isomorphic to $\mathbb{R}$. So $\mathbb{R}$ is imbedded in the constructed field $\mathbb{C}$. Work with complex numbers would be extremely cumbersome, were we not prepared to identify $\mathbb{R}^*$ with $\mathbb{R}$ (by regarding $(x, 0)$ as the same as the real number x) to give us $\mathbb{R} \subset \mathbb{C}$.

Here now is the theorem which all the foregoing explanations have been leading up to.

70.1 Let D be an arbitrary integral domain. Then there is a field F such that (1) $F \supseteq D$ and (2) each element of F is expressible as ab^{-1} for some $a, b \in D$ with $b \neq 0$.

Outline of proof. [The complete proof contains a large number of routine verifications, most of which will be left aside here.]

Stage 1 (Introduction of the appropriate set of equivalence classes.).

Let $S = D \times (D - \{0\})$, and let a relation $\sim$ be defined on S by

$$(a, b) \sim (c, d) \Leftrightarrow ad = bc \quad (a, c \in D, b, d \in D - \{0\}).$$

Then, for each $(a, b) \in S$, $(a, b) \sim (a, b)$ since $ab = ba$; and therefore $\sim$ is reflexive. Further, for $(a, b), (c, d) \in S$,

$$(a, b) \sim (c, d) \Rightarrow ad = bc \Rightarrow cb = da \Rightarrow (c, d) \sim (a, b);$$

and therefore $\sim$ is symmetric.

To prove that $\sim$ is also transitive, suppose that $(a, b) \sim (c, d)$ and $(c, d) \sim (e, f)$, where $a, c, e \in D$ and $b, d, f \in D - \{0\}$. Then

$$ad = bc \quad \text{and} \quad cf = de,$$

and hence

$$adf = (bc)f = b(cf) = bde;$$

and, from $adf = bde$, it follows (since D is an integral domain and $d \neq 0$) that $af = be$ (cf. 63.4)—i.e. $(a, b) \sim (e, f)$.

So $\sim$ is indeed transitive and is an equivalence relation on S.

Let F be the set of $\sim$-classes; and, for each $(a, b) \in S$, let $\dfrac{a}{b}$ or [same thing in one line of printing] a/b be used as notation for the $\sim$-class of (a, b). [N.B. At this stage "a/b" is merely a notation for an equivalence class. It is not to be assumed that this object is capable of being interpreted as "a divided by b", nor will any claim to that effect be made until the evidence for it is fully assembled.]

Stage 2 (Turning the set F into a field by defining addition and multiplication on F).

In the light of our knowledge of $\mathbb{Q}$, the natural definitions to propose for addition and multiplication on F are

$$\frac{a}{b} + \frac{c}{d} = \frac{ad + bc}{bd} \tag{1}$$

and
$$\frac{a}{b} \cdot \frac{c}{d} = \frac{ac}{bd} \qquad (2)$$

$(a, b, c, d \in D; b, d \neq 0)$. But, since each element of F can have several possible labels [a/b being the same as x/y iff $(a, b) \sim (x, y)$ — cf. 17.3], it is necessary to prove that these definitions are consistent. The details in the case of addition (definition (1)) are as follows. Let ξ, η be two elements of F. Let a/b and a'/b' be two possible labels for ξ ($a, b, a', b' \in D; b, b' \neq 0$), and let c/d and c'/d' be two possible labels for η ($c, d, c', d' \in D; d, d' \neq 0$). Use of the first label in each case gives, according to (1),

$$\xi + \eta = \frac{ad + bc}{bd}, \qquad (3)$$

while use of the second label in each case gives, according to (1),

$$\xi + \eta = \frac{a'd' + b'c'}{b'd'}. \qquad (4)$$

Our aim is to show that the right-hand sides of (3) and (4) necessarily coincide. Because $a/b = a'/b'$ ($= \xi$), we have $(a, b) \sim (a', b')$, i.e. $ab' = a'b$; and similarly, because $c/d = c'/d'$, $cd' = c'd$. Hence

$$(ad + bc)b'd' = ab'dd' + bb'cd'$$
$$= a'bdd' + bb'c'd \quad \text{(since } ab' = a'b \text{ and } cd' = c'd\text{)}$$
$$= bd(a'd' + b'c'):$$

so $(ad + bc, bd) \sim (a'd' + b'c', b'd')$, and therefore the right-hand sides of (3) and (4) are the same. This proves that (1) gives a consistent definition of addition on F. Similarly (2) gives a consistent definition of multiplication on F. So we end this stage of the proof knowing that addition and multiplication have been satisfactorily defined on F (by the equations (1) and (2)).

Stage 3 (Proof that the system $(F, +, .)$ that we have constructed is a field).
This is a straightforward stage — simply a matter of verifying all the relevant axioms. In the course of the verification one confirms that (a) $0/1$ is a zero in F, (b) $1/1$ is a unity (or "one") in F, (c) the element a/b of F (where $a, b \in D$ and $b \neq 0$) is nonzero iff $a \neq 0$, and, when $a \neq 0$, the element a/b has a multiplicative inverse in F — namely b/a.

As an illustration of what is entailed in all this, here is one of the less trivial verifications — that of the distributive law. Let $\frac{a}{b}, \frac{c}{d}, \frac{e}{f} \in F$ (so that $a, b, c, d, e, f \in D$ and $b, d, f \neq 0$). Then

$$\frac{a}{b} \cdot \frac{c}{d} + \frac{a}{b} \cdot \frac{e}{f} = \frac{ac}{bd} + \frac{ae}{bf} = \frac{abcf + abde}{b^2 df} = \frac{b(acf + ade)}{b^2 df}$$

(by the rules for adding and multiplying elements of F — i.e. the rules (1) and (2) in stage 2). But clearly (from the definition of $\sim$)

$$(b(acf + ade), b^2 df) \sim (acf + ade, bdf), \text{ and so } \frac{b(acf + ade)}{b^2 df} = \frac{acf + ade}{bdf}.$$

Therefore

$$\frac{a}{b} \cdot \frac{c}{d} + \frac{a}{b} \cdot \frac{e}{f} = \frac{acf + ade}{bdf} = \frac{a(cf + de)}{bdf} = \frac{a}{b} \left(\frac{cf + de}{df} \right) = \frac{a}{b} \left(\frac{c}{d} + \frac{e}{f} \right)$$

(by further use of the rules (1) and (2)). This shows that the distributive law holds for F.

Stage 4 (Demonstration that D is imbedded in F).

Let θ be the mapping from D to F defined by $\theta(x) = x/1$ ($x \in D$). Then one can easily verify that θ is a ring homomorphism. Moreover, for $x \in D$,

$$x \in \ker \theta \Leftrightarrow \theta(x) = 0/1 \quad (0/1 \text{ being the zero of } F)$$
$$\Leftrightarrow x/1 = 0/1 \Leftrightarrow (x, 1) \sim (0, 1) \Leftrightarrow x . 1 = 1 . 0 \Leftrightarrow x = 0,$$

and therefore (cf. 03.2) θ is a monomorphism. It follows that im θ (which is the set of all elements in F of the form $x/1$) is a subring of F isomorphic to D. Let us now (cf. explanatory comments in the paragraphs leading up to the statement of the theorem) identify im θ with D: we do this by identifying $x/1 \, [= \theta(x)]$ with x for each $x \in D$, and that gives us $D \subseteq F$.

Stage 5 (Proof of property (2) in the statement of the theorem).

Let $t = a/b$ ($a, b \in D$, $b \neq 0$) be an arbitrary element of F. Then

$$t = \frac{a}{1} \cdot \frac{1}{b} \quad \text{(by the definition of multiplication on } F - \text{see stage 2)}$$

$$= \frac{a}{1} \left(\frac{b}{1} \right)^{-1} \quad \text{(cf. remark on multiplicative inverses early in stage 3)}$$

$$= ab^{-1} \quad \text{(by the identification in stage 4 of } x/1 \text{ with } x \text{ for each } x \in D\text{)}.$$

This completes the proof of 70.1, bringing us in particular to the happy point of seeing that the "a/b" notation *is* appropriate: i.e. the element of F which we have (hitherto without prejudice) denoted by a/b actually *is*, after our identification of im θ with D, "a divided by b" in the sense of ab^{-1}.

The field F constructed from the given integral domain D in 70.1 is called the **field of fractions** of D. Put informally, the message of 70.1 is that if we are working with any integral domain D, we are entitled to introduce and use a field whose elements are, in the obvious sense of the words, fractions with numerator and nonzero denominator in D. As has been stressed from the beginning of this section, the extension of $\mathbb{Z}$ to $\mathbb{Q}$ is the motivating illustration. Apart from that, the other close-to-hand illustration (which will be mentioned with greater preciseness in § 72 in the next chapter) is the extension of the integral domain of polynomials with coefficients in an integral domain D to the field of all formal rational functions [objects of the form

$$\text{polynomial/(nonzero polynomial)}]$$

with coefficients in D.

EXERCISES ON CHAPTER TEN

1. Suppose that the ring R has the property that $x^2 = x$ for every $x \in R$. Prove that $\forall x \in R$, $-x = x$; and show further that R is commutative.

2. Let R be a ring. A *left unity* in R means a nonzero element $e \in R$ satisfying $ex = x$ for all $x \in R$; and similarly a *right unity* in R means a nonzero element $f \in R$ satisfying $xf = x$ for all $x \in R$.

(i) Suppose that R contains one and only one left unity e. By considering $x + e - xe$, where x is an arbitrary element of R, prove that e is in fact a unity in R (i.e. both a left unity and a right unity).

(ii) Suppose instead that R satisfies the postulate

$$xy = 0 \Rightarrow x = 0 \text{ or } y = 0 \quad (x, y \in R).$$

Could R contain a left unity which is not a right unity?

3. Let a be a fixed element of a ring R, and let $C(a)$ be the subset $\{x \in R : xa = ax\}$. Prove that $C(a)$ is a subring of R.

4. Let S and L be the subsets of $\mathbb{R}$ defined by

$$S = \{a + b\sqrt{2} : a, b \in \mathbb{Z}\} \quad \text{and} \quad L = \{a + b\sqrt{2} : a, b \in \mathbb{Q}\}.$$

Prove that S is a subring of $\mathbb{R}$ and that L is a subfield of $\mathbb{R}$.

5. A *Gaussian integer* is a complex number of the form $m + in$ with m, n integers. Use 64.2 to prove that the set of all Gaussian integers is a ring. Prove also that in this ring the group of units is $\{1, -1, i, -i\}$. Hint: $\forall z_1, z_2 \in \mathbb{C}$, $|z_1 z_2|^2 = |z_1|^2 |z_2|^2$; and $|z|^2$ is a nonnegative integer for every Gaussian integer z.

6. Prove the remark at the end of § 64 — that if D is an integral domain contained as a subring in a larger integral domain E, then the unities of D and E must coincide.

7. Suppose that F is a field and that θ is an automorphism of F (i.e. an isomorphism from F to itself). Let

$$L = \{x \in F : \theta(x) = x\},$$

i.e. the set of elements of F which are, as we say, "fixed" by θ. Prove that L is a subfield of F.

8. Let a be a fixed element in the commutative ring R, and let $A = \{x \in R : xa = 0\}$. Prove that A is an ideal of R.

9. Let R be a ring. The centre of R is the subset $Z(R)$ defined by

$$Z(R) = \{x \in R : rx = xr \text{ for all } r \in R\}.$$

Show that $Z(R)$ is a subring of R, but is not an ideal if R is non-commutative with a one.

10. Let R be a ring and F a field. Show that:

(i) the intersection of any two subrings of R is also a subring of R;

(ii) the intersection of any two ideals of R is also an ideal of R;

(iii) the intersection of any two subfields of F is also a subfield of F.

11. Suppose that D is an integral domain and that J and K are ideals of D neither of which is $\{0\}$. Show that $J \cap K \neq \{0\}$.

12. (i) Let R be a commutative ring with a one in which the only ideals are $\{0\}$ and R. Prove that R is a field.

(ii) By considering the ring of all 2×2 real matrices, show that a non-commutative ring R with a one but with no ideals except $\{0\}$ and R need not be a division ring.

13. Let D be an integral domain, and let a be a nonzero element of D which is not a unit. Show that, $\forall n \in \mathbb{N}$, $(a^{n+1}) \subset (a^n)$.

Deduce that (i) in an integral domain which is not a field there are infinitely many different ideals, and (ii) every finite integral domain is a field.

14. Let $\theta: F \to R$ be a ring homomorphism, where R is a ring and F is a field. Prove that either θ is the "zero mapping" (i.e. $\theta(x) = 0$ for all $x \in F$) or else θ is a monomorphism.

15. Let $\theta: R \to R_1$ be a ring homomorphism, where R, R_1 are rings with unities $1, 1_1$, respectively, and where R_1 is an integral domain. Suppose that $\ker \theta \neq R$. Show that $\theta(1) = 1_1$.

16. Let R be a commutative ring and J an ideal of R.

(i) Let $a, b \in R$, and suppose that a^m and b^n belong to J for certain positive integers m, n. Taking it as read that the binomial theorem (for a positive integral power) is applicable in R, show that $(a - b)^{m+n-1} \in J$.

(ii) Let $K = \{x \in R : x^m \in J \text{ for some } m \in \mathbb{N}\}$. Prove that K is an ideal of R.

17. Suppose that R and R_1 are rings and that $\theta: R \to R_1$ is a homomorphism.

(i) Let K be an ideal of R_1, and let $J = \{x \in R : \theta(x) \in K\}$. Prove that J is an ideal of R.

(ii) Now suppose that θ is an epimorphism and that K_1, K_2 are two unequal ideals of R_1. Let

$$J_1 = \{x \in R : \theta(x) \in K_1\}, \quad J_2 = \{x \in R : \theta(x) \in K_2\},$$

which are both ideals of R, by part (i). Show that $J_1 \neq J_2$.

18. Let $m, n \in \mathbb{N}$, and let $d = \text{hcf}(m, n)$. Show that the sum of the ideals (m) and (n) of $\mathbb{Z}$ is the ideal (d).

19. Let R be a ring. Let $J_1, J_2, J_3, \ldots$ be an infinite sequence of ideals of R such that $J_n \subseteq J_{n+1}$ for all $n \in \mathbb{N}$. Let K be the union of $J_1, J_2, J_3, \ldots$ Prove that K is an ideal of R.

Deduce that in a principal ideal domain there cannot exist an infinite sequence of ideals $J_1, J_2, J_3, \ldots$ with $J_n \subset J_{n+1}$ for all $n \in \mathbb{N}$.

20. Let J be an ideal of the ring R. Show that the ring R/J is commutative iff $xy - yx \in J$ for every $x, y \in R$.

Deduce that if K_1 and K_2 are ideals of R and both R/K_1 and R/K_2 are commutative, then $R/(K_1 \cap K_2)$ is also commutative.

21. Let R, S be as in the worked example in § 66. Give an alternative proof that S is an ideal of R by showing that the mapping from R to $\mathbb{R}$ given by $\begin{bmatrix} x & y \\ 0 & z \end{bmatrix} \mapsto z$ is a ring homomorphism with kernel S. Deduce the additional information that $R/S \cong \mathbb{R}$.

22. Let R denote a ring with a one. An element x of R is termed *nilpotent* iff $\exists n \in \mathbb{N}$ s.t. $x^n = 0$.

(i) Give an example of a nonzero nilpotent element in $\mathbb{Z}_{12}$. Show that if p and q are distinct prime numbers, then $\mathbb{Z}_{pq}$ contains no nonzero nilpotent elements.

(ii) Show that if J is an ideal of R and x is a nilpotent element of R, then $x + J$ is a nilpotent element of R/J.

(iii) Show that if K is an ideal of R and if all the elements of K are nilpotent and all the elements of the ring R/K are nilpotent, then all the elements of R are nilpotent.

(iv) Show that if x is a nilpotent element of R, then $1 - x$ is a unit in R.

In the remaining parts of this exercise, *suppose that R is commutative.*

(v) Using part (ii) of exercise 16, show that the set N of all nilpotent elements in R is an ideal of R and that R/N contains no nonzero nilpotent elements.

(vi) Using part (ii), show that N is contained in every maximal ideal of R.

23. Use exercises 12(i) and 17(ii) to construct an alternative proof of theorem 68.3.

24. Let R be a commutative ring with a one, and let J be an ideal of R. Show that if R/J is a field, then J is a maximal ideal of R. (This is the converse of 68.3.)

25. Give an explicit example of two different maximal ideals in the same ring.

26. Let R be a commutative ring with a one. An ideal of R is called a *prime* ideal iff the complement in R of the ideal is non-empty and closed under multiplication. Prove, for J an ideal of R, that

$$R/J \text{ is an integral domain} \Leftrightarrow J \text{ is a prime ideal.}$$

Deduce that every maximal ideal of R is prime. Deduce also that in $\mathbb{Z}$ the prime ideals are $\{0\}$ and the ideals (p) with p prime.

27. Let F be a field of characteristic p (p prime), and let σ be the mapping from F to F given by $\sigma(x) = x^p$ ($x \in F$). Show that σ is a monomorphism. Show further that if F is finite, then σ is an automorphism of F (i.e. an isomorphism from F to F).

28. Let F be a field of characteristic p (p prime), let x be an element of F and r an integer. Given that $rx = 0$ and that $p \nmid r$ in $\mathbb{Z}$, show that $x = 0$.

29. Let D be an integral domain. Show that

$$x^2 = 1 \Rightarrow x = 1 \text{ or } x = -1 \quad (x \in D).$$

Deduce that if D contains only finitely many units, then the product of those units equals -1. Hence prove that, for every prime number p, $(p - 1)! \equiv -1 \pmod{p}$. This is part of the result known as *Wilson's theorem* in number theory.

30. Let F be a field. Prove that the groups $(F, +)$ and $(F - \{0\}, .)$ are not isomorphic.

31. Supply some of the missing detail in stage 3 of the proof of 70.1 — in particular prove the assertion about multiplicative inverses, that the element a/b of F (where $a, b \in D$ and $b \neq 0$) is nonzero iff $a \neq 0$ and that, when $a \neq 0$, the element a/b has a multiplicative inverse in F (namely b/a).

32. If one starts with a *field* D and constructs its field of fractions (as in 70.1), is the resulting field F bigger than the original D?

CHAPTER ELEVEN

POLYNOMIALS AND FIELDS

71. Introduction

In the final section of Chapter 10, new light was shed on something very significant in the rudiments of mathematics — the extension of the number system from $\mathbb{Z}$ to $\mathbb{Q}$; and it was seen that this extension was but one instance of something much more general. A different but also very significant number system extension in elementary mathematics is the extension from $\mathbb{R}$ to $\mathbb{C}$, and the major climax of this chapter will be a theorem (79.1) showing that the extension from $\mathbb{R}$ to $\mathbb{C}$ is also capable of very broad (and fruitful) generalization.

When we go from $\mathbb{R}$ to $\mathbb{C}$, we are moving to a bigger field which contains a root of a polynomial equation (namely $x^2 = -1$) with no root in the original field. Theorem 79.1 will show that *whenever* one is given a field and a polynomial equation with no root in that field, it is *always* possible to extend the field to a bigger one which does contain a root of the polynomial equation in question. Not surprisingly, the road towards that theorem passes through the detailed study of polynomials, and a large proportion of this chapter is devoted to such a study. This begins in § 72 where the idea "of the ring of polynomials with coefficients in a given ring" is defined in a precise way that prepares the ground for its study in the context of abstract algebra. Quite apart from the highlight results of this chapter, rings of polynomials crop up all over mathematics and its applications; and §§73 and 75 (separated by a discussion in § 74 of some generalities concerning factorization) develop facts about polynomials (and their roots and factors) which are very widely used. After some notational matters in § 76, § 77 draws on polynomial theory to explore to good effect the structure of an arbitrary field known to be generated by a given subfield and a root of a polynomial equation with coefficients in this subfield (in the sense that $\mathbb{C}$ is generated by the subfield $\mathbb{R}$ and a root of the polynomial equation $x^2 = -1$). Together with other material from the chapter, that enables us in § 78 to prove some basic facts about finite

181

fields. Finite fields certainly deserve to be regarded as an important topic, because of the extensive use made of them in other parts of mathematics — in particular in combinatorics and its applications to such practical problems as tournament design and coding theory. (See, for example, [9] and [10].) After § 78's introduction to the theory of finite fields, § 79 features the theorem already referred to which generalizes the construction of $\mathbb{C}$ from $\mathbb{R}$, and it will be seen that, amongst other things, this generalization gives access to interesting finite fields different from the already known finite fields $\mathbb{Z}_p$ (p prime). There follows a section (§ 80) which should be ignored by students who as yet know nothing about vector space theory, but which will show those who are acquainted with that part of mathematics how usefully it can be applied to field theory.

At that point this book's systematic introduction to the world of abstract algebra comes to an end. But a final section (§ 81, entitled "Epilogue") offers a few thoughts on the benefits of studying abstract algebra and on its relation to the rest of mathematics and its applications; and, as a concrete illustration of the previous paragraph's allusion to the extensive use of finite fields in combinatorics, this epilogue gives an introduction to the subject of finite projective planes and shows how finite fields are used to construct them.

72. Polynomial rings

Throughout this section D will denote an arbitrary integral domain, and the main focus of attention will be a commutative ring S whose elements are the polynomials over D (i.e. the polynomials with coefficients in D) in a symbol X. This ring S can be more precisely described as a commutative ring with the following three properties:

(1) S contains D as a subring.

(2) S contains an element X which is transcendental over D. (X transcendental over D means that X does not satisfy any nontrivial polynomial equation with coefficients in D, i.e. that no finite sum, of whatever length, of the form $a_0 + a_1 X + a_2 X^2 + \ldots + a_n X^n$, with each $a_i \in D$ can be equal to zero unless every coefficient a_i is zero.)

(3) Each element of S is expressible as a polynomial in X over D — i.e. in the form $a_0 + a_1 X + a_2 X^2 + \ldots + a_n X^n$, for some $n \in \mathbb{N} \cup \{0\}$, with each $a_i \in D (0 \leqslant i \leqslant n)$ [and because $D \subseteq S$ and $X \in S$, S contains (by virtue of its closure properties as a ring) *every possible* polynomial in X over D].

A post-script to this section will briefly outline why we can be sure that, whatever the integral domain D is, a commutative ring S as just described does exist. Meanwhile there follows a series of notes providing a comprehensive guide to a variety of basic things that should be appreciated about the ring S.

(a) The standard notation for the ring which we have so far called "S" is $D[X]$.

(*b*) Obviously, we could have lots of rings that are just like $D[X]$ $(=S)$ except for the use of a different symbol instead of X: e.g. we could have $D[Y]$, a ring consisting of all polynomials over D in the element Y transcendental over D (or we could have $D[Z]$, etc., etc.). But, equally obviously, all such rings $(D[X], D[Y], D[Z], \ldots$, where $X, Y, Z, \ldots$, are transcendental over D) are isomorphic to one another: they are identical algebraic systems except for the superficial matter (irrelevant to their structures) of choice of notation for their member objects. It is for this reason that we call $D[X]$ *the* ring of polynomials over D, the definite article being justified by the fact that the number of possibilities is only one if isomorphic systems are regarded as the same thing.

In the extensive discussion of polynomial rings in the coming pages, we shall always take the ring of polynomials over D to be $D[X]$ (and not $D[Y]$, etc.). This accords with a widely accepted convention that the capital letter X, when it is introduced in ring theory, is automatically to be understood as an element transcendental over the given ring (i.e. over D in the present discussion). Furthermore, because the ring of polynomials over D is always to be taken to be $D[X]$, elements of this ring may be referred to simply as "polynomials over D" (rather than "polynomials in X over D") — and one can simply call them "polynomials" if "over D" is clear from the context.

(*c*) As stated above, the meaning of "X transcendental over D" is that there is no nontrivial expression for 0 as a polynomial in X over D. A consequence is that each nonzero element of $D[X]$ can be expressed in *just one* way in the form

$$a_0 + a_1 X + \ldots + a_n X^n \text{ where each } a_i \in D \text{ and } a_n \neq 0:$$

for if a polynomial $f \in D[X]$ could be expressed in two different ways in this form, then, by subtracting one of these expressions from the other, we would form a nontrivial polynomial in X over D that was equal to 0 — an impossibility, X being transcendental over D.

The fact that elements of $D[X]$ are *uniquely* expressible as polynomials in X over D has some noteworthy consequences:

(1) It means that, given a polynomial equation

$$a_0 + a_1 X + a_2 X^2 + \ldots = b_0 + b_1 X + b_2 X^2 + \ldots,$$

(where all of $a_0, a_1, a_2, \ldots, b_0, b_1, b_2, \ldots$ belong to D), we can equate corresponding coefficients to deduce

$$a_0 = b_0, \quad a_1 = b_1, \quad a_2 = b_2, \ldots .$$

(2) It also means that some nomenclature widely in use in elementary mathematics can be brought in on a sound logical footing: if a nonzero element f of $D[X]$ is equal to $a_0 + a_1 X + \ldots + a_n X^n$ (where $n \geqslant 0$, $a_n \neq 0$,

and each $a_i \in D$), we call n the **degree** of f (for which we use the notation $\deg f$), we call a_n the **leading coefficient**, we call a_0 the **constant term**, and we say that f is a **monic polynomial** iff $a_n = 1$.

(*d*) Consider two arbitrary nonzero members of $D[X]$ —

$$f = a_0 + a_1 X + \ldots + a_n X^n, \quad g = b_0 + b_1 X + \ldots + b_m X^m,$$

where $a_0, \ldots, b_m \in D$ and a_n, b_m are nonzero, so that $\deg f = n$, $\deg g = m$. On forming the product fg, we obtain

$$fg = a_n b_m X^{m+n} + \text{(terms of lower degree)},$$

and here the coefficient $a_n b_m$ of X^{m+n} is nonzero (since a_n, b_m are nonzero and D is an integral domain). From this a pair of significant conclusions can at once be drawn:

72.1 (i) The commutative ring $D[X]$ is an integral domain (for every possible integral domain D).

(ii) For all nonzero $f, g \in D[X]$, $\deg (fg) = \deg f + \deg g$.

(*e*) So far no meaning has been given to $\deg 0$ — for the simple reason that there is no such thing as "the highest power of X appearing in 0". The gap can be filled by the clever device of defining $\deg 0$ to be $-\infty$ (minus infinity). Then 72.1 extends to cover the case where f or g is 0 (or both are 0) — with the obvious convention that when $-\infty$ is added to anything, the result is still $-\infty$. Another advantage of taking $\deg 0$ to be $-\infty$ is that, for any fixed n, "the set of polynomials with degree $\leqslant n$" is the set that includes 0.

(*f*) In discussion of polynomials, elements of D (which, of course, are members of the larger ring $D[X]$) are often referred to as constant polynomials (or simply *constants*). They have degrees $\leqslant 0$ (0 in the case of a nonzero constant, $-\infty$ in the case of 0).

(*g*) The ring $D[X]$ provides a precise and usable expression of the very natural idea of "the system of polynomials over D". Achieving a satisfactory expression of that idea is a subtler matter than might at first be appreciated. For example, at first sight it might seem that the obvious approach to take would be to consider the system of all polynomial *functions* over D (i.e. of all mappings from D to D specifiable by formulae of the form

$$x \mapsto a_0 + a_1 x + \ldots + a_n x^n \ (x \in D)$$

with each coefficient $a_i \in D$). But, in fact, this approach has a number of unhappy features (which the use of $D[X]$ avoids). To give one glaring example, we would surely wish to regard X^2 and X^3 as different polynomials (and indeed, because X is transcendental over D, they are different elements of $D[X]$ — cf. above remark (*c*)): but, in the case $D = \mathbb{Z}_2$, the corresponding polynomial functions ($x \mapsto x^2$ and $x \mapsto x^3$) from D to D are equal! (They both map 0 to 0 and 1 to 1, and so have the same effect on every element of their common domain $\mathbb{Z}_2$.)

(*h*) The integral domain $D[X]$ gives rise to a field of fractions (cf. § 70), which we denote by $D(X)$. Each element of $D(X)$ is expressible (albeit in several ways) as a fraction

$$\frac{\text{element of } D[X]}{\text{nonzero element of } D[X]}.$$

We call $D(X)$ the field of *formal rational functions* in X over D, though it must be emphasized that, just as elements of $D[X]$ are not the same thing as polynomial functions from D to D (cf. note (*g*)), so elements of $D(X)$ are not to be regarded as functions in the sense of mappings.

(*i*) If the integral domain E is a subring of D, then $E[X]$ is the subring of $D[X]$ consisting of those polynomials in $D[X]$ whose coefficients all lie in E. For example, $\mathbb{Z}[X]$ is the subring of $\mathbb{Q}[X]$ consisting of the polynomials in $\mathbb{Q}[X]$ whose coefficients are all integers.

(*j*) Suppose that D is contained in a commutative ring R, that $c \in R$, and that $f \in D[X]$: say

$$f = a_0 + a_1 X + \ldots + a_n X^n$$

(where each $a_i \in D$). While it has been emphasized that f is not to be regarded as a function, it is nevertheless open to us to replace X by c in the above-displayed expression for f to produce $a_0 + a_1 c + \ldots + a_n c^n$ — which makes sense and is an element of R. We employ $f(c)$ as a natural notation for this element of R. Moreover, for each $c \in R$, we can introduce a mapping $\varepsilon_c : D[X] \to R$ specified by

$$\varepsilon_c(f) = f(c) \quad (f \in D[X]).$$

So ε_c is the mapping whose action is "putting X equal to c" or, as one might say, "evaluation at c" in elements of $D[X]$. A useful fact is:

72.2 For each $c \in R$, the mapping $\varepsilon_c : D[X] \to R$, as just defined, is a homomorphism.

Proof. Consider an arbitrary $c \in R$.

For every possible $f = a_0 + a_1 X + \ldots + a_n X^n$, $g = b_0 + b_1 X + \ldots + b_m X^m \in D[X]$,

$$\begin{aligned}
\varepsilon_c(f + g) &= \varepsilon_c[(a_0 + b_0) + (a_1 + b_1)X + (a_2 + b_2)X^2 + \ldots] \\
&= (a_0 + b_0) + (a_1 + b_1)c + (a_2 + b_2)c^2 + \ldots \\
&= (a_0 + a_1 c + a_2 c^2 + \ldots) + (b_0 + b_1 c + b_2 c^2 + \ldots) \\
&= \varepsilon_c(f) + \varepsilon_c(g),
\end{aligned}$$

and similarly (though it may seem much more awkward to write out in detail) $\varepsilon_c(fg) = \varepsilon_c(f)\varepsilon_c(g)$. So ε_c is indeed a homomorphism.

In view of 72.2 we refer to each mapping ε_c as an **evaluation homomorphism**. While it is often very helpful to think in terms of evaluation homomorphisms, the relative sophistication of this mode of thought should not be allowed to hide the simple fact that, in practice, applying ε_c to an expression made out of polynomials is really just a matter of "putting $X = c$" throughout the expression. For example, in a later section of the chapter we shall have occasion to apply ε_c to an expression $qm + r$ (where q, m, r denote polynomials). The result will be

$$\varepsilon_c(qm + r) = \varepsilon_c(qm) + \varepsilon_c(r) \left.\right\} \text{(since } \varepsilon_c \text{ is a homomorphism)}$$
$$= \varepsilon_c(q)\varepsilon_c(m) + \varepsilon_c(r)$$
$$= q(c)m(c) + r(c)$$

—which is just what one gets by going through the expression $qm + r$ and simple-mindedly putting $X = c$ in every component part of it

(k) As in note (j) let c denote an element of a commutative ring containing D. For $f \in D[X]$, we say that c is a **root** of f iff $f(c) = 0$. Observe that the polynomials in $D[X]$ which have the given element c as a root form the kernel of the homomorphism $\varepsilon_c : D[X] \to R$. Therefore, by 66.5:

72.3 For each fixed c as just described, the polynomials in $D[X]$ that have c as a root form an ideal of $D[X]$.

(l) The perceptive student will probably have realized already that it would be possible to embark on a more general discussion of polynomial rings by considering at the outset the polynomial ring $R[X]$ over a ring R which need not be an integral domain. But that more general discussion will not be undertaken here. The fact is that all the applications of polynomials to be pursued in this book are concerned with polynomials over integral domains, while the theory of polynomials over rings that are not integral domains is fraught with extra difficulties. (e.g. 72.1 will break down if the ring D contains divisors of zero).

Post-script. As promised at the beginning of § 72, here is a brief outline of how to demonstrate that, whatever the given integral domain D is, a ring $D[X]$, with X transcendental over D, does exist. Full details of such a proof would be extremely long and for the most part very tedious, and so it is strongly suggested that students content themselves with an appreciation of the basic idea. The crux of that idea is that, if a ring $D[X]$ does exist, its typical member $a_0 + a_1 X + \ldots + a_n X^n$ corresponds to, and is completely determined by, its coefficient sequence $(a_0, a_1, \ldots, a_n, 0, 0, 0, \ldots)$.

In the light of this observation, to construct $D[X]$ we start with the set S of all infinite sequence of elements of D that have [like the coefficient sequence of a polynomial] the property of being zero from some point onwards. Then we define addition and multiplication on S so as to mimic the way the coefficients of polynomials behave in addition and multiplication: i.e., for $f = (a_0, a_1, a_2, \ldots)$ and $g = (b_0, b_1, b_2, \ldots) \in S$ [where (cf. definition of S) $a_i = 0$ for all i greater than some n and $b_i = 0$ for all i greater than some m], we define $f + g$ and fg to be respectively the sequences $(c_0, c_1, c_2, \ldots)$ and $(d_0, d_1, d_2, \ldots)$ where, for each $i \geq 0$,

$$c_i = a_i + b_i \quad \text{and} \quad d_i = \sum_{j=0}^{i} a_j b_{i-j}.$$

After this start, the procedure is to prove that: (1) the system $(S, +, .)$ is a ring; (2) the sequences in S of the form $(a_0, 0, 0, 0, \ldots)$ form a subring $\hat{D}$ of S isomorphic to D, so that D is imbedded in S (cf. explanation of "imbedding" in § 70); (3) the sequence $X = (0, 1, 0, 0, 0, \ldots)$ is transcendental over $\hat{D}$; (4) every element of S is expressible as a polynomial in X with coefficients in $\hat{D}$. When all that has been done, we know that we have a ring $\hat{D}[X]$ with $\hat{D} \cong D$. If we now agree to identify the element $(a_0, 0, 0, 0, \ldots)$ of $\hat{D}$ with the corresponding element a_0 of D, so that $\hat{D}$ is identified with D, then the ring which has been constructed is $D[X]$ as required.

73. Some properties of $F[X]$, where F is a field

Throughout this section F will denote an arbitrary field.

The first proposition of the section records a straightforward answer to the question of what the units in $F[X]$ are.

73.1 The units in $F[X]$ (where F is any field) are the nonzero constant polynomials (i.e. the elements of $F - \{0\}$).

Proof. First suppose that u is a unit in $F[X]$. Then $uv = 1$ for some $v \in F[X]$, and hence (using 72.1(ii)) we have

$$0 = \deg 1 = \deg(uv) = \deg u + \deg v.$$

From this equation it is apparent (the possible values of $\deg u$ and $\deg v$ being $-\infty$ and nonnegative integers) that $\deg u$ must be 0—i.e. u must be a nonzero constant.

This shows that every unit in $F[X]$ is a nonzero constant. It is obvious that, conversely, every nonzero constant $a\, [\in F]$ has an inverse $a^{-1} (\in F \subseteq F[X])$ and so is a unit in $F[X]$. So the stated result is true.

The next theorem is a key result in opening up the properties of $F[X]$. It is referred to as the *division algorithm* for $F[X]$ and is very obviously an analogue of the division algorithm for integers (11.1 in Chapter 2).

73.2 Let d be a nonzero polynomial of degree n in $F[X]$. Then, for every $f \in F[X]$, there exist polynomials $q, r \in F[X]$ such that

$$f = qd + r \quad \text{and} \quad \deg r < n.$$

Proof. As the equation $0 = 0d + 0$ shows, the theorem is true in the trivial case $f = 0$. So each f that is a counterexample to the stated result (if there is any such counterexample f) will have a degree belonging to $\mathbb{N} \cup \{0\}$.

Suppose, with a view to obtaining a contradiction, that there is at least one polynomial f that is a counterexample. Because of the well-ordering principle (10.1) it is legitimate to introduce the lowest degree m [$\in \mathbb{N} \cup \{0\}$] possessed by a counterexample. Henceforth in this proof let f denote a counterexample of this lowest possible degree m.

Clearly $m \geqslant n$: for, if m were less than n, the equation $f = 0d + f$ would show that f was not after all a counterexample. Suppose that $f = a_0 + a_1 X + \ldots + a_m X^m$, $d = b_0 + b_1 X + \ldots + b_n Y^n$, where a_m, b_n are nonzero and [as just explained] $m \geqslant n$. The polynomial $(a_m b_n^{-1} X^{m-n})d$ [where X^{m-n} is to be interpreted as 1 if $m = n$] has the same degree (m) and the same leading coefficient (a_m) as f, and hence the difference

$$f - (a_m b_n^{-1} X^{m-n})d \quad (= f_1, \text{ say})$$

is a polynomial of degree less than m. Since there is no counterexample to the stated result of degree less than m, it follows that $f_1 = q_1 d + r$ for some $q_1, r \in F[X]$ with $\deg r < n$. Hence we have

$$f = f_1 + (a_m b_n^{-1} X^{m-n})d$$
$$= (q_1 + a_m b_n^{-1} X^{m-n})d + r$$

—which is an expression of the form in question for f and therefore shows that f is in fact not a counterexample.

As this is a contradiction, the truth of the stated result follows.

Note. The division algorithm result for $\mathbb{Z}$ (11.1) included the statement that the quotient and remainder arising in each possible case are uniquely determined. The same is true for $F[X]$: for any given $f, d \in F[X]$ with $d \neq 0$, there is just one possibility for the quotient and remainder polynomials q, r such that $f = qd + r$ and $\deg r < \deg d$. Proof of this is set as one of the exercises at the end of this chapter.

The remaining theorems in this section are results about roots of polynomials over F — important results which are quoted and used in many branches of mathematics.

73.3 [A result often called the *factor theorem*] Suppose that $f \in F[X]$ and that the element c of F is a root of f. Then

$$f = (X - c)g$$

for some $g \in F[X]$.

Proof. By the division algorithm for $F[X]$,

$$f = (X - c)g + r \qquad (*)$$

for some g, $r \in F[X]$ with $\deg r < \deg(X - c) = 1$, i.e. r constant. By evaluating at c in (*) [i.e. applying the evaluation homomorphism ε_c to both sides], we deduce

$$f(c) = (c - c)g(c) + r, \quad \text{i.e. } 0 = 0 + r, \quad \text{i.e. } r = 0.$$

So $f = (X - c)g$, and thus the theorem is proved.

Note. The converse of 73.3 is almost too trivial to be worth mentioning: visibly it is true that if $f = (X - c)g$, where $c \in F$ and $g \in F[X]$, then $f(c) = 0$.

A consequence of 73.3 is that a polynomial of positive degree n in $F[X]$ can be completely factorized into a product of linear (i.e. degree 1) factors if we know n different roots of it. In detail:

73.4 Suppose that f is a polynomial in $F[X]$ of degree $n \geqslant 1$ and that f has n different roots $c_1, c_2, \ldots, c_n$ in F. Then

$$f = a_n(X - c_1)(X - c_2)\ldots(X - c_n)$$

where a_n is the leading coefficient of f.

Proof. Since c_1 is a root of f, 73.3 tells us that

$$f = (X - c_1)g_1$$

for some $g_1 \in F[X]$. Evaluating both sides of this equation at c_2 gives (since c_2 is also a root of f)

$$0 = (c_2 - c_1)g_1(c_2),$$

and hence (the factor $c_2 - c_1$ being nonzero because $c_1, c_2, \ldots, c_n$ are all different) $g_1(c_2) = 0$. So (by another application of 73.3) $g_1 = (X - c_2)g_2$ for some $g_2 \in F[X]$, and hence

$$f = (X - c_1)(X - c_2)g_2.$$

In a further similar step, evaluation at c_3 (along with use of the fact that c_1, c_2, c_3 are all different) yields $g_2(c_3) = 0$: hence $g_2 = (X - c_3)g_3$ for some $g_3 \in F[X]$, and therefore

$$f = (X - c_1)(X - c_2)(X - c_3)g_3.$$

After n such steps the outcome is

$$f = (X - c_1)(X - c_2)\ldots(X - c_n)g_n \tag{α}$$

for some $g_n \in F[X]$. Taking the degree of both sides of this equation gives (cf. 72.1(ii))

$$n\,[= \deg f] = \deg\,[(X - c_1)(X - c_2)\ldots(X - c_n)] + \deg g_n$$
$$= n + \deg g_n;$$

and hence $\deg g_n = 0$—i.e. g_n is constant. Knowing this, we finally compare the coefficients of X^n on the two sides of the equation (α) to find that g_n equals the leading coefficient of f. This completes the proof of the result.

A simple but very widely used theorem follows quickly from 73.4, namely:

73.5 A polynomial of degree n in $F[X]$ (where $n \in \mathbb{N}$) cannot have more than n roots in the field F.

Proof. Suppose, with a view to obtaining a contradiction, that some polynomial $f \in F[X]$ of degree n ($\in \mathbb{N}$) has $n+1$ **different roots** $c_1, c_2, \ldots, c_n,$ c_{n+1} [and maybe more] in F. By 73.4,

$$f = a_n(X - c_1)(X - c_2)\ldots(X - c_n)$$

where a_n is the leading coefficient of f. Hence (by evaluation at c_{n+1})

$$f(c_{n+1}) = a_n(c_{n+1} - c_1)(c_{n+1} - c_2)\ldots(c_{n+1} - c_n)$$
$$\neq 0 \quad (\text{since } a_n \neq 0 \text{ and } c_1, \ldots, c_{n+1} \text{ all different})$$

—a contradiction since c_{n+1} is a root of f. From this contradiction it follows that the theorem is true.

74. Generalities on factorization

In the general discussion in this section D will always denote an integral domain. Some of the more interesting theorems of the section relate only to the special case where D is a principal ideal domain. Here, as in § 75, we shall note applications of the general theory to the case $D = F[X]$, where F is an arbitrary field. [As will be seen at 75.1, $F[X]$ is a principal ideal domain for every field F.]

For elements a, b of the integral domain D, we say that b **divides** a (or that b is a **factor** of a) in D iff $a = kb$ for some $k \in D$; and we write "$b|a$" as a shorthand for "b divides a".

In any integral domain D some territory is virtually devoid of interest so far as factorization is concerned. For example, every element x of D divides

0 [since $0 = 0x$], and factorization statements involving units are not worth making since every unit u of D divides every element $x \in D$ [as is seen from the equation $x = (xu^{-1})u$]. So discussion of factorization is usually restricted to factorization of nonzero non-units. Furthermore we describe a factorization $a = bc$ of an element $a \in D$ as **trivial** if either b or c is a unit in D: a **nontrivial** factorization of the element a ($\in D$) means a factorization $a = bc$ where neither of the elements b, c of D is a unit in D. Thus, in particular:

(1) In $\mathbb{Z}$ factorizations of the form $n = 1n$ or $n = (-1)(-n)$ count as trivial; a nontrivial factorization of integer n (into 2 factors) is one where both factors have modulus greater than 1 and less than that of n.

(2) In $F[X]$, where F is a field, a factorization $f = gh$ ($f, g, h \in F[X]$) of a polynomial f of degree $\geqslant 1$ is trivial if either g or h is a constant (cf. 73.1); the factorization is nontrivial if both the factors g, h have degrees $\geqslant 1$ and $< \deg f$.

The distinction we have drawn between trivial and nontrivial factorizations enables us to make the following definition: in an integral domain D, an element p is said to be an **irreducible** element of D iff (1) p is a nonzero non-unit and (2) p has no nontrivial factorization in D.

Obviously, the idea of an irreducible element is a generalization of the idea of a prime number. It is easily seen that in $\mathbb{Z}$ the irreducible elements are the prime numbers and their negatives. The first result of this section presents some very simple facts about irreducible elements in the ring of polynomials over a field:

74.1 Let F be a field.

(1) Every polynomial of degree 1 in $F[X]$ is an irreducible element of $F[X]$.

(2) A polynomial of degree 2 or 3 in $F[X]$ is irreducible iff it has no root in F.

Proof. (1) If a polynomial $f \in F[X]$ has a nontrivial factorization $f = gh$ (where $g, h \in F[X]$), then $\deg f = \deg g + \deg h \geqslant 1 + 1 = 2$. Therefore a polynomial of degree 1 (which is certainly a nonzero non-unit) must be irreducible in $F[X]$.

(2) [The case of a polynomial of degree 2 will be dealt with here. The student should have no difficulty in adapting the analysis to deal with a polynomial of degree 3.]

Let f be a polynomial of degree 2 in $F[X]$. Then certainly f is a nonzero non-unit in $F[X]$. If f has a nontrivial factorization $f = gh$ in $F[X]$, then clearly both the factors g, h must have degree 1, so that we have

$$f = (aX + b)(cX + d)$$

for some $a, b, c, d \in F$ with $a, c \neq 0$—from which it is apparent that $-b/a$ is a root of f in F. Thus

$$f \text{ not irreducible in } F[X] \Rightarrow f \text{ has a root in } F.$$

It is easily seen that the converse of this is true too: for if f has a root $-c$, say — in F, then (by 73.3) f has a nontrivial factorization $f = (X - c)g$ for some polynomial $g \in F[X]$ necessarily of degree 1.

Therefore f is irreducible in $F[X]$ iff f has no root in F.

Further remarks on irreducible polynomials. (*a*) When referring to a polynomial $f \in F[X]$, people usually say "*f* is irreducible over F" rather than "*f* is an irreducible element of $F[X]$".

(*b*) Beware of imagining that 74.1(2) can be extended to polynomials of degree greater than 3. Notice, for example, that in $\mathbb{R}[X]$ the polynomial $(X^2 + 1)(X^2 + 4)$ is plainly not irreducible, even though it has no root in $\mathbb{R}$.

(*c*) A statement to the effect that a particular element is irreducible is always made relative to some containing ring D as the "universe". (The element is irreducible *in D*.) A switch of view as to what the containing "universe" is can change whether or not an element is irreducible. For example, consider the polynomial $X^2 + 1$. Regarded as an element of the integral domain $\mathbb{R}[X]$, $X^2 + 1$ is irreducible (cf. 74.1(2)); but, in the context of the larger system $\mathbb{C}[X]$, $X^2 + 1$ has a nontrivial factorization, namely $X^2 + 1 = (X - i)(X + i)$, and so is not irreducible.

(*d*) It should be noted that 74.1 cannot be extended to polynomials in $D[X]$ where D is an integral domain but not a field. E.g. $2X - 2$ is polynomial in $\mathbb{Z}[X]$ of degree 1, but it is *not* irreducible in $\mathbb{Z}[X]$ because in $\mathbb{Z}[X]$ the factorization $2X - 2 = 2(X - 1)$ is a nontrivial factorization (2 being a non-unit in $\mathbb{Z}[X]$).

(*e*) For the purpose of doing problems on the contents of this chapter, it is useful to know the fact that all polynomials of the form $X^n - p$, with $n \in \mathbb{N}$ and p a prime number in $\mathbb{N}$, are irreducible in $\mathbb{Q}[X]$. Students should assume this to be true when attempting exercises. There is a nice piece of algebra (including the results known as Gauss's lemma and Eisenstein's irreducibility criterion) which can be used to establish the irreducibility of polynomials of the form $X^n - p$ and other polynomials with integer coefficients. For an account of this subject-matter, see, for example, Birkhoff and MacLane's classic book [7].

After these remarks on polynomials, let us return to the discussion of factorization in a general integral domain D. It turns out to be very illuminating to bring principal ideals into the discussion. A basic result that signals their relevance is:

74.2 For $a, b \in D$, the following three statements are equivalent to one another:

(1) $b \mid a$ in D;
(2) $a \in (b)$;
(3) $(a) \subseteq (b)$.

[In (2) and (3), (a) and (b) signify the principal ideals of D generated by a and b, respectively — cf. § 67.]

Proof. Let $a, b \in D$.

That (1) and (2) are equivalent is immediately clear because both these statements mean "$a = kb$ for some $k \in D$".

That $(3) \Rightarrow (2)$ is clear because, of course, $a \in (a)$; and $(2) \Rightarrow (3)$ follows at once from the fact (cf. 67.1) that (a) is the smallest ideal of D to which a belongs.

So the conditions (1), (2), (3) are indeed equivalent.

A further relevant basic fact about principal ideals is:

74.3 For $a, b \in D$, $(a) = (b)$ iff a, b are unit multiples of each other (i.e. iff $a = ub$ for some unit u in D and correspondingly $b = u^{-1}a$).

Proof. Let $a, b \in D$.

If a, b are unit multiples of each other — so that $b \mid a$ and $a \mid b$ are both true — then (by 74.2) $(a) \subseteq (b)$ and $(b) \subseteq (a)$, and therefore $(a) = (b)$.

Conversely, suppose that $(a) = (b)$, so that $a = sb$ and $b = ta$ for some $s, t \in D$.

If a or b is zero, then clearly both must be zero and certainly they are unit multiples of each other ($a = 1b$ and $b = 1a$).

Henceforth we deal with the remaining case where a and b are nonzero. In this case, because

$$1a = a = sb = s(ta) = (st)a,$$

it follows (by 63.4) that $st = 1$; and this reveals that s and t are units (each being the inverse of the other). Hence a, b are unit multiples of each other.

The proof is now complete.

Not surprisingly, the connection between factorization ideals and principal ideals can be exploited with particular effectiveness in principal ideal domains. The remaining theorems in this section all concern principal ideal domains, and students should be warned that these results do not extend to arbitrary integral domains. First in this sequence of results is a very significant fact about the principal ideal generated by an irreducible element in a principal ideal domain.

74.4 Suppose that D is a principal ideal domain and that p is an irreducible element of D. Then (p) is a maximal ideal of D.

Proof. First note that $1 \notin (p)$: for $1 \in (p)$ would imply that $pt = 1$ for some $t \in D$ and hence that p is a unit in D — which is not the case since p is irreducible. Therefore (p) is a proper ideal of D.

Suppose (with a view to obtaining a contradiction) that (p) is not a maximal ideal of D. Then there is an ideal J of D such that $(p) \subset J \subset D$; and, since D is a principal ideal domain, $J = (x)$ for some $x \in D$. So we have $(p) \subset (x) \subset D$. From $(p) \subset (x)$, it follows that $x \mid p$ (cf. 74.2) i.e. that $p = xy$ for some $y \in D$; and, since p is irreducible, either x or y must be a unit in D. If y were a unit, then (since $p = xy$) x and p would be unit multiples of each other and therefore (by 74.3) (x) and (p) would coincide—whereas in fact $(p) \subset (x)$. Therefore y is not a unit, and so x is a unit. Hence the ideal J contains a unit and is therefore the whole of D (by 66.2)—a contradiction since $J \subset D$.

It now follows that (p) must indeed be a maximal ideal of D.

From 74.4 along with 68.3 it follows at once that:

74.5 If D is a principal ideal domain and p is an irreducible element of D, then $D/(p)$ is a field.

To see an illustration of 74.5, turn first to the case where $D = \mathbb{Z}$ and p is a prime number (and therefore an irreducible element in $\mathbb{Z}$). By 74.5, $\mathbb{Z}/(p)$ is a field. As pointed out in § 68, for any positive integer m, $\mathbb{Z}/(m)$ is the same thing as $\mathbb{Z}_m$: so, by applying 74.5, we have re-discovered the fact that $\mathbb{Z}_p$ is a field for every prime number p. But 74.5 also enables us to construct new fields, and in § 79 we shall make a very significant use of it—starting with an arbitrary given field F and a polynomial f over F with no root in F, taking D to be $F[X]$ (which turns out to be a principal ideal domain), and showing that F can be extended to a larger field in which f has a root.

The next theorem is easily recognized as a generalization of a property of integers that follows immediately from the euclidean algorithm as presented in § 12. The theorem is in fact one special, but extremely useful, corner of a general theory of highest common factors (including an analogue of the euclidean algorithm for integers) which can be developed for an arbitrary principal ideal domain. (Some of the exercises at the end of the chapter guide the student to construct interesting parts of that general theory.)

74.6 Let D be a principal ideal domain, and suppose that the elements p and a of D are such that p is irreducible and does not divide a in D. Then there exist elements $x, y \in D$ such that $xp + ya = 1$.

Proof. Consider the ideal $J = (p) + (a)$ of D. This ideal contains (p) and also contains the element a, which does not lie in (p) since p does not divide a (cf. 74.2). Therefore the ideal J strictly contains (p), which, however, is a maximal ideal (by 74.4). So J must be the whole of D, and hence, in particular,

$1 \in J$ $[= (p) + (a)]$. This gives us exactly what is asserted — that $1 = xp + ya$ for some $x, y \in D$.

The last two results of § 74 are corollaries of 74.6 and are important generalizations of the results 12.5 and 12.6 about factorization of integers.

74.7 Let D be a principal ideal domain, and let p be an irreducible element of D. Suppose that $p|ab$ in D (where $a, b \in D$). Then $p|a$ or $p|b$ [or possibly both] in D.

Proof. Since $p|ab$, $ab = cp$ for some $c \in D$.

Suppose that p does not divide a. (It will suffice to prove that $p|b$ follows from this supposition.)

By 74.6, there are elements $x, y \in D$ such that $xp + ya = 1$. Hence

$$
\begin{aligned}
b = b1 &= b(xp + ya) \\
&= bxp + bya \\
&= bxp + ycp \text{ (since } ab = cp) \\
&= (bx + yc)p.
\end{aligned}
$$

Thus $p|b$; and the result follows.

By repeated application of 74.7 (cf. the deduction of 12.6 from 12.5), one easily obtains the more general result:

74.8 If $p, a_1, a_2, \ldots, a_k$ are elements of the principal ideal domain D, p is irreducible in D and p divides the product $a_1 a_2 \ldots a_k$ in D, then p divides at least one of $a_1, a_2, \ldots, a_k$ in D.

75. Further properties of $F[X]$, where F is a field

Throughout this section F will denote an arbitrary field.

The opening theorem of the section (75.1) presents, in particular, the cardinally important fact that $F[X]$ is a principal ideal domain. It gives also some technical detail about generators of ideals in $F[X]$ — detail which will prove useful when ideals of $F[X]$ enter the discussion in other places.

In this theorem we shall be considering a nonzero ideal J of $F[X]$. It should be appreciated that the set

$$\{\deg f : f \in J - \{0\}\},$$

i.e. the set of integers occurring as degrees of nonzero polynomials in J, is a nonempty set of integers bounded below by 0, and accordingly, by virtue of the well-ordering principle (10.1), this set has a least member. That is why it is legitimate to introduce the "minimum of the degrees of polynomials in $J - \{0\}$".

75.1 (i) Let J be a nonzero ideal in $F[X]$, and let n be the minimum of the degrees of polynomials in $J - \{0\}$. Then:

(a) for any polynomial g of degree n in J, $J = (g)$;

(b) there is precisely one monic polynomial h of degree n in J, and $J = (h)$.

(ii) $F[X]$ is a principal ideal domain.

Proof. (i) (a) Let g be any polynomial of degree n in J, and let f be an arbitrary polynomial in J. By the division algorithm for $F[X]$ (73.2),

$$f = qg + r \text{ for some } q, r \in F[X] \text{ with } \deg r < \deg g = n.$$

Since J is an ideal to which g belongs, qg must also belong to J; and since $f \in J$ and J is closed under subtraction, $f - qg \in J$: i.e. $r \in J$. But $\deg r < n$, and no nonzero polynomial in J has degree less than n. Therefore r must be 0; and hence $f = qg$, so that $f \in (g)$. This proves that $J \subseteq (g)$. But, since J is an ideal to which g belongs, $(g) \subseteq J$ (cf. 67.1). Therefore $J = (g)$ — which proves part (i)(a) of the theorem.

(b) Let $g = a_0 + a_1 X + \ldots + a_n X^n$, where $a_n \neq 0$, be any polynomial of degree n in J. By part (a), $J = (g)$. Let $h = a_n^{-1} g$. Then clearly h is a monic polynomial of degree n in the ideal J, and (cf. part (a) again) $(h) = J$.

Suppose that there is a second monic polynomial h_1 of degree n in J. By a further application of part (a), $J = (h_1)$, and so $(h) = (h_1)$. Hence, by 74.3 and 73.1, $h = bh_1$ for some $b \in F$. Since h and h_1 are monic, comparison of the coefficients of X^n on the two sides of the equation $h = bh_1$ yields $1 = b1$, i.e. $b = 1$. So $h = h_1$.

This proves that h is the one and only monic polynomial of degree n in J. The fact that $J = (h)$ has already been recorded, and so part (i)(b) of the theorem is now established.

(ii) We already know (cf. 72.1) that $F[X]$ is an integral domain. Part (i)(a) of the present theorem shows that every nonzero ideal of $F[X]$ is a principal ideal, while of course $\{0\}$ $[= (0)]$ is also a principal ideal. Therefore $F[X]$ is a principal ideal domain.

Because of 75.1(ii) we can apply to $F[X]$ the potent results about principal ideal domains proved towards the end of § 74, and by this means we shall be able to show that $F[X]$ has unique factorization properties very similar to those of positive integers (cf. 13.1 and 13.2). To achieve a sharply clear statement of these properties, let us agree to restrict attention to factorizations of polynomials that have been tidied up to the form

$$\text{(leading coefficient)} \times \text{(product of monic polynomials)}. \qquad (*)$$

It is easily seen that any factorization whatever of a given polynomial can be adjusted to the tidied form (*) simply by expressing each non-monic factor as (a constant) × (a monic polynomial) and then multiplying together all the constant factors: e.g. in the factorization

$$6X^4 - 3X^3 - 22X^2 + 7X + 2 = (X - 2)(2X - 1)(3X^2 + 6X + 1)$$

we can take the constant factors 2 and 3 out of the non-monic factors of the right-hand side to produce a revised version

$$6X^4 - 3X^3 - 22X^2 + 7X + 2 = 6(X - 2)(X - \tfrac{1}{2})(X^2 + 2X + \tfrac{1}{3}),$$

which is of the form (*).

Here now are the theorems (analogous to 13.1 and 13.2, respectively) which give details of the nice factorization properties of $F[X]$.

75.2 Every polynomial of degree $\geqslant 1$ in $F[X]$ can be expressed in the form

$$\text{(leading coefficient)} \times \text{(a product of monic irreducible polynomials)}.$$

Proof. Suppose the result false. Then (as a consequence of the well-ordering principle) we can legitimately consider a counterexample (f, say) of lowest possible degree. Let f have leading coefficient a_n, so that $f = a_n g$ for some monic polynomial g of the same degree as f. Since a single monic irreducible polynomial is, in a trivial way, expressible as a product of monic irreducible polynomials, it is clear (from the fact that f is a counterexample) that g cannot be irreducible. Therefore there is a nontrivial factorization $g = h_1 h_2$ of g, in which, we can assume, the factors h_1, h_2 are monic (cf. above remarks on adjusting factorizations by taking constant factors out of non-monic factors). Here $1 \leqslant \deg h_1, \deg h_2 < \deg g = \deg f$, and so (since f is of minimum possible degree for a counterexample) each of h_1, h_2 is expressible as a product of monic irreducible factors — say $h_1 = p_1 p_2 \ldots p_r$, $h_2 = q_1 q_2 \ldots q_s$, where each of the polynomials $p_1, \ldots, q_s$ is a monic irreducible polynomial. Hence we have

$$f = a_n p_1 \ldots p_r q_1 \ldots q_s,$$

which shows that f is not a counterexample. That is a contradiction, and from it the result follows.

75.3 Apart from the possibility of varying the order of the irreducible monic factors, there is *only one way* to express any given polynomial of degree $\geqslant 1$

in $F[X]$ in the form

(leading coefficient) $\times$ (product of irreducible monic polynomials).   (†)

Proof. Suppose the result false and (as in the proof of 75.2) consider a counterexample polynomial f of minimum possible degree. Since f is a counterexample, there are two genuinely different factorizations

$$f = a_n p_1 p_2 \ldots p_r \quad \text{and} \quad f = a_n q_1 q_2 \ldots q_s,$$

where a_n is the leading coefficient of f and $p_1, \ldots, q_s$ are monic irreducible polynomials in $F[X]$; in saying that these are "genuinely different" factorizations of f, we mean that the products $p_1 \ldots p_r$ and $q_1 \ldots q_s$ cannot be transformed into each other by changing the order of the factors. On equating these two factorizations of f to each other and cancelling the nonzero a_n factor, we obtain

$$p_1 p_2 \ldots p_r = q_1 q_2 \ldots q_s \ (= g, \text{ say}). \tag{A}$$

Obviously g, with its two genuinely different factorizations given here, cannot itself be irreducible, and so r and s are certainly $\geqslant 2$.

From (A), p_1 divides $q_1 q_2 \ldots q_s$. So, by 74.8, the irreducible p_1 must divide at least one of $q_1, q_2, \ldots, q_s$. Without loss of generality, we can suppose the q_is to be numbered so that p_1 divides q_1. Accordingly, $q_1 = bp_1$ for some $b \in F[X]$. But since q_1 is irreducible, the factorization $q_1 = bp_1$ must be trivial, while p_1 (as an irreducible polynomial) is not a unit. Therefore b must be a constant; and, since p_1 and q_1 are monic, comparison of leading coefficients in the equation $q_1 = bp_1$ gives $b = 1$. So $q_1 = p_1$. Since $F[X]$ is an integral domain, we can cancel the (nonzero) factor p_1 $[= q_1]$ in (A) to deduce

$$p_2 p_3 \ldots p_r = q_2 q_3 \ldots q_s \ (= h, \text{ say}). \tag{B}$$

By 72.1, $\deg h = \deg g - \deg p_1 = \deg f - \deg p_1 < \deg f$. So, because f is of the minimum possible degree for a counterexample to the theorem, the two factorizations of h in (B) cannot be genuinely different: i.e. the sequence of factors $q_2, q_3, \ldots, q_s$ must be just a rearrangement of the sequence $p_2, p_3, \ldots, p_r$. Therefore (p_1 being the same as q_1) the products $p_1 p_2 \ldots p_r$ and $q_1 q_2 \ldots q_s$ *can* be transformed into each other by changing the order of the factors. That is a contradiction.

It follows that the stated result is true.

Remarks. (*a*) It follows from 75.3 that when a given polynomial f is factorized in the form (†) (as in the statement of 75.3), the number of times a linear factor $X - c$ (c constant) appears is uniquely determined. This number is called the **multiplicity** of c as a root of f. The fact that multiplicities of roots of polynomials are uniquely determined is one of several consequences of

75.3 that are often used when polynomials crop up in different parts of mathematics and its applications.

(b) Results analogous to 75.2 and 75.3 can be proved (by a very similar approach, in fact) for an arbitrary principal ideal domain once one has dealt with the problem of when, in the general context, two factorizations of an element should be regarded as genuinely different. E.g., in $\mathbb{Z}$ one would surely regard the factorizations $6 = 2 \times 3$ and $6 = (-2) \times (-3)$ as essentially the same. More generally, one takes the view that factorizations should not be regarded as different if they can be transformed into each other by replacing factors by unit multiples of themselves (e.g. 2 by -2, 3 by -3 in the illustration) and/or by changing the order of the factors.

76. Some matters of notation

Suppose that D is an integral domain contained in a larger integral domain R, and let c be an element of R. We denote by $D[c]$ the set of all elements expressible in the form $f(c)$ for some polynomial $f \in D[X]$, i.e. in the form $a_0 + a_1c + \ldots + a_nc^n$ with each $a_i \in D$. It is easily discovered that:

76.1 In the context just described, $D[c]$ is a subring of R and is the smallest subring of R containing D as a subset and c as an element [smallest in the sense that every such subring must contain $D[c]$].

Proof. By its definition $D[c]$ is the set $\{\varepsilon_c(f) : f \in D[X]\}$ and so is the image set of the homomorphism $\varepsilon_c : D[X] \to R$. Therefore, by 65.3(i), $D[c]$ is a subring of R. Clearly $c \in D[c]$ and $D \subseteq D[c]$. Further, if S is any subring of R containing $D \cup \{c\}$, then from the fact that S is closed under addition and multiplication it is apparent that S must contain all the members of $D[c]$. So $D[c]$ is indeed the smallest subring of R that contains both D and c.

Illustration. In an exercise on Chapter 10, it is proved that the set

$$S = \{a + b\sqrt{2} : a, b \in \mathbb{Z}\}$$

is a subring of $\mathbb{R}$. Clearly it is the smallest subring of $\mathbb{R}$ that contains $\mathbb{Z}$ as a subset and $\sqrt{2}$ as an element. So S is $\mathbb{Z}[\sqrt{2}]$.

Had we not been in a position to quote the result of the exercise on Chapter 10, we would have described $\mathbb{Z}[\sqrt{2}]$ as the set of all numbers expressible as polynomials in $\sqrt{2}$ with integer coefficients, i.e. expressible in the form

$$a_0 + a_1\sqrt{2} + a_2(\sqrt{2})^2 + \ldots + a_n(\sqrt{2})^n \text{ with each } a_i \in \mathbb{Z}.$$

Of course (since $(\sqrt{2})^2 = 2$, $(\sqrt{2})^3 = 2\sqrt{2}$, etc.) every such expression can be tidied to the form $a + b\sqrt{2}$ with $a, b \in \mathbb{Z}$; and hence it is apparent that $\mathbb{Z}[\sqrt{2}]$

is nothing more elaborate than S as defined at the beginning of this illustration.

Remarks on the general context $(D, c, R$ as before)

(*a*) The subring $D[c]$ of R, consisting as it does of all the elements that can be constructed via ring operations from c and elements of D, is often referred to as the subring of R **generated by adjoining** c to D.

(*b*) In the discussion so far, the emphasis has been on picking out $D[c]$ as a subring of R. Often in future discussions, however, $D[c]$ appears in statements of the form "$R = D[c]$" intended to reveal something of the structure of an entire ring R in terms of a subring D and a member element c. (Consider, for example, the observation that $\mathbb{C} = \mathbb{R}[i]$.) In digesting such a statement "$R = D[c]$" we should appreciate that it is telling us that the subring $D[c]$ of R is the *whole* of R, i.e. that *every* element of R is expressible as a polynomial in c with coefficients in D.

(*c*) The use of square brackets introduced in this section is easily seen to be in harmony with — and indeed to be a generalization of — their use in the notation $D[X]$ introduced in § 72.

(*d*) Use of the square bracket can be extended further to cover the adjunction of more than one generating element. For example, if $x, y \in R$ $[\supseteq D]$, then $D[x, y]$ means the set of all elements of R expressible as polynomials in x and y with coefficients in D—i.e. as finite sums $\sum_{i,j \geqslant 0} a_{ij} x^i y^j$ with each coefficient $a_{ij} \in D$; and it is easily seen that this set is the smallest subring of R that contains D as a subring and contains the elements x and y.

A relatively trivial thing that perhaps needs to be pointed out explicitly to newcomers to this notation is that a subring $D[x_1, x_2, \ldots, x_n]$ can be built up from D by adjoining the generators $x_1, x_2, \ldots, x_n$ one by one in any order: e.g. the ring $(D[x])[y]$ (i.e. the ring obtained by first forming $D[x]$ and then adjoining y to it) is no different from $D[x, y]$, and similarly $(D[y])[x]$ is the same as $D[x, y]$. The student may wish to think through the details of why exactly this is so.

From the above considerations of rings and subrings, let us now turn to an analogous discussion of fields and subfields. It has been seen that square brackets are used to build a notation signifying "sub*ring* generated by adjoining the listed elements to a given sub*ring*". In a similar way, as will now be explained, round brackets are used to build a notation signifying "sub*field* generated by adjoining listed elements to a given subfield". To set the scene for a full explanation of this round brackets notation, suppose that the field F is a subfield of a larger field K and that $c \in K$. Then, for a start, $F(c)$ is defined to mean the set of all elements of K expressible in the form

$$\frac{f(c)}{g(c)} \text{ for some } f, g \in F[X] \text{ such that } g(c) \neq 0,$$

i.e., so to speak, as permissible "rational functions" of c with coefficients in F. It is easily verified that:

76.2 $F(c)$, as just defined, is a subfield of K and is the smallest subfield of K that contains F as a subset and c as an element.

In the light of 76.2, we refer to $F(c)$ as the subfield of K **generated by adjoining** c **to** F.

Illustration. In an exercise on Chapter 10 it was shown that the set

$$L = \{a + b\sqrt{2} : a, b \in \mathbb{Q}\}$$

is a subfield of $\mathbb{R}$. Rather obviously it is the smallest subfield of $\mathbb{R}$ containing $\mathbb{Q}$ and $\sqrt{2}$: so $L = \mathbb{Q}(\sqrt{2})$.

This simple conclusion might at first sight puzzle a student who, having read the general definition of "$F(c)$", might expect that a much more complicated story would be necessary to describe $\mathbb{Q}(\sqrt{2})$ — namely that it is the set of all possible quotients of the form

$$\frac{a_0 + a_1\sqrt{2} + \ldots + a_n(\sqrt{2})^n}{b_0 + b_1\sqrt{2} + \ldots + b_m(\sqrt{2})^m}$$

with all the coefficients $a_0, \ldots, b_m \in \mathbb{Q}$ (and with nonzero denominator). However, it is not too difficult to see that, by elementary algebra, all such quotients can be simplified to the form $a + b\sqrt{2}$ with $a, b \in \mathbb{Q}$, and that is why $\mathbb{Q}(\sqrt{2})$ can be described as simply as it was above. An interesting supplementary detail that students may notice is that, whereas $F(c)$ always contains $F[c]$ as a subset and looks likely to contain $F[c]$ strictly, it is apparent in the present illustration that $\mathbb{Q}(\sqrt{2})$ is just the same as $\mathbb{Q}[\sqrt{2}]$! Surprising though that may seem at first sight, it will be seen in § 77 that $F(c)$ always coincides with $F[c]$ when c is not transcendental over F.

The round brackets notation can be used to cover the adjunction of more than one element to generate a bigger field: if (with F, K as in the above general discussion) $c_1, c_2, \ldots, c_n$ are elements of K, then $F(c_1, c_2, \ldots, c_n)$ means the subset of K consisting of all elements expressible in the form

$$\frac{\text{element of } F[c_1, c_2, \ldots, c_n]}{\text{nonzero element of } F[c_1, c_2, \ldots, c_n]}$$

It can, without too much bother, be verified that this definition makes $F(c_1, \ldots, c_n)$ a field, and indeed the smallest subfield of K that contains F as a subset and contains the elements $c_1, \ldots, c_n$. It therefore makes sense to refer to $F(c_1, \ldots, c_n)$ as the subfield generated by adjoining $c_1, \ldots, c_n$ to F. It should be appreciated that $F(c_1, \ldots, c_n)$ can be built up from F by adjoining the generators $c_1, \ldots, c_n$ one by one in any order (cf. remark (d) on the square brackets notation): e.g. $(F(x))(y)$ is just the same thing as $F(x, y)$, as is $(F(y))(x)$.

Other points can be made exactly parallel to remarks made on the square brackets notation. In particular: (1) the round bracket notation often appears in statements of the form "$K = F(c)$" where the point being made is that $F(c)$ is the *whole* of K, i.e. that *every* element of K is expressible as $f(c)/g(c)$ for some $f, g \in F[X]$ with $g(c) \neq 0$; and (2) the use of round brackets introduced here is in harmony with (and indeed is a generalisation of) their use in the notation $D(X)$ as introduced in § 72.

Worked example (on the theme that $F(c_1, c_2, \ldots)$ is the *smallest* subfield containing $c_1, c_2, \ldots$). In the context of subfields of $\mathbb{C}$, show that $\mathbb{Q}(\sqrt{2}, \sqrt{3}) = \mathbb{Q}(d)$ where $d = \sqrt{3} + \sqrt{2}$.

Note. Every subfield of $\mathbb{C}$ contains $\mathbb{Q}$ (cf. 69.4). Therefore in the solution given below (and in all similar discussions) we can describe a subfield $\mathbb{Q}(c_1, c_2, \ldots)$ of $\mathbb{C}$ simply as the smallest subfield of $\mathbb{C}$ containing $c_1, c_2, \ldots$; to mention that it contains $\mathbb{Q}$ would be superfluous.

Solution. $\mathbb{Q}(\sqrt{2}, \sqrt{3})$ contains $\sqrt{2}$ and $\sqrt{3}$, and so [since it is closed under addition, being a field] it contains $\sqrt{3} + \sqrt{2} = d$. Therefore, since $\mathbb{Q}(d)$ is the smallest subfield of $\mathbb{C}$ that contains d,

$$\mathbb{Q}(d) \subseteq \mathbb{Q}(\sqrt{2}, \sqrt{3}). \tag{A}$$

Next observe that $d(\sqrt{3} - \sqrt{2}) = 1$, so that

$$\sqrt{3} - \sqrt{2} = 1/d \quad \text{while} \quad \sqrt{3} + \sqrt{2} = d.$$

Hence

$$\sqrt{3} = \tfrac{1}{2}[d + (1/d)] = (d^2 + 1)/2d \quad \text{and} \quad \sqrt{2} = \tfrac{1}{2}[d - (1/d)] = (d^2 - 1)/2d$$

—from which it is apparent that $\sqrt{3}$ and $\sqrt{2}$ belong to $\mathbb{Q}(d)$. Therefore, since $\mathbb{Q}(\sqrt{2}, \sqrt{3})$ is the smallest subfield to which $\sqrt{2}$ and $\sqrt{3}$ belong,

$$\mathbb{Q}(\sqrt{2}, \sqrt{3}) \subseteq \mathbb{Q}(d). \tag{B}$$

The containments (A) and (B) together give the desired result.

77. Minimal polynomials and the structure of $F(c)$

In this heading and throughout § 77 F denotes an arbitrary given field contained as a subfield in a larger field K, and c denotes an element of K.

In § 72 the meaning of "transcendental" was explained: to rephrase the definition in terms now available to us, c is transcendental over F iff there is no nonzero polynomial in $F[X]$ that has c as a root. In the contrary case, where there *is* a nonzero polynomial $f \in F[X]$ such that $f(c) = 0$, we describe c as **algebraic** over F.

It is easily seen that many of the irrational numbers one encounters in elementary mathematics (e.g. $\sqrt{2}$, $\sqrt{3}$, $2^{1/3}$) are algebraic over $\mathbb{Q}$: e.g. $\sqrt{2}$ is algebraic over $\mathbb{Q}$ because there is a nonzero polynomial (e.g. $X^2 - 2$) in $\mathbb{Q}[X]$ that has $\sqrt{2}$ as a root. It is known on the other hand that many real numbers are transcendental over $\mathbb{Q}$: such numbers are often called "transcendental numbers", and it is known that they include the frequently used numbers π and e. (For a proof that π and e are transcendental numbers, see for example the famous textbook by Hardy and Wright [8].)

There is not much to say about the structure of the field $F(c)$ [$\subseteq K$] in the case where c is transcendental over F. In this case $F(c)$ is isomorphic to the field $F(X)$ of all formal rational functions over F as defined in § 72. But a number of illuminating things can be said about the field $F(c)$ in the case where c is algebraic over F. For the rest of § 77 we shall concentrate on that case: so henceforth we suppose that the element c is algebraic over F.

Consider the set J_c of polynomials in $F[X]$ that have c as a root. By 72.3, J_c is an ideal of $F[X]$; and, because c is algebraic over F, $J_c \neq \{0\}$. We can, therefore, apply 75.1(i): it tells us that, if n is the lowest degree possessed by a nonzero polynomial in J_c, J_c contains a unique monic polynomial $- m_c$, say $-$ of degree n and $J_c = (m_c)$. This polynomial m_c is called the **minimal polynomial** of c over F, and we shall use m_c as the standard notation for it (or else $m_{c/F}$ if we wish to emphasize the role of F in the considerations).

As an illustration, consider the case $F = \mathbb{Q}$, $c = \sqrt{2}$. The polynomials in $\mathbb{Q}[X]$ with $\sqrt{2}$ as a root form an ideal $J_{\sqrt{2}}$ of $\mathbb{Q}[X]$. For simplicity, let us call this ideal simply J. Clearly J contains no polynomial of degree 0 or 1, but it does contain $X^2 - 2$: so 2 is the lowest degree possessed by a nonzero polynomial in J, and $X^2 - 2$ must be the unique monic polynomial in J of this minimum possible degree (cf. 75.1(i)(b)). That tells us that $m_{\sqrt{2}/\mathbb{Q}}$ is $X^2 - 2$.

The above general discussion has introduced $m_{c/F}$ in a sophisticated way that gives immediate access to its properties. To guard against the possibility that the sophistication of approach might make $m_{c/F}$ seem a difficult or arcane notion, the following proposition 77.1 provides a user-friendly summary of its key properties. With the aid of this summary the student should rapidly begin to feel at home with the idea of minimal polynomials.

77.1 $m_{c/F}$ has the following properties.
    (1) It is a monic polynomial in $F[X]$ with c as a root [in the larger field K].
    (2) No nonzero polynomial of lower degree in $F[X]$ has c as a root.

(3) $m_{c/F}$ generates the ideal of $F[X]$ consisting of all the polynomials in $F[X]$ with c as a root, and consequently, for $g \in F[X]$,

$$g(c) = 0 \Leftrightarrow m_{c/F} \text{ divides } g \text{ in } F[X].$$

A further important property of $m_{c/F}$ is:

77.2 In all cases $m_{c/F}$ is irreducible over F.

Proof (in which m_c will be used for $m_{c/F}$). By its definition m_c is nonzero and clearly (since it has c as a root) m_c is non-constant. So m_c is a nonzero non-unit in $F[X]$.

Suppose the result to be false. Then $m_c = gh$ for some $g, h \in F[X]$ with $1 \leqslant \deg g, \deg h < \deg m_c$. Evaluating both sides of the equation $m_c = gh$ at c gives

$$m_c(c) = g(c)h(c), \quad \text{i.e. } 0 = g(c)h(c);$$

and so either $g(c) = 0$ or $h(c) = 0$—a contradiction (cf. 77.1(2)) since g, h are polynomials of degree less than $\deg m_c$.

It follows that m_c must be irreducible over F.

As a sort of converse to 77.3, the next result provides a useful tool for identifying $m_{c/F}$.

77.3 Suppose that q is a monic irreducible polynomial in $F[X]$ with c as a root [in the larger field K]. Then $m_{c/F} = q$.

Proof (in which m_c will be used for $m_{c/F}$). Since $q \in F[X]$ and $q(c) = 0$, m_c divides q in $F[X]$ (cf. 77.1(3)): i.e. $q = bm_c$ for some $b \in F[X]$. Since q is irreducible and m_c is certainly not constant, b must be constant. Further, since m_c and q are both monic, comparison of the leading coefficients in the equation $q = bm_c$ yields $b = 1$. So $q = m_c$, as asserted.

Remark. Implicit in 77.3 is a four-point checklist:

(1) q is monic;
(2) $q \in F[X]$;
(3) q has c as a root;
(4) q is irreducible over F;

and 77.3 says that if q satisfies all four points in this list, then q is $m_{c/F}$. When it comes to the application of this in problems, (1) and (2) are often visibly true, and, that being the case, to make the claim that $q = m_{c/F}$, it suffices to verify that (3) and (4) hold.

Illustrations. (1) $X^2 - 2$ is irreducible in $\mathbb{Q}[X]$ (cf. 74.1(2)), and it has $\sqrt{2}$ as a root. So (by a much shorter analysis than that which led to the same conclusion immediately after minimal polynomials were introduced) $m_{\sqrt{2}/\mathbb{Q}} = X^2 - 2$.

(2) Since 2 has no rational cube root, $X^3 - 2$ is irreducible in $\mathbb{Q}[X]$ (cf. 74.1(2)) while it has $2^{1/3}$ as a root in $\mathbb{C}$. So $m_{2^{1/3}/\mathbb{Q}} = X^3 - 2$.

(3) The polynomial $X^2 + X + 1$ is irreducible over $\mathbb{Q}$ (having no root in $\mathbb{R}\ (\supset \mathbb{Q})$) and has $\omega = \exp(2\pi i/3)$ as a root in $\mathbb{C}$. So $m_{\omega/\mathbb{Q}} = X^2 + X + 1$. [The student should recognise ω as one of the complex roots of unity.]

In an illustration in § 76 it was noted that $\mathbb{Q}(\sqrt{2})$ is just the same as $\mathbb{Q}[\sqrt{2}]$, and a remark was made that this is no fluke but is, rather, one instance of a more general story. That more general story is revealed in 77.4, after which we go on to a very interesting theorem, 77.5, which tells us a great deal about the nature and structure of any field of the form $F(c)$ with c algebraic over F.

77.4 When c is algebraic over F,
(1) $F[c]$ is a field, and
(2) $F(c)$ and $F[c]$ coincide.

Proof. (1) Since $F[c]$ is certainly a commutative ring with a one, it will suffice to prove that each nonzero element of $F[c]$ has an inverse [i.e. a multiplicative inverse] in $F[c]$. For this purpose, let x be an arbitrary nonzero element of $F[c]$. Then (by the meaning of $F[c]$) $x = f(c)$ for some $f \in F[X]$; and, since $f(c)\ (= x)$ is nonzero, 77.1(3) tells us that the minimal polynomial m_c of c over F does not divide f. So, since (by 77.2) m_c is irreducible, it follows by 74.6 that there are polynomials $s,\ t \in F[X]$ such that $sm_c + tf = 1$. Evaluating both sides of this equation at c (and using the fact that $m_c(c) = 0$), we get $t(c)f(c) = 1$, i.e. $t(c)x = 1$; and therefore x has an inverse—namely $t(c)$—in $F[c]$. Part (1) of the result follows.

(2) By part (1), $F[c]$ is a field, and it certainly contains F and c. So, as $F(c)$ is the smallest such field, $F(c) \subseteq F[c]$. But (in view of the meaning of the notations) the reverse containment is a triviality. Therefore $F(c)$ and $F[c]$ do indeed coincide.

77.5 Suppose that m_c, the minimal polynomial of c over F, has degree n. Then each element of $F(c)$ can be uniquely expressed in the form

$$a_0 + a_1 c + \ldots + a_{n-1}c^{n-1} \text{ with } a_0, a_1, \ldots, a_{n-1} \in F.$$

Proof. Let x be an arbitrary element of $F(c)$. In view of 77.4(2), $x = f(c)$ for some $f \in F[X]$. By the division algorithm for $F[X]$, $f = qm_c + r$ for some $q, r \in F[X]$ with $\deg r < n$. Evaluating both sides of the equation $f = qm_c + r$ at c gives

$$f(c) = q(c)m_c(c) + r(c) = q(c).0 + r(c) = r(c),$$

i.e. $x = r(c)$.

So, since $\deg r < n$,

$$x = a_0 + a_1 c + \ldots + a_{n-1}c^{n-1} \text{ for some } a_0, a_1, \ldots, a_{n-1} \in F.$$

If there were a second, different expression of this form for x, then subtraction would give

$$0[= x - x] = b_0 + b_1 c + \ldots + b_{n-1}c^{n-1}$$

for some $b_0, b_1, \ldots, b_{n-1} \in F$ which are not all zero, and consequently c would be a root of a nonzero polynomial in $F[X]$ with degree $< n (= \deg m_c)$. That would be a contradiction (cf. 77.1(2)): therefore there can only be one expression of the kind in question for x.

The stated result is now fully proved.

Many simple illustrations of 77.5 can be given. E.g.
(1) elements of $\mathbb{Q}(\sqrt{2})$ are uniquely expressible in the form $a + b\sqrt{2}$ with $a, b \in \mathbb{Q}$;
(2) elements of $\mathbb{Q}(2^{1/3})$ are uniquely expressible in the form

$$a + b2^{1/3} + c2^{2/3} \text{ with } a, b, c \in \mathbb{Q}.$$

In § 78 it will be seen that 77.5 provides almost immediate proof of one of the most important facts about finite fields.

Post-script. Most students who have advanced to this stage in their study of abstract algebra will also have learned the rudiments of vector space theory. Those who have done so should note that, just as $\mathbb{C}$ is obviously a vector space over $\mathbb{R}$ (with basis $(1, i)$), so any field that contains F as a subfield is a vector space over F. When that point has been noted, one sees that 77.5 could be re-stated to say: when c is algebraic over F and $\deg(m_{c/F}) = n$, $F(c)$ is an n-dimensional vector space over F and $(1, c, c^2, \ldots, c^{n-1})$ is a basis of this vector space.

Further remarks on the application of vector space theory to field theory will be found in § 80.

78. Some elementary properties of finite fields

Finite fields are of great intrinsic interest. Moreover, as was stated in the introduction to this chapter and as will be illustrated in § 81, there are many applications of finite fields to other parts of mathematics (to combinatorics, in particular). We are already familiar with one family of finite fields — the fields $\mathbb{Z}_p$ with p prime. The main result of § 79 will enable us to construct and explore less obvious examples.

The first theorem in this section (78.1) presents a striking fact about all finite fields. In proving it, we shall draw on the result of the worked example at the end of Chapter 6—the fact that if d ($\in \mathbb{N}$) divides the order of a finite cyclic group G, then G contains a subgroup of order d.

78.1 In any finite field F, the multiplicative group (i.e. the group $(F - \{0\}, .)$) is cyclic.

Proof. Let F be an arbitrary finite field, let F^* stand for $F - \{0\}$, let $n = |F - \{0\}|$, and let G be a multiplicative cyclic group of order n, generated, say, by t. [Such a group G certainly exists—cf. 38.5 and the explanation leading to it.]

For each $r \in N$ with $1 \leqslant r \leqslant n$, let

a_r = number of elements of period r in the multiplicative group F^*,
b_r = number of elements of period r in the multiplicative group G.

Clearly, since the period of an element of a finite group divides the order of the group, no element in F^* can have period exceeding n. Therefore

$$\left(\begin{array}{c}\text{number of elements} \\ \text{of period 1 in } F^*\end{array}\right) + \left(\begin{array}{c}\text{number of elements} \\ \text{of period 2 in } F^*\end{array}\right) + \cdots$$

$$+ \left(\begin{array}{c}\text{number of elements} \\ \text{of period } n \text{ in } F^*\end{array}\right)$$

will be the total number of elements in F^*: i.e. $\sum\limits_{r=1}^{n} a_r = n$. Similarly $\sum\limits_{r=1}^{n} b_r = n$; and therefore

$$\sum_{r=1}^{n} a_r = \sum_{r=1}^{n} b_r. \tag{*}$$

Suppose now that the integer r is such that $a_r > 0$, so that we can introduce an element y of period r in F^*. By 44.1, $r|n$, and so (by the result of the worked example at the end of Chapter 6) G contains a (cyclic) subgroup H of order r. Since cyclic groups of the same order are isomorphic [cf. worked example in §46], $H \cong \langle y \rangle$. Notice that [although not all elements of $\langle y \rangle$ need have period r precisely] all the r elements of $\langle y \rangle$ have periods dividing r [by 44.1], and therefore all these r elements are roots of the polynomial $X^r - 1$. By 73.5, it follows that the r elements of $\langle y \rangle$ must be *all* the roots that $X^r - 1$ has in F. Hence, in particular, *all* of the a_r elements of period precisely r in F must lie in $\langle y \rangle$. Therefore a_r can be re-described as the number of elements of period r in $\langle y \rangle$, and hence

a_r = number of elements of period r in H (since $H \cong \langle y \rangle$)

$\leqslant$ number of elements of period r in G (since $H \subseteq G$)

$= b_r$.

This proves that $a_r \leqslant b_r$ whenever $a_r > 0$ $(1 \leqslant r \leqslant n)$. But, clearly, $a_r \leqslant b_r$ is true whenever $a_r = 0$. Therefore

$$a_r \leqslant b_r, \text{ for } every \text{ integer } r \text{ in the range } [1, n].$$

From (*) it now follows that a_r must equal b_r for every integer $r \in [1, n]$; and, in particular, $a_n = b_n$. But $b_n > 0$ (since G, a cyclic group of order n, certainly must contain an element of period n). Hence $a_n > 0$: i.e. F^* [which is a group of order n] contains at least one element of period n. Therefore, F^* is, as asserted, cyclic.

Worked example. Let F be a finite field of order q. Show that F contains a cube root of unity other than 1 itself $\Leftrightarrow q \equiv 1$ (mod 3).

Solution. (1) [The "$\Rightarrow$" half] Suppose that F contains an element $x \neq 1$ such that $x^3 = 1$. From these properties of x it is clear that the period of x in the multiplicative group $(F - \{0\}, .)$ is 3. Therefore [cf. 44.1] $3 \mid |F - \{0\}|$, i.e. $3 \mid (q - 1)$, i.e. $q \equiv 1$ (mod 3).

(2) [The "$\Leftarrow$" half] Suppose that $q \equiv 1$ (mod 3), so that $q = 3k + 1$ for some $k \in \mathbb{N}$. Let t be a generator of the cyclic group $(F - \{0\}, .)$, so that t is an element with period $q - 1$ $(= 3k)$ in that group. Let $u = t^k$. Since $1 \leqslant k < 3k =$ period of t, $t^k \neq 1$, i.e. $u \neq 1$. But $u^3 = t^{3k} = 1$ [$3k$ being the period of t]. Therefore F contains a cube root of 1 other than 1 itself, namely u.

Parts (1) and (2) together establish the fact to be proved.

Students who have gone some distance into number theory will recognize that a particular case of 78.1 is important there — namely the fact that the multiplicative group of $\mathbb{Z}_p$ is cyclic (for every prime p). In number theory this fact is usually expressed by saying that, for every prime p, there exists a *primitive root* (mod p): a primitive root (mod p) means an integer whose congruence class (mod p) generates the cyclic group $(\mathbb{Z}_p - \{0\}, .)$.

Certain simple facts about finite fields are immediate consequences of points made in §69 in Chapter 10. E.g. a finite field must have characteristic p for some prime p. Furthermore, every finite field F of characteristic p must contain (as its prime subfield) a subfield P isomorphic to $\mathbb{Z}_p$; and, unless there is some special risk of confusion, we normally identify the prime subfield P with $\mathbb{Z}_p$ to give us (for an arbitrary finite field F of characteristic p) $F \supseteq \mathbb{Z}_p$. This state of affairs is assumed in the statements and proofs of the results which now follow, beginning with a very simple observation:

78.2 If F is a finite field of characteristic p (p a prime number), then $F = \mathbb{Z}_p(t)$ where t is any generator of the cyclic multiplicative group of F.

(This is obvious because the list $0, 1, t, t^2, \ldots, t^{q-1}$ (where $q = |F|$) contains all the elements of F, and therefore every element of F is expressible as a polynomial in t with coefficients in $\mathbb{Z}_p$ [to which, of course, 1 belongs].)

An extremely important corollary is the next theorem, most especially its second part:

78.3 (i) If t is algebraic over $\mathbb{Z}_p$ the order of the field $\mathbb{Z}_p(t)$ is p^n where $n = \deg(m_{t/\mathbb{Z}_p})$.

(ii) The order of every finite field is an integral power of its characteristic (and, in particular, is a prime power).

Proof. (i) Suppose that the field K is $\mathbb{Z}_p(t)$, where t is algebraic over $\mathbb{Z}_p$ and $\deg(m_{t/\mathbb{Z}_p}) = n$. Then, by 77.5, each element of K is uniquely expressible in the form

$$a_0 + a_1 t + a_2 t^2 + \ldots + a_{n-1} t^{n-1} \quad \text{with each } a_i \in \mathbb{Z}_p.$$

So $|K|$ is the number of possibilities for a coefficient sequence $(a_0, a_1, \ldots, a_{n-1})$ with each $a_i \in \mathbb{Z}_p$. But there are p possibilities for a_0 (it can be any of the p elements of $\mathbb{Z}_p$), and similarly p possibilities for a_1, p possibilities for $a_2, \ldots$, and p possibilities for a_{n-1}, so that the number of possibilities for the sequence $(a_0, a_1, \ldots, a_{n-1})$ is $p \times p \times p \times \ldots$ to n factors, i.e. p^n. Therefore $|K| = p^n$ — which establishes part (i).

(ii) Now let K be an arbitrary finite field of characteristic p. By 78.2, $K = \mathbb{Z}_p(t)$, where t is a generator of the cyclic multiplicative group of K. The element t is certainly algebraic over the prime subfield $\mathbb{Z}_p$ of K (since [cf. 44.1] it is a root of $X^{q-1} - 1$, where $q = |K|$. So by part (i) it follows that $|K| = p^n$ for some $n \in \mathbb{N}$ (namely $n = \deg(m_{t/\mathbb{Z}_p})$); and this establishes part (ii). ∎

Remarks. (*a*) The result 78.3(ii) reveals a big contrast between finite group theory and finite field theory. At 38.5 it was pointed out that, for every positive integer n, there exists at least one group of order n; but now we see, as one of the most basic facts of finite field theory, that, for an arbitrary $n \in \mathbb{N}$, there may not exist a field of order n—and indeed there won't exist such a field unless n is a prime power.

(*b*) It can be shown that, in fact, for every integer of the form p^n (where p is prime and $n \in \mathbb{N}$) there does exist a finite field of order p^n and that any two finite fields of the same order p^n are isomorphic to each other. So if we regard isomorphic systems as the same thing, we can sum the situation up in the simple statement that, for every possible p^n (p prime, $n \in \mathbb{N}$) there is *precisely one* field of order p^n. This field is often called the *Galois field* of order p^n and denoted by $GF(p^n)$—a fitting nomenclature since it was Evariste Galois

(1811–1832) who first discovered the principal facts about finite fields. Note that $GF(p)$ is $\mathbb{Z}_p$.

(c) Proof of the facts just stated in remark (b) lies beyond the scope of this textbook. However, it will be seen that the main theorem of §79 enables us to show that certain particular finite fields do exist and to reveal enough details of their structures to be able to use them in applications.

79. Construction of fields by root adjunction

Throughout this section F will denote an arbitrary field.

In §77, and especially in theorem 77.5, we gained understanding of the structure of fields of the form $F(c)$ where c is a root of some nonzero polynomial (f, say) in $F[X]$. But the discussion in §77 started with the supposition that we were given a field K containing F as a subfield and containing also the element c, so that the existence of a field of the form $F(c)$ with c a root of f *was being assumed*. So an obvious question which arises is: if we are merely given the field F and a non-constant polynomial $f \in F[X]$ with no root in F, does there necessarily exist a field K which is an extension of F (i.e. contains F as a subfield) and contains a root of f? The following theorem tells us that the answer to this question is yes in all cases.

79.1 Let F be a field and f a polynomial of degree $\geqslant 1$ in $F[X]$. Then there exists a field K which contains F as a subfield and also contains an element which is a root of f.

Foreword to proof. As always when we set ourselves the task of constructing an extension of a given system F, we shall be content to construct a new system K containing a subsystem $\hat{F}$ which is isomorphic to F and can be identified with F to give $K \supseteq F$.

Proof. Let g be an irreducible factor of f in $F[X]$. [In the case f irreducible, we would take $g = f$.] By 74.5 and the fact (cf. 75.1) that $F[X]$ is a principal ideal domain, $F[X]/(g)$ is a field. Let K denote this field, and let v be the natural epimorphism [given by $h \mapsto h + (g)$] from $F[X]$ to K.

With a view to showing that $v|_F$ (the restriction of v to F as domain) is injective, consider $\ker(v|_F)$, which is equal to $F \cap \ker v$, i.e. $F \cap (g)$. Apart from 0, all elements of (g) [$= \{hg : h \in F[X]\}$] have degrees $\geqslant \deg g$ ($\geqslant 1$) and so lie outside F. Therefore $F \cap (g)$ consists of 0 only: i.e. $\ker(v|_F) = \{0\}$, and hence $v|_F$ is a monomorphism.

It follows that $\mathrm{im}(v|_F)$, i.e. $v(F)$, is an isomorphic copy $\hat{F}$ of F. By identifying $v(x)$ with x for each $x \in F$, we identify $\hat{F}$ with F, and that gives us $K \supseteq F$. [This is the identification step mentioned in the foreword to the proof.]

We now have a field K containing F as a subfield. Let c be the element $v(X)$ of K. Then c is a root of g (and therefore of f, since g is a factor of f): for, if $g = a_0 + a_1 X + \ldots + a_n X^n$ (with each $a_i \in F$), then

$$g(c) = a_0 + a_1 v(X) + \ldots + a_n [v(X)]^n \quad \text{(since } c = v(X)\text{)}$$
$$= v(a_0) + v(a_1)v(X) + \ldots + v(a_n)[v(X)]^n \quad \text{(by the identification of } v(x)$$
$$\text{with } x \text{ for each } x \in F)$$
$$= v(a_0 + a_1 X + \ldots + a_n X^n) \quad \text{(since } v \text{ is a homomorphism)}$$
$$= v(g)$$
$$= 0 \quad \text{(since } g \in (g) = \ker v).$$

So the existence of a field K as described in the statement of the theorem has been demonstrated.

Remarks. (a) The theorem 79.1 has a number of very pleasing features. One is its universality: it tells us that an extension field of the kind in question *always* exists. Further, it provides us with the satisfaction of seeing the generality of which something very familiar and important in elementary mathematics — namely the construction of $\mathbb{C}$ from $\mathbb{R}$ — is one special case (the case where the given field F is $\mathbb{R}$ and one constructs an extension field containing a root of the polynomial $X^2 + 1$).

(b) The fact that we have constructed a field K containing F and containing the root c of F means that a field $F(c)$, generated by adjoining a root of f to F, certainly exists (as a subfield of K). In fact, as is easily shown, the subfield $F(c)$ of the field K constructed in the proof of 79.1 is the whole of K. A further supplementary point that can be proved as a post-script to the proof of 79.1 is that (in the notation of that proof) the minimal polynomial of c over F is g if g is monic (and is the monic constant multiple of g, obtained by dividing g by its leading coefficient, if g is not monic).

(c) For any given polynomial $f \in F[X]$, we can, by one or more applications of 79.1, build up a field L, generated by adjoining roots of f to F, with the property that, as mathematical parlance puts it, f *splits completely* over L: that means f factorizes into a product of linear factors in $L[X]$. Such a field L is termed a **splitting field** for f over F. To give an illustration from inside the complex number universe, take F to be $\mathbb{Q}$ and f to be $X^4 - X^2 - 2$ $[= (X^2 - 2)(X^2 + 1)]$. By adjoining the roots $\sqrt{2}$ and i of f to $\mathbb{Q}$ we produce the field $L = \mathbb{Q}(\sqrt{2}, i)$. Since L is generated by adjoining roots of f to $\mathbb{Q}$ and since f splits completely over L [because

$$f = (X - \sqrt{2})(X + \sqrt{2})(X - i)(X + i)],$$

L is a splitting field for f over $\mathbb{Q}$.

(*d*) As promised in §78, we are now in a position to show that certain finite fields exist and at the same time to supply plenty of information about the structure of these fields. Here are two simple illustrations, which will introduce the student to GF(9) and GF(4), finite fields of orders 9 and 4, respectively.

(1) In $\mathbb{Z}_3[X]$, $X^2 + 1$ is irreducible (since, by inspection, none of the 3 elements 0, 1, -1 of $\mathbb{Z}_3$ is a root of $X^2 + 1$). By 79.1 [see also remark (*b*) above] there is a field $K = \mathbb{Z}_3(i)$ obtained by adjoining to $\mathbb{Z}_3$ an element i which is a root of $X^2 + 1$ and which therefore has the property that $i^2 = -1$. Clearly the minimal polynomial of the adjoined element i over $\mathbb{Z}_3$ is $X^2 + 1$ (of degree 2), and so, by 78.3(i), the field K has order 9. So K is "GF(9)". Moreover, by 77.5, each element of K is uniquely expressible in the form $a + bi$ with $a, b \in \mathbb{Z}_3$, and therefore a complete repetition-free list of the elements of K is

$$0,\ 1,\ -1,\ i,\ 1 + i,\ -1 + i,\ -i,\ 1 - i,\ -1 - i.$$

The facts that K must (as an extension of $\mathbb{Z}_3$) have characteristic 3 and that $i^2 = -1$ enable full addition and multiplication tables for K to be easily constructed and any calculation involving elements of K to be easily performed. Possibly the simplest way of describing K is to say that its elements are like complex numbers with coefficients which are integers mod 3.

(2) In $\mathbb{Z}_2[X]$, $X^2 + X + 1$ is irreducible (since, obviously, neither of the elements 0 and 1 of $\mathbb{Z}_2$ is a root of it). By 79.1 there is a field $L = \mathbb{Z}_2(\omega)$ obtained by adjoining to $\mathbb{Z}_2$ an element ω which is a root of $X^2 + X + 1$ and therefore satisfies $\omega^2 + \omega + 1 = 0$, i.e. (since the characteristic is 2) $\omega^2 = \omega + 1$. Clearly the minimal polynomial of the adjoined element ω over $\mathbb{Z}_2$ is $X^2 + X + 1$ (of degree 2), and so, by 78.3(i), the field L has order 4. So L is "GF(4)". Moreover, by 77.5, each element of L is uniquely expressible in the form $a + b\omega$ with $a, b \in \mathbb{Z}_2$, and therefore a complete repetition-free list of the elements of L is

$$0, 1, \omega, 1 + \omega,$$

the last, as already pointed out, being the same thing as ω^2. The facts that L has characteristic 2 and that $\omega^2 = \omega + 1$ enable calculations involving elements of L to be easily performed, and in particular it is a very easy task to construct full addition and multiplication tables for L. Notice incidentally that, when the elements of L are listed as 0, 1, ω, ω^2, the cyclic nature of the multiplicative group of L is clearly apparent. [Of course, it is a consequence of $\omega^2 + \omega + 1 = 0$ that $\omega^3 = 1$ since $\omega^3 - 1 = (\omega - 1)(\omega^2 + \omega + 1)$].

This section concludes with a worked example which shows 79.1 in use to establish an impressively nontrivial fact. In the solution reference will be made to the result of one of the exercises at the end of this chapter — the fact that in a finite field of order q every element is a root of the polynomial $X^q - X$.

Worked example. Show that every irreducible polynomial of degree n in $\mathbb{Z}_p[X]$ is a factor (in $\mathbb{Z}_p[X]$) of $X^{p^n} - X$.

Solution. Let f be an arbitrary irreducible polynomial of degree n in $\mathbb{Z}_p[X]$. By 79.1 we may consider a field $K = \mathbb{Z}_p(c)$, where c is a root of f. Since f is irreducible, it is readily seen (cf. 77.3) that the minimal polynomial of c over $\mathbb{Z}_p$ is $a_n^{-1}f$, where a_n is the leading coefficient of f. Since this minimal polynomial has degree n, it follows from 78.3(i) that $|K| = p^n$. So, by the result of one of the exercises at the end of the chapter, the element c of K must be a root of the polynomial $X^{p^n} - X$ (which belongs to $\mathbb{Z}_p[X]$), and therefore $m_{c/\mathbb{Z}_p}$ must divide $X^{p^n} - X$ in $\mathbb{Z}_p[X]$. So $a_n^{-1}f$ divides $X^{p^n} - X$, and therefore f itself divides $X^{p^n} - X$ in $\mathbb{Z}_p[X]$. This proves the result.

80. Degrees of field extensions

This section of the chapter should be read only by students acquainted with the rudiments of vector space theory (at least to the extent of knowing what one means by a basis of a vector space and by the dimension of a vector space).

Suppose throughout the coming general discussion that the field K is an extension of the field F—which is an alternative way of saying that F is a subfield of K. As already briefly indicated in §77, vector space theory is relevant here because, in this situation, K is a vector space over F. (To see this, regard the elements of F as "scalars" and the elements of K as "vectors": then, quite simply, all the vector space axioms are satisfied.) Cases can be exhibited where K is an infinite-dimensional vector space of F. But in many cases that we explore, the dimension (n, say) of K as a vector space over F is finite; and when that is so, we describe K as a **finite extension** of F and call n the **degree** of K over F, for which we use the notation $[K:F]$. For example, clearly $\mathbb{C}$ is a 2-dimensional vector space over $\mathbb{R}$: so $\mathbb{C}$ is a finite extension of $\mathbb{R}$, and $[\mathbb{C}:\mathbb{R}] = 2$.

Since the field K, whatever it is, certainly contains more than its zero, the minimum possible value for $[K:F]$ is 1. Note that:

80.1 (In the notation of the discussion so far) $[K:F] = 1 \Leftrightarrow K = F$.

Proof. First suppose $[K:F] = 1$. Then any nonzero element of K by itself is a basis for K over F. In particular 1 by itself is a basis for K over F. So every element of K is expressible as $x1$ ($= x$) for some $x \in F$ and thus lies in F. Therefore F is the whole of K.

This proves that $[K:F] = 1 \Rightarrow K = F$. The converse is easily seen to be true too since F as a vector space over itself is obviously 1-dimensional (1 by itself, or indeed any other nonzero element by itself, being a basis). So the stated result is true.

As remarked in passing in §77, the very significant theorem 77.5 can be rephrased as follows:

80.2 If c is algebraic over F and $\deg(m_{c/F}) = n$, then $(1, c, c^2, \ldots, c^{n-1})$ is a basis for $F(c)$ over F and $[F(c):F] = n$.

A further very simple observation is:

80.3 If K is a finite extension of F, then every element of K is algebraic over F.

Proof. Suppose that K is a finite extension of F. Let $n = [K:F]$, and let c be an arbitrary element of F.

In an n-dimensional vector space, no sequence of more than n elements can be linearly independent. Therefore $(1, c, c^2, \ldots, c^n)$ must be linearly dependent over F. Hence there exist elements $a_0, a_1, \ldots, a_n \in F$, not all zero, such that $a_0 + a_1 c + \ldots + a_n c^n = 0$, and thus c is algebraic over F. The stated result follows.

Last in this sequence of results comes a very useful fact usually referred to as the *multiplicative property* of field extension degrees.

80.4 Suppose that F, L, K are fields such that $F \subseteq L \subseteq K$, L is a finite extension of F and K is a finite extension of L. Then K is a finite extension of F, and

$$[K:F] = [K:L][L:F].$$

Proof. Let $(\ell_1, \ell_2, \ldots, \ell_m)$, where $m = [L:F]$, be a basis for L over F, and let $(k_1, k_2, \ldots, k_n)$, where $n = [K:L]$, be a basis for K over F.

Let x be an arbitrary element of K. Then, since $k_1, k_2, \ldots, k_n$ span K as a vector space over L, $x = \sum_{i=1}^{n} a_i k_i$ for some $a_1, \ldots, a_n \in L$; and, moreover, since $\ell_1, \ell_2, \ldots, \ell_m$ span L as a vector space over F, each $a_i = \sum_{j=1}^{m} b_{ij} \ell_j$ for some $b_{i1}, \ldots, b_{im} \in F$. Hence

$$x = \sum_{i=1}^{n} \left(\sum_{j=1}^{m} b_{ij} \ell_j \right) k_i = x = \sum_{i=1}^{n} \sum_{j=1}^{m} b_{ij}(k_i \ell_j).$$

As x was arbitrary in K, this demonstrates (since all the coefficients b_{ij} belong to F) that the mn elements $k_i \ell_j$ ($1 \leq i \leq n$, $1 \leq j \leq m$) span K as a vector space over F.

Further, these mn elements are linearly independent over F. To see this, observe that, for elements $c_{ij} \in F$ $(1 \leqslant i \leqslant n, 1 \leqslant j \leqslant m)$,

$$\sum_{i,j} c_{ij} k_i \ell_j = 0 \Rightarrow \sum_{i=1}^{n} \left(\sum_{j=1}^{m} c_{ij} \ell_j \right) k_i = 0$$

$$\Rightarrow \sum_{j=1}^{m} c_{ij} \ell_j = 0 \text{ for each } i$$

(since $k_1, \ldots, k_n$ linearly independent over L)

$$\Rightarrow c_{ij} = 0 \text{ for all } i, j$$

(since $\ell_1, \ldots, \ell_m$ linearly independent over F).

It now follows that the mn elements $k_i \ell_j$ form a basis for K as a vector space over F. Therefore K is a finite extension of F, and

$$[K:F] = mn = [K:L][L:F]$$

—which completes proof of the stated result.

The result 80.4 is extremely useful. One obvious use is in determining the value of a field extension degree of the form $[F(a, b):F]$ via the equation

$$[F(a, b):F] = [F(a, b):F(a)][F(a):F].$$

As a worked example, the value of $[\mathbb{Q}(\sqrt{2}, i):\mathbb{Q}]$ will now be found.

Since $X^2 - 2$ is irreducible over $\mathbb{Q}$ and has $\sqrt{2}$ as a root, it is clear that $m_{\sqrt{2}/\mathbb{Q}} = X^2 - 2$; and therefore, by 80.2,

$$[\mathbb{Q}(\sqrt{2}):\mathbb{Q}] = \deg(X^2 - 2) = 2.$$

Next note that $X^2 + 1$ is irreducible over $\mathbb{Q}(\sqrt{2})$ (since $X^2 + 1$ has no root in $\mathbb{R}$, which contains $\mathbb{Q}(\sqrt{2})$). So $X^2 + 1$, which has i as a root, is the minimal polynomial of i over $\mathbb{Q}(\sqrt{2})$, and therefore (by 80.2 again)

$$[\mathbb{Q}(\sqrt{2}, i):\mathbb{Q}(\sqrt{2})] = \deg(X^2 + 1) = 2.$$

It now follows that

$$[\mathbb{Q}(\sqrt{2}, i):\mathbb{Q}] = [\mathbb{Q}(\sqrt{2}, i):\mathbb{Q}(\sqrt{2})][\mathbb{Q}(\sqrt{2}):\mathbb{Q}] \quad \text{(by 80.4)}$$
$$= 2 \times 2$$
$$= 4.$$

The final result concerns finite fields and may be viewed as a clarificatory post-script to 78.3.

80.5 Let K be a finite field. Then $|K| = p^n$, where p is the characteristic of K and n is the degree of K over its prime subfield.

Proof. Let p be the characteristic of K, so that the prime subfield P of K has order p, and let $n = [K:P]$. Then K is an n-dimensional vector space over P and so is in one-to-one correspondence with the set of all n-tuples of elements of P—a set whose order is $|P|^n = p^n$. Therefore $|K| = p^n$—which proves the result.

81. Epilogue

Those who have explored abstract algebra with the aid of this textbook will, it is hoped, realize that here is a subject where one meets appealing mathematical ideas, often of great elegance. But students should recognize that there is very much more to the subject than just its aesthetic appeal!

Perhaps the most important benefit obtained through studying abstract algebra is the rewarding insight that comes from understanding mathematical phenomena at the appropriate level of generality. Of course, the achieving of generality of insight is a theme which permeates all scientific enquiry: the most significant scientific discoveries include those which expose the general principles that underlie and govern hosts of diverse phenomena, and scientific understanding is gained when one is able to see a range of phenomena not as separate unconnected things but as manifestations of a single law or principle. This is true no less in mathematics than in physical science. And it will surely be clear to students that abstract algebra not only contributes to efficiency in mathematical work by repeatedly enabling several problems to be discussed and solved by a single analysis but also by giving that genuine extra depth of insight which is achieved when one knows one has penetrated to the real reason why something is true. Moreover, it is to be expected that those who have studied abstract algebra at length will thereby have been educated into an alertness to the importance of tackling mathematical and other problems at a suitable level of generality.

However, it certainly should not be supposed that abstract algebra exists as an isolated branch of mathematics unconnected with other parts of the subject. The truth is that the language, insights and results of abstract algebra are very widely employed in the rest of mathematics, not only in obviously "algebraic" territory, but in fact much more widely (e.g. to a considerable extent in some important parts of analysis). One example is that the use of simple ideas about polynomials and of relatively subtle facts about them (such as their unique factorization properties) is widespread in both pure and applied mathematics. To give a more technical example alluded to earlier in the chapter, the fact that the multiplicative group of $\mathbb{Z}_p$ (p a prime number) is cyclic is the basis of significant results in number theory. And, under the heading of the use of abstract algebra in other parts of mathematics, mention must definitely be made of Galois theory. Galois theory is an outstandingly attractive part of mathematics, in which group theory and field theory (of a slightly more advanced kind than has been reached in this book) can be applied to answer an obvious question about elementary mathematics — namely, can one produce a formula (or something equivalent to a formula) enabling the roots of polynomial equations to be calculated (from the coefficients) by use of the basic arithmetical operations (addition, subtraction, multiplication, division) supplemented only by the taking of square roots, cube roots, etc.? It is a prominent result in school algebra that an arbitrary quadratic equation $ax^2 + bx + c = 0$ can be solved in this way, by the formula

$$x = \frac{-b \pm \sqrt{b^2 - 4ac}}{2a};$$

and, although the details are more elaborate and very much less widely known, polynomial equations of degrees 3 and 4 can be similarly solved. But in Galois theory, through addressing related questions about the structure of fields and the structure of groups, one can show that there is no possibility of such a formula-based procedure for solving polynomial equations of degree $\geqslant 5$. Students who have derived some enjoyment from studying abstract algebra to the level covered in this book will undoubtedly get pleasure from exploring Galois theory and so should look out for the opportunity to learn about it through lecture-courses or books.

After the emphases of the last paragraph, it is necessary to say further that significant application of abstract algebra is by no means confined to neighbouring parts of mathematics. Several instances could be quoted of the real-life use of mathematics where abstract algebraic ideas are of central importance. This is so, for example, in coding theory. Coding theory is concerned with the translation of a message into a "coded" form (generally a string of 0s and 1s), not so much to hide the content of the message from third parties but rather to enable the message to be satisfactorily transmitted and received by modern technology (e.g. transmitted from an unmanned space probe and received on Earth). Interference may cause errors to appear in the received form of the message: every now and again what should have been a 0 is received as a 1, and vice versa. In coding theory a number of algebraic ideas have been cleverly deployed with the result that the recipient of a suitably coded message can in many cases not only detect where there are errors but also correct these errors.

Finite fields are among the algebraic tools used in coding theory; and they and their properties are widely used also in other applications of abstract algebra. Several illustrations can be found in the highly applicable subject of combinatorics — a subject which is too wide and varied to define in a single phrase but which includes the study of highly symmetrical configurations (such as might be used in the efficient design of statistical experiments or in the design of tournaments). One example of a "highly symmetrical configuration" is the following collection of 7 subsets of the set $\{1, 2, 3, 4, 5, 6, 7\}$:

$$\{1, 2, 6\}, \quad \{1, 3, 5\}, \quad \{1, 4, 7\}, \quad \{2, 3, 4\}, \quad \{2, 5, 7\}, \quad \{3, 6, 7\}, \quad \{4, 5, 6\}.$$

This collection of subsets has the following remarkable properties:

(*a*) each subset contains 3 objects, and each of the objects 1, 2, 3, 4, 5, 6, 7 belongs to precisely 3 of the subsets;

(*b*) any two of the of the subsets have precisely one object in common;

(*c*) any pair of (different) objects from 1, 2, 3, 4, 5, 6, 7 can be found in precisely one of the subsets.

In order to exhibit in detail one application of finite fields, the rest of this section will converge on a demonstration of how finite fields can be used to construct not only the above symmetrical configuration but also an infinite family of configurations with similar properties. The motivatory background for the construction lies in a branch of geometry termed *projective* geometry, and so there now follows a short digression to explain just a little of what projective geometry is about.

Two-dimensional projective geometry could be said to arise from the study of what happens when plane geometrical figures undergo the kind of transformation that occurs in photography (see Figure 1), and it is concerned with what is invariant under such transformations (projective transformations, let us call them). For example, under projective transformation points go to points, lines go to lines, collinear points go to collinear points, and concurrent lines go to concurrent lines: so points, lines, collinear points, concurrent lines

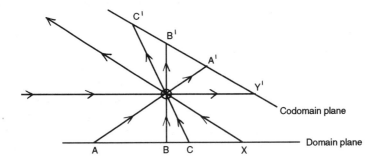

Figure 1   The domain plane is the plane being photographed; ○ is the camera lens; the codomain plane is the photographic plate on which the image is formed. Under this transformation (as the arrows indicate) the points *A*, *B*, *C* produce images *A′*, *B′*, *C′*, respectively; the image of the point *X* is a point at infinity in the codomain plane; the point *Y′* (an ordinary point of the photographic plate) is the image of a point at infinity in the domain plane.

are acceptable topics for discussion in projective geometry. On the other hand, projective transformations distort distances and ratios of distances, and so those ideas are left aside in projective geometry. Moreover, it is familiar to everyone who has seen a photograph of railway lines that a projective transformation can transform parallel lines into non-parallel lines. Therefore, parallelism of lines is another notion that has no place in projective geometry. A newcomer to this branch of geometry may find it helpful to imagine a "projective plane" (i.e. a plane as studied in projective geometry) to be like a familiar geometrical plane with an added "line at infinity", at each point of which a family of parallel lines in the "familiar plane" meet; but it must be understood that the line at infinity has the same status as every other line in the plane (cf. the fact apparent in Figure 1, that points at infinity can be mapped under a projective transformation to ordinary points and vice versa).

Such an ultra-brief description of projective geometry cannot hope to do justice to the richness to be found in the subject when it is properly explored. But out of what has been said the reader should latch on to one cardinal thing — the fact that in 2-dimensional projective geometry there is symmetry in what we term the basic incidence properties of points and lines, namely:

(I1) any two distinct points are incident with (i.e. lie on) precisely one line (the line joining them); and

(I2) any two distinct lines are incident with (i.e. contain) precisely one point (their point of intersection).

This symmetry does not exist in the familiar world of 2-dimensional euclidean geometry, where (I1) holds but where (I2) does not because parallel lines exist with no point of intersection.

A second matter vital to the theory about to be unfolded is the distinctive way in which coordinates can be used to great effect in projective geometry. The story is rather different from the use of coordinates in euclidean geometry. In 2-dimensional projective geometry one uses triples $[x, y, z]$ with x, y, z not all zero as coordinates of points, but it is the ratios $x:y:z$ in a triple $[x, y, z]$ that matter rather than the actual values of x, y, z: e.g. $[1, 3, 4]$ and $[2, 6, 8]$ define the same point because the ratios $1:3:4$ are the same as the ratios $2:6:8$. Such coordinates are called homogeneous coordinates. A more sophisticated way of describing all this is that each point of the projective plane corresponds not to a single triple in $\mathbb{R}^3 - \{(0, 0, 0)\}$, but to an equivalence class of triples in $\mathbb{R}^3 - \{(0, 0, 0)\}$ — an equivalence class under the equivalence relation $\sim$ defined on $\mathbb{R}^3 - \{(0, 0, 0)\}$ by

$$(x_1, y_1, z_1) \sim (x_2, y_2, z_2) \Leftrightarrow \exists \text{ nonzero } \lambda \in \mathbb{R}$$
$$\text{such that } x_2 = \lambda x_1, y_2 = \lambda y_1 \text{ and } z_2 = \lambda z_1.$$

When such a coordinate system is in use for points of a projective plane, each line in the plane is the set of points with homogeneous coordinates (x, y, z) satisfying a homogeneous linear equation, i.e. an equation of the form $ax + by + cz = 0$ (a, b, c being real coefficients, not all zero).

This method of coordinatizing a projective plane clearly reveals the possibility of generalization: quite simply we can consider the analogous coordinate story with an arbitrary field F in place of $\mathbb{R}$. Our interest is going to focus on the case where F is a finite field.

At this point we shift attention back towards combinatorics and introduce a definition of a *finite projective plane* — a definition that will seem very natural in the light of our observation above that in the traditional projective plane points and lines have the incidence properties (I1) and (I2).

A *finite projective plane* is defined to be a finite set of objects called points, among which certain subsets termed lines satisfy the following axioms:

(P1) For any two distinct points A and B, there is a unique line ℓ that passes through A and B (i.e. is such that $A \in \ell$ and $B \in \ell$).

(P2) For any two distinct lines ℓ and m, there is a unique point A on both ℓ and m (i.e. such that $A \in \ell$ and $A \in m$).

(P3) There exist among the points four with the property that no three of the four are collinear. (Points $A, B, C, \ldots$ are termed collinear iff there is a line to which they all belong.)

(Axiom (P3), it should be explained, is included to eliminate degenerate cases.)

This simple definition has some elementary consequences that may at first sight seem surprising — e.g. the fact that in any given finite projective plane every line must contain the same number of points. The full story surrounding this fact is given in the following proposition.

81.1 Let S be a finite projective plane. Then there is an integer $n \geqslant 2$ such that:

 (i) each line of the plane contains exactly $n + 1$ points,
 (ii) each point of the plane belongs to exactly $n + 1$ lines, and
 (iii) $|S|$ (the total number of points in the plane) is $n^2 + n + 1$, and the total number of lines in the plane is also $n^2 + n + 1$.

Proof. From (P3) it is clear that S contains more than one line. Consider two arbitrary distinct lines ℓ, m in S. By (P2) there is a point (P_0, say) common to ℓ and m.

Assume (with a view to obtaining a contradiction) that every point of S belongs to the set $\ell \cup m$. By (P3) we can introduce 4 points no 3 of which are collinear; and by our assumption all these points lie on ℓ or m. Since no 3 of them are collinear, it is clear that precisely 2 of them (say A and B) belong to ℓ, while the other two (C and D, say) belong to m, and that none of A, B, C, D can coincide with P_0. There is (Figure 2) a line 'AC' containing A and C [(P1) ensures this]; and similarly there is a line 'BD' containing B

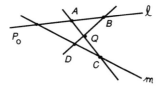

Figure 2

and D; and, by (P2), the lines AC and BD have a point Q in common. By our assumption, Q belongs to either ℓ or m. But this yields a contradiction: for if $Q \in \ell$, then ℓ and AC are two different lines containing Q and A (an impossibility by (P1)), and if $Q \in m$, then m and AC are two different lines containing Q and C. From this contradiction it follows that $\ell \cup m$ is not the whole of S, and so we may legitimately introduce into the argument a point R belonging to neither ℓ nor m.

Let the points of ℓ, other than P_0, be $P_1, P_2, \ldots, P_n$ (so that ℓ consists of $n + 1$ points). Then (see Figure 3) the lines $RP_0, RP_1, RP_2, \ldots, RP_n$ (where RP_i denotes the unique line on which R and P_i lie) are $n + 1$ different lines containing R. (They are different since, by (P1), ℓ is the only line containing two different members of the set $\{P_0, P_1, \ldots, P_n\}$.) Moreover, it can be seen (because in particular each line through R must, by (P2), have a point of intersection with ℓ) that these are all the lines through R. Hence

number of lines containing R = number of points belonging to $\ell = n + 1$.

Similarly,

number of lines containing R = number of points belonging to $m = n + 1$.

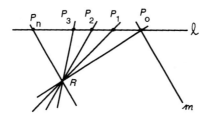

Figure 3

Therefore , the lines ℓ, m contain the same number of points $(n + 1)$.

Since ℓ, m were arbitrary lines, it follows that all lines of S contain the same number of points, namely $n + 1$.

In view of the symmetry of the axioms (P1), (P2), (P3), it can equally be shown that through any two points of S there pass the same number of lines and this number must also be $n + 1$, since there are $n + 1$ lines through the point R.

Next let $k_1, k_2, \ldots, k_{n+1}$ denote the $n + 1$ different lines through R. Then each of $k_1 - \{R\}, k_2 - \{R\}, \ldots, k_{n+1} - \{R\}$ is a set of n points. Therefore these sets together contain $(n + 1)n$ points. But these are all the points of $S - \{R\}$, since every point of $S - \{R\}$ belongs (by (P1)) to a line through R. Hence $|S - \{R\}| = (n + 1)n$, and therefore

$$|S| = n(n + 1) + 1 = n^2 + n + 1.$$

In view of the symmetry between the roles of points and lines, it can equally be proved that the total number of lines in S is also $n^2 + n + 1$.

Finally, from (P3) it follows now that $n^2 + n + 1 \geqslant 4$; and hence $n \geqslant 2$. This completes the proof of the proposition.

The integer n in 81.1 is called the **order** of the finite projective plane S. So, in particular, the order of a finite projective plane is one less than the number of points on each line of the plane.

We now approach the climax of our consideration of finite projective planes by indicating how to prove that:

81.2 If there exists a finite field of order q, then there exists a finite projective plane of order q.

Outline of proof. Suppose there is a finite field F of order q. (As foreshadowed in an earlier remark, we proceed via generalization of the homogeneous coordinates idea.) On the set $F^3 - \{(0,0,0)\}$, let a relation $\sim$ be defined by

$$(x_1, y_1, z_1) \sim (x_2, y_2, z_2) \Leftrightarrow \exists \lambda \in F - \{0\}$$
$$\text{such that } x_2 = \lambda x_1, y_2 = \lambda y_1 \text{ and } z_2 = \lambda z_1.$$

It is easily veried that $\sim$ is an equivalence relation.

Let the equivalence class of an arbitrary triple $(x, y, z) \in F^3 - \{(0,0,0)\}$ be denoted by $[x, y, z]$, and let S be the set of all such equivalence classes. We call the elements of S 'points'. For each possible homogeneous linear equation $ax + by + cz = 0$ (where $a, b, c \in F$ and are not all zero) the set of points $[x, y, z]$ satisfying the equation is termed a 'line'. (Needless to say, equivalent equations like $x + 2y + 3z = 0$ and $2x + 4y + 6z = 0$ are deemed to define the same line.) We show that points and lines in S, as just introduced, satisfy the axioms (P1), (P2) and (P3).

First let $A[a_1, a_2, a_3]$ and $B[b_1, b_2, b_3]$ be any two distinct points in S. By elementary algebraic calculation, one can show that there is precisely one line of S to which both A and B belong — namely the line $px + qy + rz = 0$ where

$$p = a_2 b_3 - a_3 b_2, \quad q = a_3 b_1 - a_1 b_3, \quad r = a_1 b_2 - a_2 b_1.$$

Moreover, (by a formally identical analysis) one can show that two arbitrary distinct lines $a_1 x + a_2 y + a_3 z = 0$ and $b_1 x + b_2 y + b_3 z = 0$ of S have precisely one common point, namely the point $[p, q, r]$ where

$$p = a_2 b_3 - a_3 b_2, \quad q = a_3 b_1 - a_1 b_3, \quad r = a_1 b_2 - a_2 b_1.$$

So (P1) and (P2) hold for S. And (P3) holds for S because, as detailed checking confirms, $[1, 0, 0]$, $[0, 1, 0]$, $[0, 0, 1]$, $[1, 1, 1]$ provide an example of 4 points in S no 3 of which are collinear. Therefore S is a projective plane.

One of the lines of S is the line $z = 0$, consisting of all points with zero third coordinate. It is easily seen that this line consists of $q + 1$ different points, namely the points $[1, \alpha, 0]$ where α can be any of the q elements of F and the point $[0, 1, 0]$. So (cf. 81.1) every line of S consists of $q + 1$ points, and thus S is a projective plane of order q. The stated proposition is, therefore, true.

Since in fact there is a finite field of order p^n for every prime p and positive integer n, it follows that there exists a projective plane of order p^n for every integer of the form p^n (p prime, $n \in \mathbb{N}$). That this is an interesting result (and that its proof is a good illustration of how finite fields are an effective tool in combinatorics) becomes clear when the news is broken that no-one has ever succeeded in demonstrating the existence of a finite projective plane which is not of prime-power order. It is known that finite projective planes of certain orders cannot exist — and in particular that finite projective planes of orders 6 and 10 do not exist. But, for example, it is not known whether a finite projective plane of order 12 exists.

It will be recognized that the illustration given earlier of a highly symmetrical configuration, namely the set of 7 subsets

$$\{1, 2, 6\}, \quad \{1, 3, 5\}, \quad \{1, 4, 7\}, \quad \{2, 3, 4\}, \quad \{2, 5, 7\}, \quad \{3, 6, 7\}, \quad \{4, 5, 6\}$$

of $\{1, 2, 3, 4, 5, 6, 7\}$, is the set of lines of a projective plane of order 2. This plane is constructible by following through the proof of 81.2 with $F = \mathbb{Z}_2$ and suitably labelling the points as 1, 2, 3, 4, 5, 6, 7. This projective plane, which is the smallest possible projective plane, is often called the *Fano plane* and (in contrast to finite projective planes of higher orders) can be illustrated by a simple diagram — namely that given in Figure 4: here the points of the

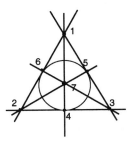

Figure 4

projective plane are represented by the points marked 1, 2, 3, 4, 5, 6, 7, and the 7 lines of the projective plane are represented by the 6 straight lines and also the circle in the diagram.

EXERCISES ON CHAPTER ELEVEN

1. Let D be an integral domain and R a commutative ring containing D. Show that if the element Y of R is transcendental over D, then Y^2 is also transcendental over D.

2. Let D be an integral domain. By applying the first isomorphism theorem for rings to a suitably selected evaluation homomorphism, show that $D[X]/(X) \cong D$.

3. In the ring $\mathbb{Z}[X]$ let J be the set of all polynomials with even constant term. Show that J is an ideal of $\mathbb{Z}[X]$ but not a principal ideal.

4. Let f be a polynomial in $\mathbb{R}[X]$. Show that if the complex number α is a root of f, then so also is its complex conjugate $\bar{\alpha}$.

5. Taking it as read that if X and Y are both transcendental over $\mathbb{Q}$, then $\mathbb{Q}(X) \cong \mathbb{Q}(Y)$, show that it is possible for a field to be isomorphic to a proper subfield of itself.

6. (Cf. note after the proof of the division algorithm result 73.2) Let F be a field, and let f, d be polynomials in $F[X]$ with $\deg d = n \geqslant 0$. Show that there is just one possibility for polynomials $q, r \in F[X]$ such that $f = qd + r$ and $\deg r < n$.

7. By applying 73.4 to the polynomial $X^5 - 1$ ($\in \mathbb{C}[X]$), show that, in $\mathbb{C}[X]$,

$$X^4 + X^3 + X^2 + X + 1 = [X^2 - (2\cos(2\pi/5))X + 1][X^2 - (2\cos(4\pi/5))X + 1].$$

On the way to this result, you will have cancelled $X - 1$ from both sides of an equation. How would you reply to someone who complained, "The result may therefore not be valid for $X = 1$"?

By comparing the coefficients of X^2 in the two sides of the above result, show that $\cos(\pi/5)\cos(2\pi/5) = \frac{1}{4}$.

8. Show that in $\mathbb{Z}_3[X]$ the polynomial $X^3 - X + 1$ is irreducible.

9. Let D be an arbitrary integral domain.

(i) Show that X is irreducible in the ring $D[X]$.

(ii) Using the result of exercise 2, deduce that if D is not a field, then $D[X]$ is not a principal ideal domain.

10. Let D be an integral domain. A nonzero non-unit element $p \in D$ is called a *prime* element of D iff it satisfies the condition

$$p|ab \Rightarrow p|a \text{ or } p|b \quad (a, b \in D).$$

(i) Show that every prime element of D is irreducible in D.

(ii) Deduce, using 74.7, that in a principal ideal domain "prime element" and "irreducible element" are synonymous. (See exercise 21 below for an example, in the context of an integral domain which is not a principal ideal domain, of an irreducible element which is not a prime element.)

(iii) Show, for $q \in D - \{0\}$, that the ideal (q) of D is a prime ideal iff q is a prime element. (See exercise 26 on Chapter 10 for the definition of "prime ideal".)

(iv) Deduce that if D is a principal ideal domain, the prime ideals in D are its maximal ideals and zero.

11. (This exercise indicates how the theory of highest common factors and the euclidean algorithm result can be extended to an arbitrary principal ideal domain.) Let D be a principal ideal domain, and let $a, b \in D$.

(i) A highest common factor of a, b means an element $d \in D$ such that (1) $d|a$ and $d|b$ in D and (2) every element c that divides both a and b in D is a factor of d. By considering the ideal $(a) + (b)$ of D, prove that at least one such highest common factor of a, b exists.

(ii) Several such highest common factors of a, b may exist. Show that (1) any two highest common factors of a, b are unit multiples of each other, and (2) each highest common factor of a, b is expressible as $xa + yb$ for some $x, y \in D$.

12. Let R be the ring of Gaussian integers (as introduced in exercise 5 on Chapter 10). Prove a division algorithm result for R — namely that if $a, d \in R$ and $d \neq 0$ then there exist $q, r \in R$ such that $a = qd + r$ and $|r| < |d|$. Deduce that R is a principal ideal domain.

13. Let D be an arbitrary principal ideal domain, and let x be an arbitrary nonzero non-unit in D. Prove that x is expressible as a finite product of irreducible elements of D. (You will find the result of exercise 19 on Chapter 10 helpful.)

14. Let K be a finite field of order q. Show that the number of irreducible monic polynomials of degree 2 in $K[X]$ is $\frac{1}{2}q(q - 1)$.

15. Describe in simple terms a typical element of the ring $\mathbb{Z}[X, 1/X]$.

16. Let $F = \mathbb{Q}(2^{1/2}, 2^{1/3})$, and let $K = \mathbb{Q}(2^{1/6})$. Is it true that $K = F$?

17. Let ϕ be the mapping from $\mathbb{Q}(\sqrt{2})$ to $\mathbb{Q}$ given by $\phi(a + b\sqrt{2}) = a^2 - 2b^2$ $(a, b \in \mathbb{Q})$. Show that, $\forall t \in \mathbb{Q}(\sqrt{2})$, $\phi(t^2) = [\phi(t)]^2$. Hence prove that $(2 + \sqrt{2})^{1/2}$ does not belong to $\mathbb{Q}(\sqrt{2})$, and deduce that $\mathbb{Q}(\sqrt{2})$ is strictly contained in $\mathbb{Q}((2 + \sqrt{2})^{1/2})$.

10. Show that $\mathbb{Q}(i, \sqrt{2}) = \mathbb{Q}(i + \sqrt{2})$.

19. Suppose that L is a subfield of $\mathbb{C}$ with the property that in $L[X]$ the polynomial $X^3 - 2$ factorizes into a product of three linear factors. Show that $L \supseteq \mathbb{Q}(2^{1/3}, \omega)$, where $\omega = \exp(2\pi i/3)$.

20. Let $c = 2^{1/4} + 2^{1/2}$ and $d = 2^{1/4}$. By considering $(c - \sqrt{2})^2$, show that

$$c^2 + 2 = (2c + 1)\sqrt{2}.$$

Hence show that $d \; [= c - \sqrt{2}]$ belongs to $\mathbb{Q}(c)$, and deduce that $\mathbb{Q}(c) = \mathbb{Q}(d)$.

21. Let $\theta = i\sqrt{5}$, and let $D = \mathbb{Z}[\theta] = \{a + b\theta : a, b \in \mathbb{Z}\}$. For $\alpha = a + b\theta \in D$ (where $a, b \in \mathbb{Z}$), we call the integer $|\alpha|^2 (= a^2 + 5b^2)$ the *norm* of α, and we write $N(\alpha)$ for the norm of α. Note that by elementary properties of complex numbers, $N(\alpha\beta) = N(\alpha)N(\beta)$ for all $\alpha, \beta \in D$.

Prove that: (i) the units in D are 1 and -1, these being the only elements of D with norm 1; (ii) there is no element of D with norm equal to 2 or 3; (iii) $2, 3, 1 + i\sqrt{5}$, and $1 - i\sqrt{5}$ are irreducible elements of D; (iv) none of these elements is a prime element of D (cf. exercise 10 above).

Let $J = \{a + b\theta : a, b \in \mathbb{Z}, a \equiv b \pmod 2\}$. Prove that J is an ideal of D, but not a principal ideal.

If K is the ideal (2) of D, is D/K an integral domain?

22. Suppose that F, K are fields with $F \subseteq K$ and that the elements a, b of K are such that $K = F(a, b)$.

(i) Suppose that the elements s, t of F are such that $s \neq t$ but $F(a + sb) = F(a + tb) (= L$, say). Show that $a, b \in L$, and deduce that L is the whole of K.

(ii) Deduce that if F is an infinite field but there exist only finitely many different fields X satisfying $F \subseteq X \subseteq K$, then K is expressible as $F(c)$ for a single generating element c.

23. What are the minimal polynomials over $\mathbb{Q}$ of (a) $3^{1/4}$, (b) $i\sqrt{3}$, (c) $1 + \sqrt{3}$?

24. Let $c = \sqrt{2} + \sqrt{3}$. Given that the minimal polynomial of c over $\mathbb{Q}$ has degree 4, find that minimal polynomial.

25. Let F, K be fields such that $F \subseteq K$ and every element of K is algebraic over F. Suppose that R is a subring of K that contains F. Use 77.4 to prove that R is in fact a subfield of K.

26. Let F be a field, K be a larger field containing F, and let c be an element of K that is algebraic over F. Prove that the field $F(c)$ is isomorphic to $F[X]/(m)$, where m is the minimal polynomial of c over F.

27. Let K be a finite field of order q.

(i) Show that every element of K is a root of the polynomial $X^q - X$.

(ii) Deduce that (in $K[X]$) $X^q - X = \prod_{c \in K} (X - c)$.

(iii) Show also that no finite field can be algebraically closed. (A field F is algebraically closed iff every non-constant polynomial in $F[X]$ has at least one root in F. It is known, for example, that $\mathbb{C}$ is an algebraically closed field.)

28. (i) Let K be a finite field of odd order q. Show that K contains a square root of -1 (i.e. an element i satisfying $i^2 = -1$) iff $q = 1 \pmod 4$.

(ii) What number theory result follows from part (i)?

(iii) Deduce from part (i) that, for every odd prime p, $GF(p^2)$ does contain a square root of -1.

29. Let K be a finite field of order q. Building on the worked example after 78.1, show that the polynomial $X^2 + X + 1$ is irreducible in $K[X]$ iff $q \equiv -1 \pmod 3$.

30. Let K be a finite field of characteristic p and order p^n (p prime, $n \in \mathbb{N}$). By exercise 27 on Chapter 10, the mapping $\sigma : K \to K$ given by $\sigma(x) = x^p$ ($x \in K$) is an automorphism of K.

(i) Let σ^r denote $\sigma_{\circ}, \sigma_{\circ}, \sigma_{\circ} \ldots$ to r factors. Show that the lowest positive value of r such that $\sigma^r = i_K$ is $r = n$. (In other words, the period of σ in the group of automorphisms of K is n.)

(ii) By exercise 7 on Chapter 10, the set of elements of K fixed (i.e. mapped to themselves) by σ is a subfield of K. Show that this subfield is the prime subfield of K.

31. (i) Show that $X^3 + X + 1$ is irreducible in $\mathbb{Z}_2[X]$.

(ii) Deduce that there exists a field K of order 8 whose elements can be represented as

$$0, 1, t, t^2, 1 + t, 1 + t^2, t + t^2, 1 + t + t^2$$

where t satisifes $t^3 = t + 1$.

(iii) The multiplicative group formed by the nonzero elements of K is a cyclic group of order 7, and obviously t is a generator of that cyclic group. Find out what powers of t the last four elements in the above list of elements of K are equal to.

32. Let f be the polynomial $X^4 - 3X^2 + 4$ in $\mathbb{Q}[X]$. Observing that $f = (X^2 + 2)^2 - 7X^2$, prove that $\mathbb{Q}(i, \sqrt{7})$ is a splitting field for f over $\mathbb{Q}$.

Note: The remaining exercises depend on the contents of §80 and should therefore be ignored by students who have omitted §80 because they don't know about vector spaces.

33. Suppose that K, F are fields with $K \supset F$ and that there are subfields L, M of K that contain F and are such that $[L:F] = 2$ and $[M:F] = 3$. Assuming that K is a finite extension of F, show that $[K:F] \geqslant 6$.

34. Suppose that K, F are fields with $K \supset F$ and $[K:F] = 2$. Let $c \in K - F$. Show that $K = F(c)$.

35. Show that any subfield of $GF(p^n)$ (p prime, $n \in \mathbb{N}$) has order p^m for some positive integer m dividing n.

36. Show that $[\mathbb{Q}(2^{1/3}, \omega):\mathbb{Q}] = 6$, where ω denotes $\exp(2\pi i/3)$.

37. Let $c = 2^{1/4} + 2^{1/2}$, as in exercise 20. Noting the result of exercise 20, show that $[\mathbb{Q}(c):\mathbb{Q}] = 4$, and find the minimal polynomial of c over $\mathbb{Q}$.

38. (i) Suppose that c_1, c_2 are both algebraic over the field F. Prove that $F(c_1, c_2)$ is a finite extension of F, and deduce that $c_1 - c_2$ and $c_1 c_2$ are both algebraic over F.

(ii) Let the field K be an extension of the field F. Prove that the set of all elements of K that are algebraic over F is a subfield of K.

39. (i) Prove that $\sqrt{3} \notin \mathbb{Q}(\sqrt{2})$.

(ii) Find a basis for $\mathbb{Q}(\sqrt{3}, \sqrt{2})$ over $\mathbb{Q}$.

BIBLIOGRAPHY

[1] P.R. Halmos, *Naive Set Theory* (Van Nostrand, 1960, reprinted by Springer-Verlag, 1974).

[2] J. Hunter, D. Monk *et al.*, *Algebra and Number System* (Blackie-Chambers, 1971).

[3] B. Baumslag and B. Chandler, *Theory and Problems of Group Theory* (McGraw-Hill, 1968).

[4] I.D. Macdonald, *The Theory of Groups* (Oxford, 1968).

[5] W.R. Scott, *Group Theory* (Prentice-Hall, 1964).

[6] J.S. Rose, *A Course on Group Theory* (Cambridge University Press, 1978).

[7] G. Birkhoff and S. MacLane, *A Survey of Modern Algebra* (Macmillan, 1953).

[8] G.H. Hardy and E.N. Wright, *An Introduction to the Theory of Numbers*, 4th Edition (Oxford, 1960).

[9] I. Anderson, *Combinatorial Designs* (Ellis Horwood, 1990).

[10] J.H. van Lint and R.M. Wilson, *A Course in Combinatorics* (Cambridge University Press, 1992).

APPENDIX TO EXERCISES

Some answers, some hints, some fragments of solutions

Chapter 1 (p. 20)

1. Negations are: (i) $\exists x \in S$ s.t. $x > 3$, (ii) $\forall x \in S$, $x > 3$, (i) is false in case (a), true in case (b); (ii) is true in both cases. Final answer: (i, a), (ii, a and b).

2. (a) (ii); (b) (iii); (c) (i) and (iii) are true, (ii) and (iv) false.

3. Prove by contradiction. Suppose x/y, y/z, z/x are all equal. Then, since their product is 1, their common value must be 1; and hence $x = y = z$ — contradiction.

4. 3, 5, 2, 10, 1.

5. For $x \in S$, $x \in A - B \Leftrightarrow x \in A$ and $x \notin B \Leftrightarrow x \in A \cap \mathscr{C}B$.
$A - (A - B) = A \cap \mathscr{C}(A \cap \mathscr{C}B) = A \cap (\mathscr{C}A \cup B)$ [7.1] $= (A \cap \mathscr{C}A) \cup (A \cap B)$ [7.4]
$\qquad = \varnothing \cup (A \cap B) = A \cap B$, etc.

6. Negation is: $\exists x \in A$ s.t. $x \in B$. This is clearly equivalent to $A \cap B \neq \varnothing$. Hence $A \cap B \neq \varnothing$ is equivalent to $\neg(A \subseteq \mathscr{C}B)$.

Last part: (i) Suppose $(A \cap X) \cup (B \cap X') = \varnothing$ for some $X \subseteq S$. Then $A \cap X = \varnothing$ and $B \cap X' = \varnothing$. Hence $A \subseteq X'$ and $X' \subseteq B$. So $A \subseteq B'$, i.e. $A \cap B = \varnothing$. (ii) Conversely, if $A \cap B = \varnothing$, then $X = A'$ satisfies $(A \cap X) \cup (B \cap X') = \varnothing$.

7. (i) Suppose $X \in$ left-side. Then $X' \subseteq A$ or $X \subseteq B$. Hence, in any event, $X \subseteq A \cup B$, i.e. $X \in$ right-side. (ii) Prove right-side $\subseteq$ left-side by the approach adopted in part (i): use 7.2(i) to prove left-side $\subseteq$ right-side. Consider $C = [0, 2]$, $D = [2, 4]$: note that $[1, 3] \in \mathscr{P}(C \cup D)$ but....

8. (a) $|A \cup B| = |A| + |B|$, (b) $|A \cup B| = |A| + |B| - |A \cap B| < |A| + |B|$.

9. (i) $A, \mathscr{C}_s A, \varnothing, S$. (ii) One method is to show, for $x \in S$, that $x \in$ left-side $\Leftrightarrow x$ belongs to an odd number of A, B, C.

10. Clearly 0 belongs to the intersection. Prove by contradiction that no nonzero number belongs to it. (If $x (\neq 0)$ belongs, then $\forall n \in \mathbb{N}, 0 < x \leqslant 1/n$, and so, $\forall n \in \mathbb{N}, n \leqslant 1/x$.)

11. $P = X \cup Y$, where $X = A \cap B \cap C$, $Y = A \cap B \cap D$. Clearly, $X \subseteq B \cap C \subseteq Q$, and $Y \subseteq B \cap D \subseteq Q$: hence $P \subseteq Q$. If $P = Q$, then $B \cap C \subseteq Q = P \subseteq A$, etc.; and if $B \cap C \subseteq A$ and $A \cap D \subseteq B$, then $X = B \cap C$ and $Y = A \cap D$, etc.

12. Suppose (1) sastisfied for some set X. Since $D = (...) \cap B$, $D \subseteq B$; and, since $D = (A \cap B) \cup (X \cap B)$ [7.4]. $A \cap B \subseteq D$. Moreover, $A \cup D = A \cup (A \cap B) \cup (X \cap B)$ $= A \cup (X \cap B) = C$. Conversely suppose (2) satisfied, and let $X = D - A$. Then $X \cap B = B \cap D \cap A' = D - A$; and from this it is easy to show (1) satisfied.

13. Since $A \neq \varnothing$, introduce $a \in A$. Then, if $A \times B = B \times A$, $x \in B \Rightarrow (a, x) \in A \times B \Rightarrow (a, x) \in B \times A \Rightarrow x \in A$, and hence $B \subseteq A$; etc.

14. Prove $(x, y) \in$ left-side $\Leftrightarrow x \in A$ and $y \in B$ and $x \in C$ and $y \in D \Leftrightarrow (x, y) \in$ right-side.

15. Show $(x, y) \in$ left-side $\Rightarrow (x, y) \in$ right-side, and conversely. Using first part, we

230

have: $(A \cup C) \times (B \cup D) = ((A \cup C) \times B) \cup (A \cup C) \times D) = (A \times B) \cup (C \times B) \cup (A \times D) \cup (C \times D)$, whence second result. In the case $A = B = \{1\}$ and $C = D = \{2\}$, left-side = $\{(1, 1), (2, 2)\}$, while right-side = $\{(1, 1), (2, 2), (1, 2), (2, 1)\}$.

16. (i) 2, (ii) 5, (iii) 15.

Chapter 2 (p. 34)

1. Proof for $x \notin \mathbb{Z}$: Write $x = n + t$, where $n \in \mathbb{Z}$, $t \in (0, 1)$. Then $-x = (-n - 1) + (1 - t)$, and $1 - t \in (0, 1)$. So $[-x] = -n - 1$, while $[x] = n$.

2. $d|c$ and $d|a \Rightarrow d|(a + b)$ and $d|a$ [11.3] $\Rightarrow d|((a + b) - a)$ and $d|a \Rightarrow d|b$ and $d|a \Rightarrow d = 1$: etc.

3. If d is a common factor of $9n + 8$ and $6n + 5$, then $d|(2(9n + 8) - 3(6n + 5))$, i.e. $d|1$.

4. By 12.3, $d = au + bv$ for some $u, v \in \mathbb{Z}$. So if $d|c$, we have $ax + by = c$, where $x = (c/d)u$, $y = (c/d)v$. Converse follows from 11.4.

5. Imitate proof of 12.5.

6. $ud_1 + vd_2 = 1$ for some $u, v \in \mathbb{Z}$. And $a = kd_1 = ld_2$ for some $k, l \in \mathbb{Z}$. Hence $a = (ud_1 + vd_2)a = ud_1 ld_2 + vd_2 kd_1 = (ul + vk)d_1 d_2$.

7. $n \equiv 7 \pmod{12} \Rightarrow n = 12k + 7$ (for some $k \in \mathbb{Z}$) $\Rightarrow n = 4(3k + 1) + 3$, etc.

8. Consider positive integer $a_n a_{n-1} \ldots a_2 a_1 a_0$, meaning $a_n 10^n + a_{n-1} 10^{n-1} + \ldots + a_2 10^2 + a_1 10 + a_0$. Since $10 \equiv 1 \pmod 9$, $10^r \equiv 1 \pmod 9$ for every $r \geq 0$; and hence our integer is congruent (mod 9) to $a_n + \ldots + a_2 + a_1 + a_0$. It is divisible by 9 iff the sum of its digits is divisible by 9.

9. If either $3 \nmid x$ or $3 \nmid y$, then $x^2 + y^2 \equiv 0 + 1$ or $1 + 0$ or $1 + 1 \pmod 3$, and so $3 \nmid (x^2 + y^2)$.

Last part: Show first that if $x = u$, $y = v$, $z = w$ is an integral solution, then all of u, v, w are divisible by 3 and $x = \frac{1}{3}u$, $y = \frac{1}{3}v$, $z = \frac{1}{3}w$ is also an integral solution. If the stated result is false, we can (by 10.1) consider a solution with least possible positive value of $|z|$, and then the observation of the last sentence provides a contradiction.

10. Modulo 11: $3^3 \equiv 5$; hence $3^{3n} \equiv 5^n$; hence $3^{3n+1} (= 3.3^{3n}) \equiv 3.5^n$. Also $2^{4n+3} \equiv 8.5^n$. So $3^{3n+1} + 2^{4n+3} \equiv 11.5^n \equiv 0$.

11. $x = -9$ is the answer one obtains by using Euclid's algorithm.

12. Suppose $a \equiv b \pmod n$. Let $d_1 = \text{hcf}(a, m)$, $d_2 = \text{hcf}(b, m)$. Using "$b = a + km$", one sees that $d_1|b$. Hence d_1 is a common factor of b, m. Hence $d_1 \leq d_2$. Prove similarly that $d_2 \leq d_1$.

13. Consider prime decomposition: integer $m \geq 2$ is a perfect square iff every prime appearing in the decomposition of m does so to an even power.

14. Suppose $\sqrt{n} = u/v$, where $u, v \in \mathbb{N}$. Then $u^2 = nv^2$. Now consider prime decompositions to deduce that n must be a perfect square.

15. Every prime is a factor of $p_1 p_2 \ldots p_n$, so not a factor of $p_1 \ldots p_n + 1$. This contradicts 13.1. *Variant*: suppose that 3, $p_1 p_2 \ldots p_n$ is a complete list of the primes congruent to 3 (mod 4) and obtain a contradiction by considering $4p_1 p_2 \ldots p_n + 3$.

Chapter 3 (p. 46) (Proofs that relations are equivalence relations will not be mentioned.)

1. $x \in C_n \Leftrightarrow [x] = [n] = n \Leftrightarrow x = n + t$ for some $t \in [0, 1)$, etc.

2. (ii) $X \in \alpha$-class of $\varnothing \Leftrightarrow X \cap A = \varnothing \cap A \Leftrightarrow X \cap A = \varnothing \Leftrightarrow X \subseteq T - A \Leftrightarrow X \in \mathscr{P}(T - A)$. (iii) 2^n classes.

3. For example, the relations p, σ, τ defined by: $x\rho y \Leftrightarrow |x - y| < 5$; $x\sigma y \Leftrightarrow x \leq y$; $x\tau y \Leftrightarrow x = y = 0$.

4. The ρ-classes are the straight lines of gradient 1.

5. Note that each R-class can be expressed as $\{2^n t : n \geqslant 0\}$, where t is its smallest member. Answer to last part is 9.

7. C_0, C_1, C_3, C_4 all satisfy $x^2 = x$ in $\mathbb{Z}_6$.

8. By Euclid's algorithm, etc., find $1 = 11.97 - 26.41$. So $C_1 = C_{-26} C_{41} = C_{71} C_{41}$; i.e. inverse is C_{71}. Modulo 97: $41x \equiv 2 \Rightarrow C_{41} C_x = C_2 \Rightarrow C_{71} C_{41} C_x = C_{71} C_2 \Rightarrow C_x = C_{142} = C_{45} \Rightarrow x \in C_{45}$; and conversely.

9. 3 β-classes: $B_0, B_1 (= B_4), B_2 (= B_3)$. Definition of multiplication is consistent because $x^2 \equiv (x')^2$ and $y^2 \equiv (y')^2 \pmod 5 \Rightarrow (xy)^2 \equiv (x'y')^2 \pmod 5$. Definition of addition is not consistent: it implies that $B_0 = B_5 = B_1 + B_4 = B_1 + B_1 = B_2$.

Chapter 4 (p. 59)

1. (i) Applied values of $f(1), \ldots, f(7)$ do not belong to the codomain. (ii) Is $g(0)$ defined? (iii) Contradictory information about $h(0)$. (iv) Definition is inconsistent: $\theta(\frac{1}{2}) = 3$ and $\theta(\frac{2}{4}) = 6$, while $\frac{1}{2} = \frac{2}{4}$.

2. Yes in both cases.

3. (i) 15. (ii) $[-1, 80]$. (iii) $[-1, \infty)$; no.

4. (i) n^m. (ii) (a) 0, (b) $n(n-1)(n-2)\ldots(n-m+1)$. (c) There are $\binom{n+1}{2}$. n (i.e. $\frac{1}{2}n(n+1).n$) ways of arranging for two members of S to be mapped to the same element of T, and, for each of these, $(n-1)!$ ways to complete the specification of a surjection. Last answer: $\frac{1}{24}n(3n+1)\{(n+2)!\}$.

5. f: Use 22.1 to prove injective; not surjective, im f being the set of even integers, g: not injection (e.g. $g(1) = g(2)$); surjective since $\forall y \in \mathbb{N}$, $y = g(2y - 1)$. Examples: (a) $x \mapsto (x-2)^2 + 1$; (b) $x \mapsto (-1)^x[\frac{1}{2}x]$.

6. For $A \subseteq S$, $f(A) \cup f(S - A) = f(A \cup (S - A)) = f(S) = T$, etc.

7. Let $y \in$ right-side. Then $y = f(a) = f(b)$ for some $a \in A$, $b \in B$. Since f injective, $a = b$; and hence $a \in A \cap B$. So $y \in f(A \cap B)$; etc.

8. "f injective $\Rightarrow \ldots$" follows from question 7. Conversely suppose that $\forall A \subseteq S$, $f(A) \cap f(S - A) = \emptyset$. Let x, y be unequal members of S. Since $f(\{x\}) \cap f(S - \{x\}) = \emptyset$ $f(x) \neq f(y)$, etc.

9. (i) $x \in A \Rightarrow f(x) \in f(A) \Rightarrow x \in B$. (ii) Suppose f injective. Then $x \in B \Rightarrow f(x) = f(a)$ for some $a \in A \Rightarrow x = a$ [since f injective] $\Rightarrow x \in A$. So $B \subseteq A$, and hence, by (i), $A = B$. (iii) $[-3, 3]$.

10. Definition of θ consistent because $C_{(a,b)} = C_{(c,d)} \Rightarrow b - a = d - c$. θ is injective because $b - a = d - c \Rightarrow C_{(a,b)} = C_{(c,d)}$. θ is surjective because $\forall n \in \mathbb{Z}$, $n = \theta(C_{(x,y)})$ where $x = |n| + 1 - n$, $y = |n| + 1$.

11. (i) (a) Number is $|S|$, (b) number is $|T|$. (ii) Take $\beta(x) = C_x$, and define α (consistently!) by $\alpha(C_x) = f(x)$.

12. Let $t = f^{-1}(T)$, so that $f(t) = T$. The definition of T now tells us that if $t \in T$, then $t \notin T$, and conversely — contradiction.

13. (ii) $[1, \infty)$, (iii) $c = 2$, (iii) $f_1^{-1}(y) = 2 - \sqrt{(y-1)}$, $f_2^{-1}(y) = 2 + \sqrt{(y-1)}$.

14. Let y be an arbitrary element of U. Then $y = (g \circ f)(x)$ for some $x \in S$; and so y is the image under g of something, viz. $f(x)$.

15. $\forall x \in S$, $h(f(x)) = (h \circ f)(x) = (h \circ g)(x) = h(g(x))$, and so, since h injective, $f(x) = g(x)$, etc.

16. $h = h \circ i_s = h \circ (f \circ g) = (h \circ f) \circ g = i_s \circ g = g$. Result now follows by 25.1.

17. Let $g_1(x) = 0$ for all $x \in T$; $g_2(x) = 0$ if $x \in$ im f, $g_2(x) = 1$ if $x \in T -$ im f.

18. Let z be an arbitrary chosen fixed element of S. Define $g: T \to S$ as follows. If $x \in T - \text{im } f$, $g(x) = z$; if $x \in \text{im } f$, $g(x) =$ the unique element $y \in S$ such that $f(y) = x$.

Chapter 5 (p. 68)

1. $\circ$ is associative on $\mathbb{Z}$, since, $\forall x, y, z \in \mathbb{Z}$, $(x \circ y) \circ z = x \circ (y \circ z) = x + y + z - xy - yz - zx + xyz$. $T \neq \varnothing$, and T is closed under $\circ$ since $1 - (x \circ y) = (1 - x)(1 - y)$.

2. Yes Ω and $\varnothing$, respectively.

3. (i) $\circ$ is associative because, $\forall x, y, z \in S$, $(x \circ y) \circ z = x \circ (y \circ z) = z$. (ii) Suppose that, in semigroup $(S, *)$, e and f are two unequal left identities. If there were a right identity g, then $e = e * g = g = f * g = f$—contradiction.

4. Positive integers k, l with the stated property exist because, since S finite, the list $a, a^2, a^3, \ldots$ must contain a repetition. From $a^k = a^{k+1}$, by multiplying successively by $a^l, a^{2l}, \ldots$, we deduce $a^k = a^{k+l} = a^{k+2l} = a^{k+3l} = \ldots$. In particular, $a^k = a^{k+kl}$. This gives the result: if $l = 1$; otherwise multiply through by $a^{k(l-1)}$.

5. Let $x \in S$, and let $y = xx'$. $y^2 = x(x'x)x' = x(ex') = xx' = y$. Hence $y = ey = (y'y)y = y'y^2 = y'y = e$. Further $xe = x(x'x) = (xx')x = ex = x$.

Chapter 6 (p. 87)

1. Identity is $(1, 1)$. Inverse of (x, y) is (x^{-1}, y^{-1}).

2. $\forall g \in G$, $(\lambda_x \circ \lambda_y)(g) = \lambda_x(\lambda_y(g)) = \lambda_x(yg) = xyg = \lambda_{xy}(g)$, etc.

3. Choose $a \in S$, and define $\lambda_a: S \to S$ by $\lambda_a(x) = ax$ $(x \in S)$. Because of cancellation λ_a is injective and so also surjective (S being finite). Hence $ae = a$ for some $e \in S$. Let x be an arbitrary element of S. Then $aex = ax$ and so (cancelling) $ex = x$. Hence $xex = xx$ and so (cancelling) $xe = x$. Thus e is identity. Existence of inverses now comes quickly from surjectivity of each λ_a. Last part: consider $(\mathbb{N}, .)$.

4. $y^{-2}x^{-2} = (xy)^{-2} = ((xy)^{-1})^2 = (y^{-1}x^{-1})^2$; i.e. $y^{-1}y^{-1}x^{-1}x^{-1} = y^{-1}x^{-1}y^{-1}x^{-1}$. Hence, by cancellation, $y^{-1}x^{-1} = x^{-1}y^{-1}$. Taking inverses now gives $xy = yx$.

5, 6. Straightforward applications of 36.3 and 36.4.

7. (i) $xy = yx \Leftrightarrow (xy)x^{-1} = (yx)x^{-1}$. (ii) Let f be a non-identity element of S_n, where $n \geqslant 3$. Suppose $f(i) = j \ (\neq i)$. Since $n \geqslant 3$, can introduce element g of S_n interchanging j and k (k different from i, j) and mapping everything else to itself. Find that gfg^{-1} maps i to k so that $gfg^{-1} \neq f$.

8. If $C(x) \subseteq C(y)$, then $x \in C(y)$ (since clearly $x \in C(x)$), and so $xy = yx$. Second part: Suppose $C(x) \subseteq C(y)$. Then $y \in C(x)$ by first part. Further, $\forall t \in C(x)$, $t \in C(y)$ [since $C(x) \subseteq C(y)$] and so $yt = ty$. It follows that $y \in Z(C(x))$. Conversely suppose $y \in Z(C(x))$. Then $yt = ty$ for all $t \in C(x)$; i.e. every element of $C(x)$ commutes with y; i.e. $C(x) \subseteq C(y)$.

9. Can introduce $h \in H - K$ and $k \in K - H$. If $H \cup K$ is a subgroup (so closed under .), then $hk \in H \cup K$, i.e. $hk \in H$ or $hk \in K$, and hence $h^{-1}(hk) \in H$ or $(hk)k^{-1} \in K$—a contradiction.

10. $H = \mathbb{N}$ in $\mathbb{R} - \{0\}$.

11. Period of A is 6. (Note that $A^3 = -I$.).

12. $x^2 = 1 \Leftrightarrow x^2x^{-1} = 1x^{-1}$. If all elements of $G - \{1\}$ have period 2, then, $\forall x, y \in G$, $xy = (xy)^{-1} = y^{-1}x^{-1} = yx$.

13. Suppose x has finite period k. Then $(x^{-1})^k = (x^k)^{-1} = 1$, and so period of $x^{-1} \leqslant$ period of x. Reverse inequality is obtained on replacing x by x^{-1}. Deal separately with the case where x has infinite period. *Second part*: Elements of period $\geqslant 3$ can be divided into pairs of the form $\{x, x^{-1}\}$; there remain an even number of elements, one of which is 1.

14. 2, 5, 10.

15. $(xax^{-1})^n = xax^{-1}xax^{-1}xax^{-1}\ldots xax^{-1} = xa1a1a1\ldots 1ax^{-1} = xa^nx^{-1}$. Hence deduce that, for $n \in \mathbb{N}$, $(xax^{-1})^n = 1 \Leftrightarrow a^n = 1$; etc. (i) Note that $bc = b(cb)b^{-1}$. (ii) Suppose s, t are distinct elements of period 2 in G. If s, t commute, then st is a third element of period 2; otherwise sts^{-1} is such.

16. 5, 2, 5.

17. $-I$ is the only element of period 2.

18. If x has period $2n - 1$ ($n \in \mathbb{N}$), then $x = x^{2n} = (x^n)^2$.

19. Prove that C_3 has period 6 by calculating its successive powers.

20. Suppose the contrary. Then there is a generator $x > 1$, and $\ldots, x^{-2}, x^{-1}, 1, x,$ $x^2, x^3, \ldots$ are in ascending order. Hence contradiction on observing that $\frac{1}{2}(1 + x)$ is a member of the group strictly between 1 and x.

21. For $r \in \mathbb{N}$, $(x, y)^r = (1, 1) \Leftrightarrow (x^r, y^r) = (1, 1) \Leftrightarrow x^r = 1$ and $y^r = 1 \Leftrightarrow m|r$ and $n|r$, and the lowest value of r for which this is true is the l.c.m. of m, n. Suppose G_1, G_2 cyclic of relatively prime orders m, n, with generators x, y, respectively. Then (x, y) has period mn, and so $\langle (x, y) \rangle$ is the whole of $G_1 \times G_2$.

22. If $\mathrm{hcf}\,(r, n) = d > 1$, then $(x^r)^{n/d} = 1$ and so x^r has period less than n. If $\mathrm{hcf}\,(r, n) = 1$, then, for $s \in \mathbb{N}$, $(x^r)^s = 1 \Rightarrow n|rs \Rightarrow n|s$, and so x^r has period n. Final questions: 2^{k-1}, 2.

23. If BG non-cyclic, then $\langle x \rangle$, where $x \in G - \{1\}$, is a non-trivial proper subgroup of G, etc.

Chapter 7 (p. 102)

1. 4, 5.

2. $HH \subseteq H$ because H is closed under multiplication. Conversely, $h \in H \Rightarrow 1h \in HH \Rightarrow h \in HH$.

3. Let g be an arbitrary element of G. Let $Y = \{x^{-1}g : x \in X\}$. Clearly $|Y| = |X|$; so, because $|X| > \frac{1}{2}|G|$, $X \cap Y \neq \emptyset$. So, for some $x_1, x_2 \in X$, $x_1^{-1}g = x_2$, i.e. $g = x_1x_2$, etc.

4. (i) $g \in$ left-side $\Rightarrow g = xt$ (where $x \in X$, $t \in Y$ and $t \in Z$) $\Rightarrow g \in XY$ and $g \in XZ$, etc. (ii) If $a^2 \neq 1$, $X(Y \cap Z) = X\emptyset = \emptyset$, while $1 \in XY \cap XZ$. (iii) Suppose $g \in$ right-side. Then $g = xy = xz$ for some $y \in Y$, $z \in Z$. By cancellation $y = z$; and so $g = xy$, where $y \in Y \cap Z$, etc.

5. Use 36.4. After introduction of arbitrary elements $x, y \in gHg^{-1}$, the next step is: $x = gh_1g^{-1}$, $y = gh_2g^{-1}$ for some $h_1, h_2 \in H$. By two applications of 41.2, $|gHg^{-1}| = |Hg^{-1}| = |H|$.

6. (i) Follows from 35.6. (ii) Trivial. (iii) Follows from 36.4. Now suppose H, K subgroups. If HK is a subgroup, then $HK = (HK)^{-1} = K^{-1}H^{-1} = KH$. Conversely, if $HK = KH$, then $HK \neq \emptyset$ and $(HK)(HK)^{-1} = HKK^{-1}H^{-1} = HKKH = HKH = HHHK = HK$, etc.

7. (i) $x^{-1}y \in H$. (ii) $ab^{-1} \in H$. (iii) $c \in H$. (iv) $s^{-1}t \in H$.

8. Suppose $xH \cap L \neq \emptyset$. Then there is an element $y \in xH \cap L$. Since $y \in xH$, $xH = yH$; and since $y \in L$ and $H \subseteq L$, $yH \subseteq L$ (as L closed under .), etc.

9. If $c \in aH \cap bK$, then $aH \cap bK = cH \cap cK = c(H \cap K)$.

10. Since $x = x1 \in xH$, we have $x \in yK$. So $yK = xK$, and therefore $xH \subseteq xK$. Hence obviously $x^{-1}(xH) \subseteq x^{-1}(xK)$.

11. $a_i(H \cap K) \neq a_j(H \cap K) \Rightarrow a_i^{-1}a_j \notin H \cap K \Rightarrow a_i^{-1}a_j \notin K$ [all the a's being in the subgroup H] $\Rightarrow a_iK \neq a_jK$. Hence there are at least as many left cosets of K in G as there are left cosets of $H \cap K$ in H. Last part: since $|G:K| = 9$, $|H:H \cap K|$ is finite and $\leqslant 9$. Hence, since $|H \cap K| = 5$, H is finite and $|H| \leqslant 45$. Hence, since $|G, H| = 8$,

G is finite and $|G| \leqslant 8 \times 45 = 360$. But, by Lagrange, $|G|$ is divisible by all of $|G:H|$, $|G:K|$, and $|H \cap K|$, so that $|G|$ is at least the l.c.m. of 8, 9, 5, i.e. $|G| \geqslant 360$. It follows that $|G| = 360$.

12. For $h_1, h_2 \in H$, $h_1 K = h_2 K \Leftrightarrow h_1^{-1} h_2 \in K \Leftrightarrow h_1^{-1} h_2 \in H \cap K$, etc. Note that HK is the union of all the cosets hK with $h \in H$. Unequal such cosets don't overlap; each contains $|K|$ elements; and, by previous part, the number of different such cosets is $|H:H \cap K|$. Hence formula for $|HK|$. *Last part*: note that $|HgK| = |HgKg^{-1}|$, and apply previous part with K replaced by gKg^{-1}.

13. (i) By Lagrange, $|H \cap K|$ divides $|H|$ and $|K|$. (ii) By Lagrange, both $|H|$ and $|K|$ divide $|G|$.

14. $|H \cup K| = |H| + |K| - |H \cap K| = 32 - |H \cap K|$; and, by Lagrange, $|H \cup K|$ is 1 or 2 or 4 or 8. (Can't be 16 because $H \neq K$.)

15. Argue as in worked example 2 of § 43 to prove $|G:H| = 1$.

16. $\langle x \rangle \subseteq C(x)$: hence result, by Lagrange.

17. Consider the coset xG_t. Suppose $x(t) = s$. Then $y \in xG_t \Leftrightarrow y^1 x \in G_t \Leftrightarrow y^{-1}(x(t)) = t \Leftrightarrow y^{-1}(s) = t \Leftrightarrow y(t) = s$. Hence result.

18. (i) Let x have period k. Then $|G| = qk$ for some $q \in \mathbb{N}$. So $x^{|G|} = (x^k)^q = 1^q = 1$. (ii) By (i), $C_n^{p-1} = C_1$, i.e. $C_{n^{p-1}} = C_1$, i.e. $n^{p-1} \equiv 1 \pmod{p}$.

19. $|\langle x \rangle \cap \langle y \rangle| = 1$ because $|\langle x \rangle \cap \langle y \rangle|$ divides both $|\langle x \rangle|(= m)$ and $|\langle y \rangle|(= n)$. Suppose $xy = yx$. Clearly $(xy)^{mn} = 1$ and so period of $xy \leqslant mn$. On the other hand suppose $(xy)^r = 1$. Then $x^r = y^{-r} \in \langle x \rangle \cap \langle y \rangle$, i.e. $x^r = y^r = 1$, and so r is divisible by both m and n, etc.

20. Since group non-cyclic, every non-identity element has period 2 or 4. Hence result by 44.4.

21. Each of the subgroups contains 1 and $p - 1$ elements of period p (see 44.2); and the m sets of $p - 1$ elements so arising are disjoint (see worked example 1 in § 43). Hence $m(p - 1)$ elements of period p in the subgroups; and hence result (since every element of period p lies in such a subgroup). (i) If a non-cyclic group of order 55 has m subgroups of order 5 and n or order 11, then (from counting the non-identity elements) $4m + 10n = 54$, and hence $m, n \neq 0$. (ii) Let there be m subgroups of order p. The $m(p - 1) =$ number of elements of period $p = p^2 - 1$, i.e. $m = p + 1$. There are 2 more subgroups, viz. $\{1\}$ and the whole group.

22. There are elements x, y of period 3 such that x, x^2, y all different. By 36.5 the subset $\{x^i y^j : 0 \leqslant i, j \leqslant 2\}$ is a subgroup of G. Its order is 9: so $9 \, | \, |G|$. *Generalization*: if G is an abelian group containing more than $p - 1$ elements of period p, then $p^2 \, | \, |G|$. By this result, the abelian G of order pq contains $\leqslant p - 1$ elements of period p and $\leqslant q - 1$ elements of period q. Hence the number of elements of period pq in G is at least $(pq - 1) - (p - 1) - (q - 1) = (p - 1)(q - 1)$, which is positive.

Chapter 8 (p. 121)

1. θ injective because $\theta(f) = \theta(g) \Rightarrow \alpha \circ f \circ \alpha^{-1} = \alpha \circ g \circ \alpha^{-1} \Rightarrow \alpha^{-1} \circ (\alpha \circ f \circ \alpha^{-1}) \circ \alpha = \alpha^{-1} \circ (\alpha \circ g \circ \alpha^{-1}) \circ \alpha$, etc.; θ surjective because the arbitrary element g of S_Y equals $\theta(\alpha^{-1} \circ g \circ \alpha)$. It is easy to prove that θ is a homomorphism.

2. The mapping $\begin{bmatrix} 1 & t \\ 0 & 1 \end{bmatrix} \mapsto t$ is an isomorphism.

3. (i) $\forall x, y \in G_1$, $(\phi \circ \theta)(xy) = \phi(\theta(xy)) = \phi(\theta(x)\theta(y)) = \phi(\theta(x))\phi(\theta(y))$, etc. (ii) Take aribtrary elements $x, y \in G_1 : x = \theta(s)$, $y = \theta(t)$ where $s = \theta^{-1}(x)$, $t = \theta^{-1}(y)$. Hence $xy = \theta(st)$, and $\theta^{-1}(x)\theta^{-1}(y) = st = \theta^{-1}(xy)$; etc.

4. (i) $y \in \theta(\langle x \rangle) \Leftrightarrow y = \theta(x^n)$ for some $n \in \mathbb{Z} \Leftrightarrow y = \{\theta(x)\}^n$ for some $n \in \mathbb{Z} \Leftrightarrow y \in \langle \theta(x) \rangle$.
(ii) $\{\theta(x)\}^k = \theta(x^k) = \theta(1) = 1_1$. Hence period of $\theta(x)$ divides k. If θ is a monomorphism, then, for $r \in \mathbb{N}$, $\{\theta(x)\}^r = 1_1 \Rightarrow x^r \in \ker \theta \Rightarrow x^r = 1 \Rightarrow k|r$, etc.

5. $\ker \theta = \{(1, y) : y \in G_2\}$, $y \mapsto (1, y)$ is an isomorphism from G_2 to $\ker \theta$.

6. Let y be an arbitrary element of $\theta(\mathbb{Z})$ and t an arbitrary element of G_1. $y = \theta(z)$ for some $z \in \mathbb{Z}$, and, since θ surjective, $t = \theta(s)$ for some $s \in G$. Hence $yt = \theta(z)\theta(s) = \theta(zs) = \theta(sz) = \theta(s)\theta(z) = ty$; etc.

7. Use 49.3 (along with 36.6).

8. (i) Note that (clearly) $xHx^{-1} = H$ is equivalent to $xH = Hx$ $(x \in G)$. (ii) Only one subgroup of order 2: so it is normal, by (i); subgroups of order 4 are normal by 49.2.

9. (i) Use 36.4, noting that $(k_1 h_1)^{-1}(k_2 h_2) = (h_1^{-1} k_1^{-1} k_2 h_1)(h_1^{-1} h_2)$. (ii) Note that $x(kh)x^{-1} = (xkx^{-1})(xhx^{-1})$.

10. $H \lhd G$ and $K \lhd H$ by 49.2. But $CAC^{-1} = -A \notin K$, and so $K \ntriangleleft G$.

11. Let $H = \langle x \rangle$. $H \lhd G$, by 49.2. Let $y = x^{2^{n-2}}$, which has period 2. $\forall g \in G$, gyg^{-1} must be an element of H (as $H \lhd G$) with same period as y; etc.

12. (i) Straightforward by 36.3 and definition of normality. (ii) As L, M abelian, $L \cap M \lhd L$ and $L \cap M \lhd M$. Hence L, M are subsets of $N(L \cap M)$, and so $L \cup M \subseteq N(L \cap M)$. Since $N(L \cap M)$ is a subgroup, it follows that $H \subseteq N(L \cap M)$. Hence result, by (i).

13. $\theta(a^n) = \{\theta(a)\}^m = 1_1$. So $a^m \in \ker \theta$. Hence $a^{m|\ker \theta|} = 1$ (cf. exercise 18(i) on Chapter 7). Hence result, by 38.4(ii).

14. Clearly always $(xh)^2 h^{-1}(x^{-1})^2 \in H$. For every $x \in G$, $(xH)^2 = x^2 H = H$ (by given property); and so 44.4 applies to G/H.

15. *Last part:* Suppose xH has finite period m in G/H $(x \in G)$. Then $(xH)^m = H$, i.e. $x^m H = H$, i.e. $x^m \in H$. Hence $(x^m)^n = 1$ for some $n \in \mathbb{N}$. So $x \in H$, i.e. $xH = H$.

16. Note that $xy = yx[x, y]$. $[h, k] = h^{-1}(k^{-1}hk) = (h^{-1}k^{-1}h)k$. First version shows $[h, k] \in H$ if $H \lhd G$. Second version shows $[h, k] \in K$ if $K \lhd G$, etc.

17. $[xN, yN] = (xN)^{-1}(yN)^{-1}(xN)(yN) = (x^{-1}y^{-1}xy)N = [x, y]N$. G/N abelian $\Leftrightarrow \forall x, y \in G$, $[xN, yN] = $ identity of G/N, i.e. $[x, y]N = N$, i.e. $[x, y] \in N$. G/H and G/K abelian $\Rightarrow \forall x, y \in G$, $[x, y] \in H$ and $[x, y] \in K \Rightarrow \forall x, y \in G$, $[x, y] \in H \cap K \Rightarrow G/(H \cap K)$ abelian. *Last part:* No (a counterexample is easily obtained by taking G to be a four-group).

18. $yZ = (xZ)^{-1} \Rightarrow (xZ)(yZ) = Z \Rightarrow xyZ = Z \Rightarrow xy \in Z \Rightarrow (xy)x = x(xy) \Rightarrow yx = xy$ (by cancellation of x on the left).

19. Take G_1 to be a four-group and G_2 to be a cyclic group of order 4, with K_1 and K_2 subgroups of order 2.

20. Consider a counterexample G of least order. G is abelian by 44.4. Obtain contradiction by considering order of $G/\langle x \rangle$, where x is any element of $G - \{1\}$.

21. (i) G/H exists by 49.2. And, $\forall x \in G$, $x^2 H = (xH)^2 = H$, since G/H has order 2. (ii) No: consider a subgroup of order 2 in the group of symmetries of the equilateral triangle.

22. (i) $C(x)$ contains $Z(G)$, clearly; and the containment is strict because x is an element of $C(x)$ that does not belong to $Z(G)$. (ii) Let G be a group of order p^2, and let Z be its centre. By Lagrange and 48.4, $|Z| = p$ or p^2. But the former leads to a contradiction, as follows. $|Z|$ being p, we can introduce an element $x \in G - Z$. By part (i) $|C(x)| > p$, and so, since $|C(x)| \,|\, |G|$, $|C(x)| = p^2$; i.e. $C(x) = G$; i.e. $x \in Z$ — contradiction.

23. A fairly easy consequence of 49.3, which one could re-phrase by saying that a normal subgroup of G is a subgroup closed under the taking of conjugates in G.

24. For the first part, adapt the proof of 48.4. If $|N| = p$, then, as $|N \cap Z(G)|$ is > 1 and divides $|N|$, $N \cap Z(G)$ must be the whole of N, etc.

25. Let $Z = Z(G)$, (i) By Lagrange, $|Z| = 1$ or p or p^2 or p^3. But not 1 by 48.4; not p^2 by 51.1; and not p^3 since G non-abelian. (i) Exercise 22(i) enables one to show that, $\forall x \in G - Z$, $|C(x)| \subseteq = p^2$ and therefore $|\mathrm{cl}(x)| = p$. So the $p^3 - p$ elements of $G - Z$ must partition into $(p^3 - p)/p = p^2 - 1$ conjugacy classes. And there are in addition p singleton conjugacy classes inside Z. (iii) By exercise 24, every normal subgroup of order p is contained in Z and so (by part (i)) coincides with Z. (iv) By 51.1, G/Z can contain no element of period p^2, and hence all its non-identity elements have period p. So, $\forall x \in G$, $(xZ)^p = Z$, i.e. $x^p \in Z$.

26. Follow closely the proof of 51.1. Introduce a generator tN of H/N; every element of H then is expressible in the form $t^i n$, with $i \in \mathbb{Z}$ and $n \in N$; etc.

27. (i) By Lagrange, $|Z(G)| = 1$ or p or q or pq. pq is out since G non-abelian; p, q are ruled out by 51.1. (ii) Let x be an element of period p in G. Since $x \notin Z(G)$, $C(x) \subset G$. Hence, by Lagrange and the fact that $p \| C(x) \|$, $|C(x)| = p$; hence $|\mathrm{cl}(x)| = q$ (by 48.2). Hence the elements of period p come in batches of q.

28. By exercise 27, a non-abelian group of order 15 would contain $10a$ elements of period 3 and $12b$ elements of period 5, for some $a, b \in \mathbb{N} \cup \{0\}$; hence (from counting non-identity elements) $10a + 12b = 14$—an impossibility. So every group of order 15 is abelian, etc.

29. Consider the mapping from C to G/H given by $c \mapsto cH$. *Second part*: exploit fact that there is only one element of period 2.

30. (i) Apply 52.1 with $\theta|_H$ as the homomorphism. This gives $\theta(H) \cong H/K$ where $K = \ker(\theta|_H)$. The result follows on considering orders. (ii) Let H be an arbitrary subgroup of order m. By part (i), $|v(H)| \| m$. But also $|v(H)| \| |G:K|$, by Lagrange, since $v(H)$ is a subgroup of G/K. Hence, as m and $|G:K|$ are relatively prime, $|v(H)| = 1$; i.e. $H \subseteq \ker v = K$. So, since $|H| = |K|$, $H = K$.

31. First part is generalization of worked example 1 of §44. $K \lhd G$ follows from 49.2. Let b be an arbitrary element of $G - K$. By exercise 21, $b^2 \in K$. If b had period p, then (cf. worked example 1 in §43) $K \cap \langle b \rangle$ would be $\{1\}$, in contradiction to $b^2 \in K$. Hence b must have period 2. Since also $ba \in G - K$, $(ba) = 1$; so $bab = (ba)^2 a^{-1} = a^{-1}$.

32. Use 36.3 to prove set is subgroup of group of all permutations of G.

33. θ is a homomorphism because $(\tau_g \circ \tau_h)(x) = \tau_{gh}(x)$ for all $g, h, x \in G$. $g \in \ker \theta \Leftrightarrow \theta(g) = i_G \Leftrightarrow \tau_g(x) = x$ for all $x \in G$,. etc. $I = \mathrm{im}\, \theta$;, so, by 52.1, $I \cong G/Z(G)$. $I \lhd \mathrm{Aut}\, G$ because, if $\alpha \in \mathrm{Aut}\, G$, $g \in G$, then $\alpha \tau_g \alpha^{-1} = \tau_{\alpha(g)} \in I$.

34. $x^{-1}\alpha(x) = y^{-1}\alpha(y) \Rightarrow \alpha(xy^{-1}) = xy^{-1} \Rightarrow xy^{-1} = 1 \Rightarrow x = y$. Hence the mapping $\beta : G \to G$ given by $\beta(x) = x^{-1}\alpha(x)$ is injective and so surjective (G being finite). Now suppose $\alpha^2 = i_G$. Then, for $g \in G$, $\alpha(g) = \alpha(x^{-1}\alpha(x))$ for some $x \in G = \{\alpha(x)\}^{-1}x = g^{-1}$. G is abelian since, $\forall s, t \in G$, $s^{-1}t^{-1} = \alpha(s)\alpha(t) = \alpha(st) = (st)^{-1} = t^{-1}s^{-1}$, etc. G has odd order, since otherwise (cf. exercise 13 on Chapter 6) G would contain an element of period 2 and it would be mapped to itself by α.

35. Let $G = \langle x \rangle$, and let ψ_m denote the mapping from G to G given by $x^i \mapsto x^{mi}$. Key points include: (a) every homomorphism from G to G is equal to ψ_m for some $m \in \mathbb{Z}$; (b) $\psi_m \circ \psi_n = \psi_{mn}$; (c) $\psi_m \in \mathrm{Aut}\, G \Rightarrow x^m$ generates G.

Chapter 9 (p. 149)

1. Second rows are respectively $2\,6\,4\,3\,1\,5$, $3\,4\,6\,2\,1\,5$, $3\,6\,2\,1\,4\,5$, $6\,3\,5\,4\,2\,1$, $1\,3\,5\,4\,2\,6$.

2. $(1\,5\,6\,2\,4\,3)$, $(1\,6)(2\,5\,3)(4)$, respectively.

3. $(1\,2\,5\,4\,3)$, $(1\,4\,2\,3\,5)$, respectively.

4. (a) 2.4^2, $1.2^3.3$. (b) $(1\,3\,8\,5\,4)(6\,9\,10)(2)(7)$; $(1\,4\,7\,9\,10)(2\,3\,8)(5)(6)$; $(1\,4)(2\,6\,8\,9)$ $(3\,7\,5\,10)$; $(1\,5\,7)(2\,6)(3\,9)(8\,10)(4)$; $(2\,8)(3\,5)(6\,9)(7\,10)(1)(4)$; $(1\,4)(2\,6\,8\,9)(3\,7\,5\,10)$; $(1\,5\,3\,4\,8)(2)(6)(7)(9)(10)$. (c) $4, 6, 15$.

5. (a) $(1\,3\,4\,6)(2\,8\,7)(5)$; $(1\,7\,4)(2\,5\,8)(3)(6)$; $(1\,2\,5\,7\,6)(3\,4)(8)$; $(1\,6\,7\,3)(2\,4\,5)(8)$. (b) 12, 3. (c) 1 120. (d) $(1\,4\,7)(2\,8\,5)(3\,6)$.

6. $\begin{pmatrix} 1 & 2 & 3 & 4 & 5 & 6 & 7 \\ 2 & 3 & 5 & 4 & 7 & 1 & 6 \end{pmatrix}$

7. $2^2.3$, 3.4, 2.5, and 7. Total number is 1 854.

8. (a) 350; (b) 10 640.

9. The elements of period 3 belong to the conjugacy classes consisting of elements of cycle patterns $1^3.3$ and 3^2; both these classes have order 40. The centralizers of $(a\,b\,c)(d)(e)(f)$ and $(a\,b\,c)(d\,e\,f)$ contain respectively $(a\,b\,c)(d\,e)(f)$ and $(a\,d\,b\,e\,c\,f)$.

10. Consider the fact that α^r moves *all* of the objects moved by α r places round a certain circle: if one object gets back top where it started, so do they all.

11. 2 $(1^2.3.4$ and $2.3.4)$.

12. $(1\,2\,4)(3\,9\,7)(5\,8\,6)$.

13. By 57.3, $|\mathrm{cl}_{s_n}(\alpha)| = (n-1)!$ Hence by 48.2 $|\mathrm{cl}_{s_n}(\alpha)| = n$. Hence result since $\langle a \rangle \subseteq C_{s_n}(\alpha)$ and $\langle a \rangle$ also has order n.

14. $+1$ (6 inverted pairs).

15. $(1\,5\,10\,9\,7)(2\,6)(3\,4)(8)$, $(1\,3\,10\,5\,7\,6)(2\,4)(8\,9)$, respectively; even, odd, respectively.

16. σ^2, τ^2, τ^{-1}.

17. (a) 11; (b) 13.

18. (a) $(1\,4)(2\,5)(3\,6)$; (b) $(1\,2\,3)$.

19. There are 7 conjugacy classes: $\{\iota\}$, the 45 elements of cycle pattern $1^2.2^2$, the 40 elements of cycle pattern $1^2.3$, the 40 elements of cycle pattern 3^2, the 90 elements of cycle pattern 2.4, and two classes each consisting of 72 elements of cycle pattern 1.5.

20. (a) follows at once from 54.2. (b) By (a) and 59.4, an arbitrary non-identity element of A_n can be expressed as a product of the form $(1\,c_1)(1\,c_2)(1\,c_3)(1\,c_4)\ldots (1c_{2r-1})(1c_{2r})$, where one may stipulate that adjacent c_js are different since $(1_{2j-1}c)^2$ simplifies to ι; and in this product $(1_{2j-1}c_j)(1c_{2j})$ can be re-expressed as $(1c_{2j}c_{2j-1})$. The case of ι is trivial since $\iota = (1\,2\,3)^3$.

21. $\lambda_c(x) = x \Rightarrow cx = x \Rightarrow c = 1$ — contradiction. So λ_c moves all the objects; hence must have cycle pattern 2^n. Hence, n being odd, it is an odd permutation. So, by 59.6, $\theta(G)$ contains a subgroup of index 2.

22. Otherwise G has order $2n$ with n odd and >1, and by exercise 21 G has a subgroup of index 2, a normal subgroup by 49.2; and this contradicts the simplicity of G.

23. (a) If a, b are arbitrary objects, then (i) ι is an element of A_n that maps a to b if $a = b$, and (ii) $(a\,b\,c)$, where c is a third object, is an element of A_n mapping a to b in the case $a = b$. (b) Suppose $\sigma \in G$ fixes object a (G being as described in question). Let t be an arbitrary object. Since G transitive, there is an element $\tau \in G$ such that $\tau(a) = t$; and we have: $\sigma(t) = \tau\sigma\tau^{-1}(t)$ [G being abelian] $= \tau\sigma(a) = \tau(a) = t$. Since t arbitrary, it follows that $\sigma = \iota$. Hence result.

24. (a) is straightforward. (b) Show that $\tau(\in S_n)$ fixes $\sigma(a)$ iff $\sigma^{-1}\tau\sigma \in G_a$. (c) Totalling group element by group element gives the number of pairs as described is $\sum\limits_{\gamma \in G} F(\gamma)$.

Totalling object to object gives that the same total is $\sum\limits_{a=1}^{n} |G_a|$. In this latter sum the terms corresponding to the objects in any particular orbit of size m are m terms each equal to $|G|/m$ (by the result quoted in part (a) of the question); and hence the subtotal corresponding to each orbit is $m \times |G|/m$, i.e. $|G|$. Hence result.

25. (i) A subgroup of order 56 would be normal, by 49.2, and so would be the union of some of the conjugacy classes of A_4 (cf. proof of 60.5). However, this is impossible since the conjugacy classes of A_4 have orders 1, 3, 4, and 4. A_4 is not simple because the four-group formed by ι along with the 3 permutations of cycle pattern 2^2 is a non-trivial normal subgroup. (ii) $15 \,|\, |S_5|$; and 15 and $|S_5|/15 \,(= 8)$ are relatively prime. However, S_5 contains no subgroup of order 15, since such a subgroup would be cyclic and S_5 contains no element of period 15.

Chapter 10 (p. 178)

1. $\forall x \in R$, $x + x = (x + x)^2 = x^2 + x^2 + x^2 + x^2 = x + x + x + x$, and hence $x + x = 0$. $\forall x, y \in R$, $x + y = (x + y)^2 = x^2 + xy + yx + y^2 = x + y + xy + yx$, and so $xy = -yx = yx$.

2. (i) Verify that, for all $y \in R$, $(x + e - xe)y = y$. Hence (e being the unique left unity) $x + e - xe = e$; i.e. $xe = x$ for the arbitrary x. (ii) No, by part (i): for the given postulate is easily shown to preclude the existence of two different left unities (cf. proof of 63.4).

3. Straightforward by 64.2; note $C(a) \neq \varnothing$ as $0 \in C(a)$.

4. Fairly straightforward applications of 64.2, 64.3, respectively.

5. If u is a unit (with inverse v), then $1 = |uv|^2 = |u|^2 |v|^2$, and so (since $|u|^2$, $|v|^2$ integers) $|u|^2 = 1$, etc.

6. Let 1_D, 1_E be the unities of D, E, respectively. Result follows on cancelling the nonzero 1_D from the equation $1_D 1_D = 1_E 1_D$ (cf. 63.4).

7. Let $F^* = F - \{0\}$. Since $\theta(0) = 0$ and θ is a bijection, there is a bijective restriction θ^* of θ with F^* as domain and codomain. If we temporarily forget addition, then we see that θ^* is an automorphism of the group $(F^*, .)$, and that enables items (1) and (4) of the subfield test 64.3 to be easily checked for L. The rest is very straightforward.

8. Straightforward by 66.4.

9. First part is an easy application of 64.2. If $Z(R)$ were an ideal, then, since $1 \in Z(R)$, $Z(R)$ would be the whole of R — contradiction when R non-commutative.

10. Straightforward applications of 64.2, 64.3, 64.4, respectively.

11. Introduce nonzero elements a, b of J, K, respectively. Then $ab \neq 0$ since D is an integral domain; and $ab \in J$ and $ab \in K$ since J and K are ideals.

12. (i) $\forall a \in R - \{0\}$, (a) must be whole of R and so $1 \in (a)$, etc. (ii). Let R be the ring of all 2×2 real matrices (not a division ring), and let J be an ideal of R containing a nonzero matrix A with, say, nonzero (r, s)th entry λ. Let E_{ij} be the 2×2 matrix with (i, j)th entry 1 and all other entries zero. Then $(1/\lambda)E_{ir} A E_{sj} = E_{ij}$, and $(\forall i, j)$ this belongs to the ideal J. Now it is easy to prove that J is whole of R.

13. $(a^{n+1}) \subseteq (a^n)$ is trivial. $(a^{n+1}) = (a^n) \Rightarrow a^n \in (a^{n+1}) \Rightarrow a^n = a^{n+1}x$ for some $x \in D \Rightarrow 1 = ax$ [63.4] — contradiction, a being a non-unit. (i) There is a non-unit a, and $(a), (a^2), (a^3), \dots$ are different ideals. (ii) Follows immediately from (i).

14. $\ker \theta$ is an ideal of F: hence result by 66.3.

15. If $\theta(1)$ were 0, then, $\forall x \in R$, $\theta(x) = \theta(x1) = \theta(x)\theta(1) = 0$ — contradiction. So $\theta(1) \neq 0$. Result follows now on noting that $1_1\theta(1) = \theta(1) = \theta(1^2) = \theta(1)\theta(1)$.

16. (i) Observe that every term in the expansion of $(a - b)^{m+n-1}$ contains either a factor a^m or a factor b^n, and so belongs to the ideal J. (ii) (1) $K \neq \varnothing$ since $0 \in K$. (2) Closure of K under $-$ follows from part (i). (3) Since R commutative, $x^m \in J \Rightarrow \forall r \in R$, $(rx)^m = r^m x^m \in J$.

17. (i) Apply 66.4. (ii) Suppose $J_1 = J_2$. Then, for $y \in R_1$, $y \in K_1 \Leftrightarrow y = \theta(x)$ for some $x \in J_1 = J_2 \Leftrightarrow y \in K_2$; and therefore $K_1 = K_2$ — contradiction.

18. By the euclidean algorithm, $d = xm + yn$ for some $x, y \in \mathbb{Z}$. Hence $d \in (m) + (n)$, and therefore (cf. 67.1) $(d) \subseteq (m) + (n)$. But, since m, n are multiples of d, so is every element of $(m) + (n)$; and hence $(m) + (n) \subseteq (d)$. So $(m) + (n) = (d)$.

19. Closure of K under subtraction: Let $x, y \in K$; then $x \in J_k, y \in J_l$ for some $k, l \in \mathbb{N}$, and so x and y belong to J_m, where $m = \max(k, l)$; since J_m is an ideal, $x - y \in J_m$, which is a subset of K. *Second part*: Suppose there is such a sequence of ideals, and let K be their union; K must equal (a) for some $a \in K$, and $a \in J_m$ for some $m \in \mathbb{N}$; hence $K \subseteq J_m$, and so $J_{m+1} \subseteq J_m$—contradiction.

20. R/J commutative $\Leftrightarrow \forall x, y \in R$, $(x + J)(y + J) = (y + J)(x + J) \Leftrightarrow \forall x, y \in R$, $xy + J = yx + J$, etc. *Second part*: R/K_1 and R/K_2 both commutative $\Rightarrow \forall x, y \in R$, $xy - yx \in K_1$ and K_2, etc.

21. $R/S \cong \mathbb{R}$ follows from 68.4.

22. (i) C_6 is nilpotent in $\mathbb{Z}_{12}$. C_x nilpotent in $\mathbb{Z}_{pq} \Rightarrow C_{x^r}$ is zero (for some $r \in \mathbb{N}$) $\Rightarrow pq \mid x^r \Rightarrow p$ and q both divide x (cf. 12.5), etc. (ii) If $x^r = 0$, then $(x + J)^r = x^r + J = J =$ zero of R/J. (iii) If K is as described and x is arbitrary in R, then, since $x + K$ is nilpotent in R/K, $(x + K)^r = K$ for some $r \in \mathbb{N}$; hence $x^r + K = K$, i.e. $x^r \in K$; hence $(x^r)^s = 0$ for some $s \in \mathbb{N}$—which shows that x is nilpotent. (iv) If $x^n = 0$, then $(1 - x)(1 + x + x^2 + \ldots + x^{n-1}) = 1$. (v) N is an ideal by exercise 16(ii) applied with $J = \{0\}$. $(x + N)^r = N \Rightarrow x^r \in N \Rightarrow (x^r)^s = 0$ for some $s \in \mathbb{N} \Rightarrow x \in N \Rightarrow x + N = N$. (vi) It suffices to prove that an arbitrary $x \in N$ belongs to an arbitrary maximal ideal M. By part (ii), $x + M$ is nilpotent in R/M, which is a field (by 68.3). So $x + M = M$, i.e. $x \in M$.

23. Suppose J is a maximal ideal of R, a commutative ring with a one, and let v be the natural epimorphism from R to R/J. Suppose R/J has an ideal K such that $0 \subset K \subset R/J$. Let I be the ideal of R defined by $I = \{x \in R : v(x) \in K\}$; and observe that $\{x \in R : v(x) \in \{0\}\} = \ker v = J$ and $\{x \in R : v(x) \in R/J\} = R$. Clearly, $J \subseteq I \subseteq R$. By $1'$/(ii) with $K_1 = \{0\}$ and $K_2 = K$, $J \subset I$; and by $1'$/(ii) with $K_1 = K$, $K_2 = R/J$, $I \subset R$. But J is a maximal ideal of R—contradiction. This proves that R/J has no ideals except itself and $\{0\}$. Therefore, by 12(i), it is a field.

24. Suppose R/J a field. Let K be an ideal of R with $J \subset K$. Then K contains an element x not in J. $x + J$ has an inverse $(y + J$, say) in R/J. Hence obtain $xy = j + 1$ for some $j \in J$, and use this equation to prove $1 \in K$, i.e. $K = R$, etc.

25. (As can be seen immediately from 24) (2) and (3) are both maximal ideals of $\mathbb{Z}$.

26. Suppose R/J is an integral domain. Then $J \subset R$. And $x, y \in R - J \Rightarrow x + J$, $y + J$ nonzero in $R/J \Rightarrow (x + J)(y + J) \neq J \Rightarrow xy + J \neq J \Rightarrow xy \in R - J$. So J prime. Converse must also be tackled. J maximal $\Rightarrow R/J$ a field $\Rightarrow R/J$ an integral domain $\Rightarrow J$ prime. Every ideal of $\mathbb{Z}$ is $\{0\}$ or (m) for some $m \in \mathbb{N} : \mathbb{Z}/\{0\} \cong \mathbb{Z}$, which is an integral domain; and $\mathbb{Z}/(m) = \mathbb{Z}_m$, which is an integral domain iff m is prime.

27. $\forall x, y \in F$, $\sigma(xy) = \sigma(x)\sigma(y)$, clearly, and $\sigma(x + y) = \sigma(x) + \sigma(y)$ because p divides the binomial coefficient $\binom{p}{r}$ if $1 \leqslant r \leqslant p - 1$. Prove σ injective by showing $\ker \sigma = \{0\}$. That σ is an automorphism when F is finite follows from the fairly obvious fact that an injective mapping from a finite set to itself is necessarily also surjective.

28. $rx = 0$ can be re-written as $(r1)(x) = 0$. (That is obvious by the distributive law if $r > 0$ and not too difficult to justify in the $r < 0$ case.) And we can multiply $(r1)x = 0$ through by $(r1)^{-1}$ (which exists since $r1$ is a nonzero element of F) to obtain $x = 0$. Alternatively: clearly $\mathrm{hcf}(r, p) = 1$, so that (by euclidean algorithm) there are $a, b \in \mathbb{Z}$ such that $ar + bp = 1$; and hence $x = 1x = (ar + bp)x = a(rx) + b(px)$ (by additive versions of index laws) $= a0 + b0 = 0 + 0 = 0$.

29. $x^2 = 1 \Rightarrow (x - 1)(x + 1) = 0 \Rightarrow x - 1 = 0$ or $x + 1 = 0$ (D being an integral domain). *Second part*: Consider first the product of the units that are equal to neither 1 nor -1; this product is 1 because the factors in it divide into pairs of the form $\{u, u^{-1}\}$; result now follows (though the characteristic 2 case needs separate comment). *Last part*: Apply second part with $D = \mathbb{Z}_p$.

30. Suppose that θ is an isomorphism from $(F - \{0\}, .)$ to $(F, +)$. F must have characteristic 2, since otherwise $0 = \theta(1) = \theta((-1)^2) = \theta(-1) + \theta(-1) \neq 0$. If $x \in F - \{0\}$, then $\theta(x^2) = \theta(x) + \theta(x) = 0$ and hence (as θ injective) $x^2 = 1$, so that (cf. exercise 29) $x = 1$. Hence $|F| = 2$, and θ is a bijection from a set of order 1 to a set of order 2 — contradiction.

31. Straightforward; note that (in all cases where $a, b \in D - \{0\}$) $ab/ab = 1/1$ because $(ab, ab) \sim (1, 1)$.

32. No. Each element of F equals ab^{-1} for some $a, b \in D$ with $b \neq 0$ and so (by the closure properties of the field D) belongs to D.

Chapter 11 (p. 225)

1. Y^2 not transcendental over $D \Rightarrow 0 = a_0 + a_1 Y^2 + a_2(Y^2)^2 + \ldots + a_n(Y^2)^n = a_0 + a_1 Y^2 + a_2 Y^4 + \ldots + a_n Y^{2n}$ for some $a_0, a_1, \ldots, a_n \in D$, not all zero $\Rightarrow Y$ not transcendental over D — contradiction.

2. Apply the first isomorphism theorem to the homomorphism $\varepsilon_0 : D[X] \to D$: $\ker \varepsilon_0 = (X)$, and $\operatorname{im} \varepsilon_0 = D$.

3. One can prove that J is an ideal of $\mathbb{Z}[X]$ by a fairly easy application of 66.4. Suppose J is a principal ideal — say $J = (f)$. Then, because $2 \in J$, it can be seen that $f = \pm 2$ or ± 1 — all of which possibilities obviously contradict $J = (f)$.

4. Suppose that $f = a_0 + a_1 X + a_2 X^2 + \ldots + a_n X^n$. The result follows on taking the complex conjugate of both sides of the equation $a_0 + a_1 \alpha + a_2 \alpha^2 + \ldots + a_n \alpha^n = 0$.

5. $\mathbb{Q}(X) \cong \mathbb{Q}(X^2)$; and it is not difficult to prove by contradiction that $X \notin \mathbb{Q}(X^2)$, so that $\mathbb{Q}(X^2)$ is a proper subfield of $\mathbb{Q}(X)$.

6. Suppose $f = q_1 d + r_1$ and also $f = q_2 d + r_2$, with $q_1, q_2, r_1, r_2 \in F[X]$ and $\deg r_1$ and $\deg r_2$ both $< n$. Equating the two expressions for f to each other gives $(q_1 - q_2)d = r_2 - r_1$. Here $\deg(\text{right-hand side}) < n$, whereas $\deg(\text{left-hand side}) \geq \deg d = n$ unless $q_1 = q_2$. Therefore q_1, q_2 must be equal, and hence also $r_1 = r_2$.

7. The roots in $\mathbb{C}$ of $X^5 - 1$ are 1 and the complex numbers $\exp(2k\pi i/5)$ $(k = 1, 2, 3, 4)$. Use of 73.4 gives: $X^5 - 1 = (X - 1)(X - e^{2\pi i/5})(X - e^{8\pi i/5})(X - e^{4\pi i/5})(X - e^{6\pi i/5}) = (X - 1)[X^2 - (2\cos(2\pi/5))X + 1][X^2 - (2\cos(4\pi/5))X + 1]$ after complex conjugate factors have been multiplied together to produce real quadratic factors. The factorization required now follows on observing that $X^5 - 1 = (X - 1)(X^4 + X^3 + X^2 + X + 1)$ and then cancelling the $X - 1$ factor. The person complaining about the cancellation of $X - 1$ perhaps needs to be reminded that we are talking about elements of $\mathbb{C}[X]$ and not about polynomial functions: the cancellation of $X - 1$ is unquestionably legitimate because what we are doing is cancelling a nonzero element (namely $X - 1$) in the integral domain $\mathbb{C}[X]$. *Last part*: equating coefficients of X^2 yields first of all $1 = 2 + 4\cos(2\pi/5)\cos(4\pi/5)$, i.e. $\cos(2\pi/5)\cos(4\pi/5) = -\frac{1}{4}$ — from which the result follows since $\cos(4\pi/5) = -\cos(\pi/5)$.

8. Follows from 74.1(2) once it has been checked that none of the three elements of $\mathbb{Z}_3$ $(0, 1, -1)$ is a root of $X^3 - X + 1$.

9. (i) It is easy to show that any factorization of X into two factors within $D[X]$ must be of the form $X = a(bX)$ with $a, b \in D$, and here $ab = 1$, so that a is a unit in D. (ii) $D[X]$ is a principal ideal domain $\Rightarrow (X)$ is a maximal ideal of $D[X]$ [by part (i) and 74.4] $\Rightarrow D[X]/(X)$ is a field $\Rightarrow D$ is a field (by exercise 2).

10. (i) Let p be a prime element in D and suppose that $p = ab$ where $a, b \in D$. Then (since $ab = 1p$) $p|ab$, and so (p being a prime element) $p|a$ or $p|b$ — say the former (else switch notation). So $a = cp$ for some $c \in D$, and hence $a = c(ab) = a(bc)$. As a is clearly a nonzero element of the integral domain D, it follows that $bc = 1$, and so b is a unit. Therefore p is irreducible. (ii) 74.7 showed that, in a principal ideal domain, irreducible $\Rightarrow$ prime, while part (i) shows that prime $\Rightarrow$ irreducible is true in every integral domain. (iii) Suppose (q) is a prime ideal of D (where $q \in D - \{0\}$). Since $D - (q) \neq \varnothing$, q is not a unit. Further, for $a, b \in D$, $q|ab \Rightarrow ab \in (q) \Rightarrow a \in (q)$ or $b \in (q)$ [since (q) is a prime ideal] $\Rightarrow q|a$ or $q|b$; and thus q is a prime element. The converse must also be tackled. (iv) Let D be a principal ideal domain. (As in every integral domain, clearly) $\{0\}$ is a prime ideal of D. Let J be any nonzero prime ideal of D. Since D is a principal ideal domain, $J = (q)$ for some element $q \in D$ which is prime by part (iii) and hence irreducible by part (ii). Therefore J is a maximal ideal.

11. (i) Let $J = (a) + (b)$. Since D is a principal ideal domain, $J = (d)$ for some $d \in D$. Since $a, b \in J$, $d|a$ and $d|b$. Further, for $c \in D$, $c|a$ and $c|b$ in $D \Rightarrow c$ divides every element of $(a) + (b) \Rightarrow c|d$. Therefore d is a highest common factor of a, b. (ii) (1) Let e, f be any two highest common factors of a, b. From property (2) in the definition of "highest common factor" it follows that $e|f$ and $f|e$. Hence (cf. 74.2 and 74.3) $(e) = (f)$ and e, f are unit multiples of each other. (2) Let g be any highest common factor of a, b. By part (ii, 1) g is a unit multiple of the element d of part (i), and so $g \in (d) = (a) + (b)$: so $g = xa + yb$ for some $x, y \in D$.

12. Let $a, d \in R$, with $d \neq 0$. If $|a| \geqslant |d|$, it is clear that, in the Argand diagram, one of the points $a + d$, $a - d$, $a + id$, $a - id$ is closer to the origin than a is: so one can subtract Gaussian-integer-multiples of d from a until the square of the modulus of the remainder (which can take only integral values) drops below $|d|^2$. A division algorithm result follows: there exist $q, r \in R$ such that $a = qd + r$ and $|r| < |d|$. Now let J be an arbitrary nonzero ideal of R. Choose an element d of least possible modulus in J then use the division algorithm to prove $J = (d)$.

13. Let S be the set of nonzero non-unit elements in D that are not expressible as finite products of irreducible elements. Assume (with a view to obtaining a contradiction) that $S \neq \varnothing$, and let $a_1 \in S$. Since a_1 certainly cannot be irreducible, $a_1 = a_2 b_2$, where neither a_2 nor b_2 is a unit and where, since $a_1 \in S$, at least one of $a_2, b_2 \in S$. Say $a_2 \in S$. By repetition of the same argument, $a_2 = a_3 b_3$, where neither a_3 nor b_3 is a unit and where $a_3 \in S$; and $a_3 = a_4 b_4$, where neither a_4 nor b_4 is a unit and where $a_4 \in S$, etc. It is not difficult to show that the sequence of elements $a_1, a_2, a_3, a_4, \ldots$ generated in this way satisfies $(a_1) \subset (a_2) \subset (a_3) \subset (a_4)\ldots$—a contradiction, by exercise 19 on Chapter 10.

14. The total number of monic polynomials of degree 2 in $K[X]$ is the number of possibilities for the coefficient sequence (a, b) in $X^2 + aX + b$ [with $a, b \in K$]; and that number is q^2. Among these q^2 polynomials, there are $\binom{q}{2} [= \frac{1}{2}q(q - 1)]$ of the form $(X - \alpha)(X - \beta)$ with $\alpha \neq \beta$ and q of the form $(X - \gamma)^2$, and so altogether $\frac{1}{2}q(q - 1) + q$ polynomials that are not irreducible. Hence the required number of irreducible polynomials is $q^2 - [\frac{1}{2}q(q - 1) + q] = \frac{1}{2}q(q - 1)$.

15. A finite sum of the form $a_m X^{-m} + \ldots + a_1 X^{-1} + c + b_1 X + b_2 X^2 + \ldots + b_n X^n$, with coefficients $a_m, \ldots, a_1, c, b_1, b_2, \ldots, b_n \in \mathbb{Z}$.

16. Yes. (a) $2^{1/2} = (2^{1/6})^3$ and $2^{1/3} = (2^{1/6})^2$: so $2^{1/2}$ and $2^{1/3} \in K$, and therefore $F \subseteq K$. (b) $2^{1/6} = 2^{1/2}/2^{1/3}$: so $2^{1/6} \in F$, and therefore $K \subseteq F$.

17. For arbitrary $t = a + b\sqrt{2} \in \mathbb{Q}(\sqrt{2})$ where $a, b \in \mathbb{Q}$, $\phi(t^2) = [\phi(t)]^2 = a^4 - 4a^2 b^2 + 4b^4$. Let $c = \sqrt{(2 + 2^{1/2})}$. Then $c \in \mathbb{Q}(\sqrt{2}) \Rightarrow [\phi(c)]^2 = \phi(c^2) = \phi(2 + \sqrt{2}) = 2$, while $\phi(c) \in \mathbb{Q}$—contradiction. Therefore $c \notin \mathbb{Q}(\sqrt{2})$. $\mathbb{Q}(\sqrt{2}) \subseteq \mathbb{Q}(c)$ since $\sqrt{2} = c^2 - 2 \in \mathbb{Q}(c)$; the containment is strict because c belongs to $\mathbb{Q}(c)$ but not to $\mathbb{Q}(\sqrt{2})$.

18. Very similar to the worked example at the end of § 76: use the fact that $(i + \sqrt{2})(i - \sqrt{2}) = -3$.

19. After adjustment, if necessary, to make the linear factors monic, we have $X^3 - 2 = (X - a)(X - b)(X - c)$ for some $a, b, c \in L$. By evaluating both sides of this equation at $2^{1/3}$, $2^{1/3}\omega$, $2^{1/3}\omega^2$, we see that a, b, c must be $2^{1/3}$, $2^{1/3}\omega$, $2^{1/3}\omega^2$ in some order, and so these complex numbers belong to L. Hence further L contains $2^{1/3}\omega/2^{1/3} = \omega$; and the result follows from the fact that L contains $2^{1/3}$ and ω.

20. $(c - \sqrt{2})^2 = \sqrt{2}$—from which, via expansion of the left-hand side, it quickly follows that $c^2 + 2 = (2c + 1)\sqrt{2}$. Hence $\sqrt{2} = (c^2 + 2)/(2c + 1) \in \mathbb{Q}(c)$; and so (since $\mathbb{Q}(c)$ is closed under subtraction) $d = c - \sqrt{2} \in \mathbb{Q}(c)$. It follows that $\mathbb{Q}(d) \subseteq \mathbb{Q}(c)$; and the reverse containment follows from $c = d + d^2$.

21. (i) Units in D must have norm 1 (cf. exercise 5, on Chapter 10).
(ii) Rather obviously, there are no integers satisfying $a^2 + 5b^2 = 2$ or 3.
(iii) Any non-trivial factorizations would involve factors of norms 2 or 3.
(iv) Since $2 \times 3 = 6 = (1 + i\sqrt{5})(1 - i\sqrt{5})$, 2 divides the product $(1 + i\sqrt{5})(1 - i\sqrt{5})$; but it does not divide either $1 + i\sqrt{5}$ or $1 - i\sqrt{5}$ in D, and so it is not a prime element. Similarly for the other elements listed. To prove J an ideal, use 66.4. To check condition (3), let $x + y\theta \in R$ and $a + b\theta \in J$, so that $a \equiv b \pmod{2}$. Then $(x + y\theta)(a + b\theta) = (xa - 5yb) + (xb + ya)\theta$, and this belongs to J since (mod 2) $xa - 5yb \equiv xa + yb \equiv xb + ya$ (since $a \equiv b$). Note that $2 \in J$. So if $J = (d)$, $d|2$ and hence (by parts (i) and (iii)) $d = \pm 1$ or ± 2; i.e. $J = R$ or $J = (2)$—both of which present obvious contradictions. R/K is not an integral domain since $(1 + \theta) + K$ and $(1 - \theta) + K$ are nonzero elements of R/K whose product is zero.

22. (i) Let $x = a + sb$, $y = a + tb$: these are elements of L. Solving $x = a + sb$ and $y = a + tb$ as simultaneous equations for a, b, we get $a = (sy - tx)/(s - t)$, $b = (x - y)/(s - t)$. [Note $s \neq t$ is given.] Hence, since $s, t \in F \subseteq L$, it follows by closure properties of L that $a, b \in L$. Hence L contains $F(a, b) (= K)$, so that L is the whole of K. (ii) Under the hypotheses stated, the fields $F(a + xb)$, where x varies through F, cannot all be different; and so part (i) applies and gives the result.

23. (a) $X^4 - 3$, (b) $X^2 + 3$, (c) $X^2 - 2X - 2$.

24. $(c - \sqrt{3})^2 = 2$ expands to give $c^2 + 1 = (2\sqrt{3})c$; then squaring and simplifying gives $c^4 - 10c^2 + 1 = 0$. So m_c is a monic quartic divisor of $X^4 - 10X^2 + 1$, and hence $m_c = X^4 - 10X^2 + 1$.

25. Let $c \in R - \{0\}$. Since R is a subring containing F, $F[c] \subseteq R$; i.e. (by 77.4) $F(c) \subseteq R$. So $c^{-1} \in R$.

26. Apply the first isomorphism theorem to the evaluation homomorphism ε_c; $\operatorname{im} \varepsilon_c = F[c] = F(c)$, and $\ker \varepsilon_c = (m)$.

27. (i) $\forall x \in K - \{0\}$, $x^{q-1} = 1$ (by 44.1) and so $x^q = x.x^{q-1} = x.1 = x$; and of course 0 is also a root of $X^q - X$. (ii) follows immediately from (i) by 73.4. (iii) In view of (i), the polynomial $X^q - X + 1$ has no root in K.

28. (i) Let t be a generator of the cyclic multiplicative group of F, and let $u = t^{(q-1)/2}$. [The question stipulates q odd.] Then $u^2 = 1$ but $u \neq 1$, and therefore (cf. exercise 29 on Chapter 10) $u = -1$. If $q \equiv 1 \pmod{4}$, we may legitimately consider the element $t^{(q-1)/4}$; and its square is -1. Conversely, if an element i exists in K satisfying $i^2 = -1$, then i has period 4 in the multiplicative group of K, and $q \equiv 1 \pmod{4}$ follows by 44.1. (ii) For p an odd prime, $\exists x \in \mathbb{Z}$ s.t. $x^2 \equiv -1 \pmod{p}$ iff $p \equiv 1 \pmod{4}$. (iii) For every odd prime p, $p \equiv \pm 1 \pmod{4}$ and therefore $p^2 \equiv 1 \pmod{4}$.

29. Realize that $X^3 - 1 = (X - 1)(X^2 + X + 1)$, so that any root of $X^2 + X + 1$ is a cube root of 1. First suppose $X^2 + X + 1$ is irreducible in $K[X]$. Then the characteristic of K is not 3 (since, over a field of characteristic 3, $X^2 + X + 1$ factorizes as $(X - 1)^2$). Hence q, which is a power of the characteristic of K, $\not\equiv 0 \pmod 3$. Further, if $q \equiv 1 \pmod 3$, then, by the worked example in § 78, K contains a cube root of 1 (ω, say) different from 1, and $X - \omega$ is a factor of $X^2 + X + 1$ — contradiction. So it follows that q must be congruent to $-1 \pmod 3$. Conversely, if $q \equiv -1 \pmod 3$, then $X^2 + X + 1$ has no root in K, since neither is 1 a root of it (since the characteristic of K cannot be 3) nor does K contain a cube root of 1 different from 1.

30. (i) $\sigma^r(x) = x^{p^r}$ ($x \in K$, $r \in \mathbb{N}$). That $\sigma^n = i_K$ follows from exercise 27, part (i). If $\sigma^r = i_K$, where $r \in \mathbb{N}$ but $r < n$, then all the p^n elements of K are roots of the polynomial $X^{p^r} - X$, which has degree less than p^n — contradiction. (ii) Let F (of order p) be the prime subfield of K. That σ fixes every element of F follows from exercise 27(i); and 73.5 shows that σ cannot fix any more elements than the p elements of F.

31. (i) Clearly neither of the elements 0, 1 of $\mathbb{Z}_2$ is a root of $X^3 + X + 1$. (ii) By 79.1 there is a field $\mathbb{Z}_2(t)$ with t a root of $X^3 + X + 1$, and this is a field as described. (iii) $t^3 = 1 + t$, $t^4 = t + t^2$, $t^5 = 1 + t + t^2$, $t^6 = 1 + t^2$.

32. From the hint in the question, $f = (X^2 - (\sqrt{7})X + 2)(X^2 + (\sqrt{7})X + 2)$ and so has roots $\frac{1}{2}(\sqrt{7} + i)$, $\frac{1}{2}(\sqrt{7} - i)$, $\frac{1}{2}(-\sqrt{7} + i)$, $\frac{1}{2}(-\sqrt{7} - i)$: call these a, b, c, d, respectively, and let $K = \mathbb{Q}(i, \sqrt{7})$. Since all of a, b, c, d belong to K, f splits into linear factors over K. Moreover K is generated by adjoining roots of f to $\mathbb{Q}$, since $K = \mathbb{Q}(a, b)$.

33. $[K:F] = [K:L][L:F] = 2[K:L]$, which is a multiple of 2; and similarly $[K:F]$ is a multiple of 3.

34. $2 = [K:F] = [K:F(c)][F(c):F]$ and $[F(c):F] > 1$ since $F(c) \supsetneq F$ (as $c \notin F$). Hence $[K:F(c)] = 1$, i.e. $F(c) = K$.

35. Let $K = \mathrm{GF}(p^n)$, let F be an arbitrary subfield of K, let P be the prime subfield of K, and let $[F:P] = m$. Then (cf. 80.5) $n = [K:P] = [K:F][F:P] = m[K:F]$, and so $m|n$. Moreover (cf. 80.5 again) $|F| = p^m$.

36. Let $K = \mathbb{Q}(2^{1/3}, \omega)$, $L = \mathbb{Q}(2^{1/3})$. Since $m_{2^{1/3}/\mathbb{Q}} = X^3 - 2$, $[L:\mathbb{Q}] = 3$. Further, since $X^2 + X + 1$ has no root in $\mathbb{R}$ ($\supset L$) but has ω as a root in K, $X^2 + X + 1$ is $m_{\omega/L}$: and so $[K:L] = 2$. By 80.4, $[K:\mathbb{Q}] = [K:L][L:\mathbb{Q}] = 2 \times 3 = 6$.

37. Since (in the notation of exercise 20) $\mathbb{Q}(c) = \mathbb{Q}(d)$, $\deg(m_{c/\mathbb{Q}}) = [\mathbb{Q}(c):\mathbb{Q}] = [\mathbb{Q}(d):\mathbb{Q}] = \deg(m_{d/\mathbb{Q}}) = \deg(X^4 - 2) = 4$. From $c^2 + 2 = (2c + 1)\sqrt{2}$ [see exercise 20] squaring and simplifying gives $c^4 - 4c^2 - 8c + 2 = 0$. So m_c is a monic quartic divisor of $X^4 - 4X^2 - 8X + 2$ and hence $m_c = X^4 - 4X^2 - 8X + 2$.

38. (i) $F(c_1)$ is a finite extension of F since c_1 is algebraic over F, and $F(c_1, c_2)$ is a finite extension of $F(c_1)$ since c_2 is algebraic over F and therefore over $F(c_1)$. Hence (see 80.4) $F(c_1, c_2)$ is a finite extension of F. Hence (cf. 80.3) all elements of $F(c_1, c_2)$ are algebraic over F: in particular $c_1 - c_2$ and $c_1 c_2$ are algebraic over F. (ii) Use 64.3: part (i) disposes of the awkward closure properties.

39. (i) A simple contradiction proof: start by supposing $\sqrt{3} = a + b\sqrt{2}$ with $a, b \in \mathbb{Q}$, square both sides, and use the linear independence of 1 and $\sqrt{2}$ over $\mathbb{Q}$. (ii) One easily sees that $(1, \sqrt{3})$ is a basis for $\mathbb{Q}(\sqrt{2}, \sqrt{3})$ over $\mathbb{Q}(\sqrt{2})$ and that $(1, \sqrt{2})$ is a basis for $\mathbb{Q}(\sqrt{2})$ over $\mathbb{Q}$. It follows from the detail of how 80.4 was proved that a basis for $\mathbb{Q}(\sqrt{2}, \sqrt{3})$ over $\mathbb{Q}$ is $(1, \sqrt{2}, \sqrt{3}, \sqrt{6})$.

Index